Advanced Models of Neural Networks

Gerasimos G. Rigatos

Advanced Models of Neural Networks

Nonlinear Dynamics and Stochasticity in Biological Neurons

Gerasimos G. Rigatos
Unit of Industrial Automation
Industrial Systems Institute
Rion Patras
Greece

ISBN 978-3-662-43763-6 ISBN 978-3-662-43764-3 (eBook)
DOI 10.1007/978-3-662-43764-3
Springer Heidelberg New York Dordrecht London

Library of Congress Control Number: 2014947828

Printed on acid-free paper

Springer is part of Springer Science+Business Media (www.springer.com)

To Elektra

Preface

This book provides a complete study on neural structures exhibiting nonlinear and stochastic dynamics. The book elaborates on neural dynamics by introducing advanced models of neural networks. It overviews the main findings in the modelling of neural dynamics in terms of electrical circuits and examines their stability properties with the use of dynamical systems theory. Such electric circuit models are characterized by attractors and fixed points while in certain cases they exhibit bifurcations and chaotic dynamics. Moreover, solutions of the neural dynamics in the form of travelling waves equations are derived. The dynamics of interconnected neurons is analyzed with the use of forced oscillator and coupled oscillator models. It is shown that by introducing stochastic uncertainty and variations in the previous deterministic coupled oscillator model, stochastic coupled oscillator models can be derived. Next, going into a more refined analysis level it is shown how neural dynamics can be interpreted with the use of quantum and stochastic mechanics. It is proven that the model of interacting coupled neurons becomes equivalent to the model of interacting Brownian particles that is associated with the equations and dynamics of quantum mechanics. It is shown that such neural networks with dynamics compatible to quantum mechanics principles exhibit stochastic attractors. Furthermore, the spectral characteristics of such neural networks are analyzed. Additionally, a stochastic mechanics approach to stabilization of particle systems with quantum dynamics is presented. It also shown that the eigenstates of the quantum harmonic oscillator (QHO) can be used as activation functions in neural networks. Moreover, a gradient-based approach to stabilization of particle systems with quantum dynamics is provided. There are remarkable new results in the book concerned with: (1) nonlinear synchronizing control of coupled neural oscillators, (2) neural structures based on stochastic mechanics or quantum mechanics principles, (3) nonlinear estimation of the wave-type dynamics of neurons, (4) neural and wavelet networks with basis functions localized both in space and frequency, (5) stochastic attractors in neural structures, and (6) engineering applications of advanced neural network models.

This book aims at analyzing advanced models of neural networks, starting with methods from dynamical systems theory and advancing progressively to stochasticity-based models and models compatible with principles of quantum mechanics. Advanced models of neural networks enable on the one side to understand patterns of neuronal activity seen in experiments in the area of neuroscience and on the other side to develop neurocomputing methods in the area of information sciences that remain consistent with physics principles governing biological neural structures. The first chapters of the book (Chaps. 1–6) make use of dynamical systems theory to explain the functioning of neural models. Dynamical systems and computational methods are widely used in research to study activity in a variety of neuronal systems. The dynamics of the neural structures are described with the use of linear or nonlinear ordinary differential equations or with the use of partial differential equations (PDEs). This approach to neural modelling focuses on the following issues: (1) Modelling neural networks in terms of electrical circuits (The Hodgin–Huxley equations. The FitzHugh–Nagumo equations. Ion channels) (2) Elements of dynamical systems theory: The phase plane. Stability and fixed points. Attractors, Oscillations. Bifurcations, Chaotic dynamics (3) Chaotic dynamics in neural excitation. Travelling waves solutions to models of neural dynamics (4) Neural oscillators (forced oscillators and coupled oscillator models).

The latter chapters of the book (Chaps. 7–13) analyze the significance of noise and stochasticity in modelling of neuronal dynamics. The dynamics of neural structures are no longer described by linear or nonlinear ordinary differential equations or by PDEs but are formulated in terms of stochastic differential equations. Neural computation based on principles of quantum mechanics can provide improved models of memory processes and brain functioning and is of primary importance for the realization of quantum computing machines. To this end, the book studies neural structures with weights that follow the model of the QHO. The proposed neural networks have stochastic weights which are calculated from the solution of Schrödinger's equation under the assumption of a parabolic (harmonic) potential. These weights correspond to diffusing particles, which interact with each other as the theory of Brownian motion (Wiener process) predicts. The learning of the stochastic weights (convergence of the diffusing particles to an equilibrium) is analyzed. In the case of associative memories the proposed neural model results in an exponential increase of patterns storage capacity (number of attractors). It is also shown that conventional neural networks and learning algorithms based on error gradient can be conceived as a subset of the proposed quantum neural structures. Thus, the complementarity between classical and quantum physics can be also validated in the field of neural computation. Furthermore, in continuation to modelling of neural networks as interacting Brownian particles it is shown how stabilization can be succeeded either for particle systems which are modelled as coupled stochastic oscillators or for particle systems with quantum dynamics. Finally, engineering applications of the previously described advanced models of neural dynamics are provided. The book's chapters are organized as follows: (1) Modelling biological neurons in terms of electrical circuits, (2) Systems theory for the analysis of biological neuron dynamics, (3) Bifurcations and limit cycles in biological neurons, (4) Oscillatory dynamics in biological neurons, (5) Synchronization of circadian neurons and protein synthesis control, (6) Wave dynamics in the transmission of neural signals, (7) Stochastic models of biological neuron dynamics, (8) Synchronization of stochastic neural oscillators using Lyapunov methods, (9) Synchronization in coupled stochastic neurons using differential flatness theory, (10) Attractors in associative memories with stochastic weights, (11) Spectral analysis of neural models with stochastic weights, (12) Neural networks based on the eigenstates of the QHO and engineering applications, (13) Gradient-based feedback control for dynamical systems following the model of the QHO.

A summary of the chapters' content is given in the following:

In Chap. 1, it is shown that the functioning of the cells membrane can be represented as an electric circuit. To this end: (1) the ion channels are represented as resistors, (2) the gradients of the ions concentration are represented as voltage sources, (3) the capability of the membrane for charge storage is represented as a capacitor. It is considered that the neurons have the shape of a long cylinder, or of a cable with specific radius. The Hodgkin–Huxley model is obtained from a modification of the cables PDE, which describes the change of the voltage along dendrites axis. Cable's equation comprises as inputs the currents which are developed in the ions channels. Other models of reduced dimensionality that describe voltage variations along the neuron's membrane are the FitzHugh–Nagumo model and the Morris–Lecar model. Cable's equation is also shown to be suitable for describing voltage variations along dendrites. Finally, the various types of ionic channels across the neurons' membrane are analyzed.

In Chap. 2, the main elements of systems theory are overviewed, thus providing the basis for modelling of biological neurons dynamics. To understand oscillatory phenomena and consequently the behavior of biological neurons benchmark examples of oscillators are given. Moreover, using a low order mathematical model of biological neurons the following properties are analyzed: phase diagram, isoclines, attractors, local stability, fixed points bifurcations, and chaos properties.

In Chap. 3, a systematic method is proposed for fixed point bifurcation analysis in biological neurons using interval polynomials theory. The stages for performing fixed point bifurcation analysis in biological neurons comprise (1) the computation of fixed points as functions of the bifurcation parameter and (2) the evaluation of the type of stability for each fixed point through the computation of the eigenvalues of the Jacobian matrix that is associated with the system's nonlinear dynamics model. Stage (3) requires the computation of the roots of the characteristic polynomial of the Jacobian matrix. This problem is nontrivial since the coefficients of the characteristic polynomial are functions of the bifurcation parameter and the latter varies within intervals. To obtain a clear view about the values of the roots of the characteristic polynomial and about the stability features they provide to the system, the use of interval polynomials theory and particularly of Kharitonov's stability theorem is proposed. In this approach the study of the stability of a characteristic polynomial with coefficients that vary in intervals is equivalent to the study of the stability of four polynomials with crisp coefficients computed from the boundaries of the aforementioned intervals. The efficiency of the proposed approach for the analysis of fixed point bifurcations in nonlinear models of biological neurons is tested through numerical and simulation experiments.

In Chap. 4, it is shown that the voltage of the neurons membrane exhibits oscillatory variations after receiving suitable external excitation either when the neuron is independent from neighboring neural cells or when the neuron is coupled to neighboring neural cells through synapses or gap junctions. In the latter case it is significant to analyze conditions under which synchronization between coupled neural oscillators takes place, which means that the neurons generate the same voltage variation pattern possibly subject to a phase difference. The loss of synchronism between neurons can cause several neuro-degenerative disorders. Moreover, it can affect several basic functions of the body such as gait, respiration, and hearts rhythm. For this reason synchronization of coupled neural oscillators has become a topic of significant research during the last years. It is also noted that the associated results have been used in several engineering applications, such as biomedical engineering and robotics. For example, synchronization between neural cells can result in a rhythm generator that controls joints motion in quadruped, multi-legged, and biped robots.

In Chap. 5, a new method is proposed for synchronization of coupled circadian cells and for nonlinear control of the associated protein synthesis process using differential flatness theory and the derivative-free nonlinear Kalman Filter. By proving that the dynamic model of the synthesis of the FRQ protein (protein extracted from the frq gene) is a differentially flat one its transformation to the linear canonical (Brunovsky) form becomes possible. For the transformed model one can find a state feedback control input that makes the oscillatory characteristics in the concentration of the FRQ protein vary according to desirable setpoints. To estimate the nonmeasurable elements of the state vector, a new filtering method named *Derivative-free nonlinear Kalman Filter* is used. The Derivative-free nonlinear Kalman Filter consists of the standard Kalman Filter recursion on the linearized equivalent model of the coupled circadian cells and of computation of state and disturbance estimates using the diffeomorphism (relations about state variables transformation) provided by differential flatness theory. Moreover, to cope with parametric uncertainties in the model of the FRQ protein synthesis and with stochastic disturbances in measurements, the Derivative-free nonlinear Kalman Filter is redesigned in the form of a disturbance observer. The efficiency of the proposed Kalman Filter-based control scheme is tested through simulation experiments.

In Chap. 6, an analysis is given on wave-type PDEs that describe the transmission of neural signals and proposes filtering for estimating the spatiotemporal variations of voltage in the neurons' membrane. It is shown that in specific neuron models the spatiotemporal variations of the membrane's voltage follow PDEs of the wave type while in other models such variations are associated with the propagation of solitary waves in the membrane. To compute the dynamics of the membrane PDE model without knowledge of boundary conditions and through the processing of noisy measurements, the Derivative-free nonlinear Kalman Filter is proposed. The PDE of the membrane is decomposed into a set of nonlinear ordinary differential equations with respect to time. Next, each one of the local models associated with the ordinary differential equations is transformed into a model of the linear canonical (Brunovsky) form through a change of coordinates (diffeomorphism) which is based on differential flatness theory. This transformation provides an extended model of the nonlinear dynamics of the membrane for which state estimation is possible by applying the standard Kalman Filter recursion. The proposed filtering method is tested through numerical simulation tests.

In Chap. 7, neural networks are examined in which the synaptic weights correspond to diffusing particles and are associated with a Wiener process. Each diffusing particle (stochastic weight) is subject to the following forces: (1) a spring force (drift) which is the result of the harmonic potential and tries to drive the particle to an equilibrium and (2) a random force (noise) which is the result of the interaction with neighboring particles. This interaction can be in the form of collisions or repulsive forces. It is shown that the diffusive motion of the stochastic particles (weights update) can be described by Fokker–Planck's, Ornstein–Uhlenbeck, or Langevins equation which under specific assumptions are equivalent to Schrödinger's diffusion equation. It is proven that Langevin's equation is a generalization of the conventional gradient algorithms.

In Chap. 8, a neural network with weights described by the interaction of Brownian particles is considered again. Each weight is taken to correspond to a Brownian particle. In such a case, neural learning aims at leading a set of M weights (Brownian particles) with different initial values on the 2D phase plane, to a desirable final position. A Lyapunov function describes the evolution of the phase diagram towards the equilibrium Convergence to the goal state is assured for each particle through the negative definiteness of the associated Lyapunov function. The update of each neural weight (trajectory in the phase diagram) is affected by (1) a drift force due to the harmonic potential and (2) the interaction with neighboring weights (particles). Using Lyapunov methods it is shown that the mean of the particles converges to the equilibrium while using LaSalle's theorem it is proven that the individual particles remain within a small area encircling the equilibrium.

In Chap. 9, examples of chaotic neuronal dynamics are presented first, and control of chaotic neuron model with the use of differential flatness theory is explained. Moreover, a synchronizing control method is presented for neurons described again as particle systems and modeled as coupled stochastic oscillators. The proposed synchronization approach is flatness-based control. The kinematic model of the particles is associated with the model of the QHO and stands for a differentially flat system. It is also shown that after applying flatness-based control the mean of the particle system can be steered along a desirable path with infinite accuracy, while each individual particle can track the trajectory within acceptable accuracy levels.

In Chap. 10, neural associative memories are considered in which the elements of the weight matrix are taken to be stochastic variables. The probability density function of each weight is given by the solution of Schrödinger's diffusion equation. The weights of the proposed associative memories are updated with the use of a learning algorithm that satisfies quantum mechanics postulates. The learning rule is proven to satisfy two basic postulates of quantum mechanics: (1) existence in superimposing states and (2) evolution between the superimposing states with the use of unitary operators. Therefore it can be considered as a quantum learning algorithm. Taking the elements of the weight matrix of the associative memory to be stochastic variables means that the initial weight matrix can be decomposed into a superposition of associative memories. This is equivalent to mapping the fundamental memories (attractors) of the associative memory into the vector spaces which are spanned by the eigenvectors of the superimposing matrices and which are related to each other through unitary rotations. In this way, it can be shown that the storage capacity of the associative memories with stochastic weights increases exponentially with respect to the storage capacity of conventional associative memories.

In Chap. 11, spectral analysis of neural networks with stochastic weights (stemming from the solution of Schrödinger's diffusion equation) is performed. It is shown that: (1) The Gaussian basis functions of the weights express the distribution of the energy with respect to the weights' value. The smaller the spread of the basis functions is, the larger becomes the spectral (energy) content that can be captured therein. Narrow spread of the basis functions results in wide range of frequencies of a Fourier-transformed pulse (2) The stochastic weights satisfy an equation which is analogous to the principle of uncertainty.

In Chap. 12, feedforward neural networks with orthogonal activation functions (Hermite polynomials) which come from the solution of Schrödinger's diffusion equation are considered. These neural networks have significant properties: (1) the basis functions are invariant under the Fourier transform, subject only to a change of scale, (2) the basis functions are the eigenstates of the QHO and stem from the solution of Schrödinger's harmonic equation. The proposed neural networks have performance that is equivalent to wavelet networks and belong to the general category of nonparametric estimators. They can be used for function approximation, image processing, and system fault diagnosis. The considered basis functions are also analyzed with respect to uncertainty principles and the Balian–Low theorem.

In Chap. 13, the interest is in control and manipulation of processes at molecular scale, as the ones taking place in several biological and neuronal systems. To this end, a gradient method is proposed for feedback control and stabilization of particle systems using Schrödinger's and Lindblad's descriptions. The eigenstates of the quantum system are defined by the spin model. First, a gradient-based control law is computed using Schrödinger's description. Next, an estimate of state of the quantum system is obtained using Lindblad's differential equation. In the latter case, by applying Lyapunov's stability theory and LaSalle's invariance principle one can compute a gradient control law which assures that the quantum system's state will track the desirable state within acceptable accuracy levels. The performance of the control loop is studied through simulation experiments for the case of a two-qubit quantum system.

The book contains teaching material that can be used in several undergraduate or postgraduate courses in university schools of engineering, computer science, mathematics, physics, and biology. The book can be primarily addressed to final year undergraduate students and to first years postgraduate students pursuing studies in electrical engineering, computer science as well as in physics, mathematics, and biology. The book can be also used as a main reference in upper level courses on machine learning, artificial intelligence, and computational intelligence as well as a complementary reference in upper level courses on dynamical systems theory. Moreover, engineers and researchers in the areas of nonlinear dynamical systems, artificial and computational intelligence, and machine learning can get profit from this book which analyzes advanced models of neural networks and which goes beyond the common information representation schemes with the use of artificial neural networks.

Acknowledgements

The author would like to thank research associates and colleagues who have contributed to the preparation of this book either explicitly or implicitly. Discussions, exchange of ideas, and reviewing of research work from people working in the field have given the motivation for an in-depth analysis of the topic of *Advanced Models of Neural Networks* and have provided a useful feedback while editing and revising this book.

Athens, Greece
October 2013

Gerasimos G. Rigatos

Contents

Acronyms

FHN	FitzHugh–Nagumo neurons
EEG	Electroengephalogram
CPG	Central pattern generators
GABA	Gamma aminobutyric acid
KF	Kalman filtering
EKF	Extended Kalman filtering
DKF	Derivative-free nonlinear Kalman filter
MIMO	Multi input multi output
CDK	Cycline dependent kinases
MRNA	Messenger RNA
PDE	Partial differential equations
QHO	Quantum harmonic oscillator
CLT	Central limit theorem
FNN	Feedforward neural networks
FDI	Fault detection and isolation

Chapter 1
Modelling Biological Neurons in Terms of Electrical Circuits

Abstract The functioning of the cells membrane can be represented as an electric circuit. To this end: (1) the ion channels are represented as resistors, (2) the gradients of the ions concentration are represented as voltage sources, (3) the capability of the membrane for charge storage is represented as a capacitor. It is considered that the neurons have the shape of a long cylinder, or of a cable with specific radius. The Hodgkin–Huxley model is obtained from a modification of the cable's PDE, which describes the change of the voltage along dendrites axis. Cable's equation comprises as inputs the currents which are developed in the ions channels. Other models of reduced dimensionality that describe voltage variations along the neuron's membrane are the FitzHugh–Nagumo model and the Morris–Lecar model. Cable's equation is also shown to be suitable for describing voltage variations along dendrites. Finally, the various types of ionic channels across the neurons' membrane are analyzed.

1.1 The Hodgkin–Huxley Equations

In neural cells there is a voltage difference appearing between the inner and the outer side of the membrane

$$V_M = V_{\text{in}} - V_{\text{out}} \tag{1.1}$$

where V_{in} is the voltage at the inner part of the cell, V_{out} is the voltage at the outer part of the cell, and V_M is the membrane potential. The term "resting potential" describes the membrane's voltage when the cell is in rest and has an approximate value of $-70\,\text{mV}$.

In the neural cell two types of current are developed: The incoming current is associated with the inflow of Na^+ ions and has as a result to raise the cell's potential (approaching to $0\,\text{mV}$). The outgoing current is associated with the outflow of K^+ ions, or with the inflow of Cl^- ions and has as a result to reduce significantly the cell's potential.

G.G. Rigatos, *Advanced Models of Neural Networks*,
DOI 10.1007/978-3-662-43764-3_1,

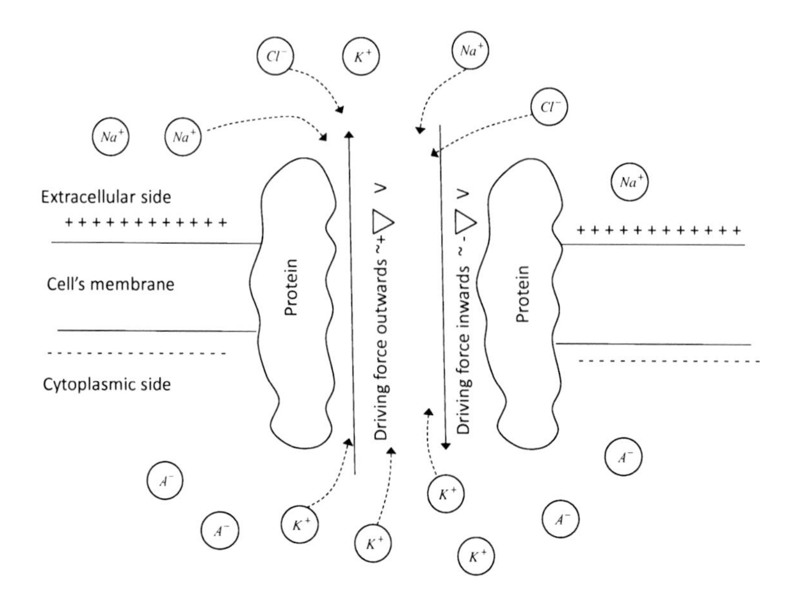

Fig. 1.1 Ionic currents in cell's membrane: inflow of Na^+ and Cl^- and outflow of K^+

The structural component of the membrane are lipids and proteins and based on these an explanation of its conductivity features can be provided (Fig. 1.1). Lipids prohibit current's conduction whereas proteins enable ions to go through them, thus forming ion channels.

There are two types of ion channels, the gated and the non-gated ones. The non-gated ion channels are always open. The gated ion channels switch from the open to the closed status according to a probability that depends on the membrane's potential (voltage-gated channels).

When the cell's membrane is in rest the gated channels remain closed and the associated resting potential of the membrane is due only to ions getting through the non-gated channels. When specific gates in the membrane are open there is an inflow of Na^+ and Cl^- and an outwards motion of K^+. The inflow of ions continues until an equilibrium condition is reached, i.e. until the concentration of positive ions in the inner part of the membrane becomes equal to the concentration of ions to the outer part of the membrane (equivalent is the equilibrium condition for negative ions).

The balance mechanism in the inflow or outflow of ions is as follows: It is assumed that there is an outflow of ions K^+ towards the outer part of the membrane. In the inner part of the membrane there is a surplus of positive charges while in the outer part of the membrane there is a surplus of negative charges (organic ions A^-). The outflow of the K^+ ions continues until there is an equalization of the concentration of positive and negative charges in the inner and outer part of the membrane.

The equilibrium in the inflow or outflow of charges is reached according to Nernst's conditions [16]:

$$E_K = -\frac{RT}{zF}\ln\frac{[K^+]_{in}}{[K^+]_{out}} \tag{1.2}$$

where E_K is the K^+ Nernst's potential, $[K^+]_{in}$ is the concentration of the K^+ ions in the inner part of the membrane, $[K^+]_{out}$ is the concentration of the K^+ ions in the outer part of the membrane, R is a gas constant, F is the Faraday constant, z is the valence of K^+, and T is the absolute temperature in Kelvin.

When the neurons are at rest their membrane is permeable to ions of the K^+, Na^+, and Cl^- type. In the equilibrium condition the number of open channels for ions of the K^+ or Cl^- type is larger than the number of the open channels for ions of the Na^+ type. The equilibrium potential of the neuron's membrane depends on the Nerst potential of the K^+ and Cl^- ions.

The exact relation that connects the membrane's potential to ions concentrations in its inner and outer part results from the Goldman–Hodgkin–Katz equation. The inflow and outflow of ions continues until the associated voltage gradients become zero (there is a voltage gradient for the positive ions Na^+ and K^+ and a voltage gradient for the negative ions Cl^-). Voltage gradients are also affected by ion pumps, that is channels through which an exchange of specific ions takes place. For example, the Na^+-K^+ pump is a protein permeable by ions which enables the exchange of $3Na^+$ with $2K^+$. As long as there is a non-zero gradient due to the concentration of K^+ in the inner and outer part of the membrane the outflow of these ions continues.

For instance, when the concentration of positive ions K^+ in the outer part of the membrane raises there is an annihilation of the gradient that forces the K^+ ions to move outwards. This has as a result, an equilibrium condition to be finally reached.

1.1.1 Nernst's Equation

The following notations are used next: $[C]_x$ is the concentration of ions in the membrane, $V(x)$ is the membrane's potential at point x, along the membrane. Then according to Fick's diffusion law one has about the diffusive flux J_{diff} [16, 65]

$$J_{diff} = -D\frac{\partial[C]}{\partial x} \tag{1.3}$$

where D is the diffusion constant (cm^2/s) and $[C]$ in the concentration of molecules (ions) N/cm^3. Moreover, there is a force that is responsible for the drift motion of the molecules [16, 65]

$$J_{drift} = -\mu z[C]\frac{\partial V}{\partial x} \tag{1.4}$$

where $E = -\frac{\partial V}{\partial x}$ is the electric field and z is the valence of the ion, that is $\pm 1, \pm 2$, etc. and parameter μ is the so-called mobility and is measured in cm^3/V s. The aggregate flow through the membrane is given by the relation

$$J_{\text{total}} = -D\frac{\partial [C]}{\partial x} - \mu z[C]\frac{\partial V}{\partial x} \tag{1.5}$$

The diffusion coefficient that appears in the aforementioned equations is given by

$$D = \frac{kT}{q} \cdot \mu \tag{1.6}$$

where k is the Boltzmann constant, T denotes the absolute temperature, and q is the electric charge (measured in Coulombs). Therefore the aggregate flux can be written as

$$J_{\text{total}} = -\frac{kT\mu}{q}\frac{\partial [C]}{\partial x} - \mu z[C]\frac{\partial V}{\partial x} \tag{1.7}$$

Multiplying this current flux with the valence and Faraday's constant and using RT/F in place of kT/q one has [16, 65]

$$I = -\left(zRT\mu\frac{\partial [C]}{\partial x} - \mu z^2 F[C]\frac{\partial V}{\partial x}\right) \tag{1.8}$$

This is the Nernst–Planck equation describing current flow through the membrane. Nernst's equation is obtained setting the current in Nernst–Planck equation to be equal to zero ($I = 0$). That is, at equilibrium condition of the diffusion and electric effects is reached. Thus, after intermediate operations one has the Nernst potential equation

$$V_{\text{eq}} = V_{\text{in}} - V_{\text{out}} = -\frac{RT}{zF}\ln\left(\frac{[C_{\text{in}}]}{[C_{\text{out}}]}\right) \tag{1.9}$$

1.1.2 The Goldman–Hodgkin–Katz Equation

Under the assumption that the electric field across the lipid membrane remains constant one has [16, 65]

$$E = -\frac{V_M}{l} \text{ and } \frac{dV}{dx} = \frac{V_M}{l} \tag{1.10}$$

The mobility of ions across the membrane is denoted by u^*. The aqueous concentration of ions is denoted by $[C]$ and the concentration of ions on the membrane is denoted as $\beta[C]$. The Nernst–Planck equation for current flow across the membrane becomes

$$I = -u^* z^2 F\beta[C]\frac{V_M}{l} - u^* zRT\beta\frac{d[C]}{dx} \quad 0 < x < l \tag{1.11}$$

with boundary conditions $[C](0) = [C]_{\text{in}}$ and $[C](l) = [C]_{\text{out}}$. Using these boundary conditions the solution for the Nernst–Planck equation becomes

$$I = \frac{u^* z^2 F V_M \beta}{l} \left(\frac{[C]_{\text{out}} e^{-\xi} - [C]_{\text{in}}}{e^{-\xi} - l} \right) \tag{1.12}$$

where $\xi = \frac{z V_M F}{RT}$. By defining the membrane's permeability as

$$P = \frac{\beta u^* RT}{lF} \tag{1.13}$$

one obtains that the current's equation through the membrane becomes

$$I = PzF\xi \left(\frac{[C]_{\text{out}} e^{-\xi} - [C]_{\text{in}}}{e^{-\xi} - 1} \right) \tag{1.14}$$

It is assumed now that there is flow of a large number of ions, e.g. K^+, Na^+, and Cl^-. At the equilibrium condition, the current through the membrane drops to zero, i.e. it holds

$$I = I_K + I_{Na} + I_{Cl} = 0 \tag{1.15}$$

The potential of the membrane for which the zeroing of the current I takes place is

$$V_M = \frac{RT}{F} \ln \left\{ \frac{P_K [K^+]_{\text{out}} + P_{Na} [Na^+]_{\text{out}} + P_{Cl} [Cl]_{\text{in}}}{P_K [K^+]_{\text{in}} + P_{Na} [Na^+]_{\text{in}} + P_{Cl} [Cl]_{\text{out}}} \right\} \tag{1.16}$$

where variables P_K, P_{Na}, and P_{Cl} denote the associated permeabilities. Equation (1.16) is the Goldman–Hodgkin–Katz and the associated voltage is measured in mV.

1.2 Equivalent Circuits of the Cell's Membrane

1.2.1 The Electric Equivalent

The functioning of the cell's membrane can be represented as an electric circuit. To this end: (1) the ion channels are represented as resistors, (2) the gradients of the ions' concentration are represented as voltage sources, (3) the capability of the membrane for charge storage is represented as a capacitor (Fig. 1.2). This is depicted in the following diagram:

For the charge that is stored in the membrane it holds

$$q = C_M V_M \tag{1.17}$$

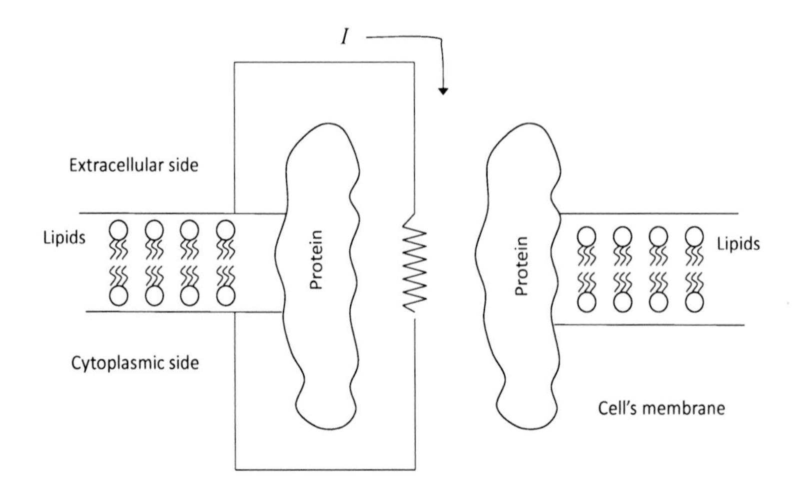

Fig. 1.2 Electric equivalent of the cell's membrane

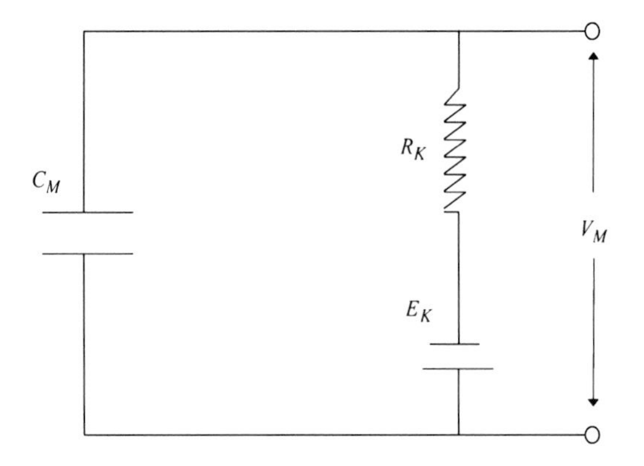

Fig. 1.3 Equivalent circuit of cell's membrane

Voltage V_M is computed from the Goldman–Hodgkin–Katz equation and C_M is the membrane's capacity. The normalized membrane capacity (capacity per cm^2) is noted as c_M. Considering the equivalent circuit of Fig. 1.3 it holds that

$$I = \frac{dq}{dt} \tag{1.18}$$

and

$$\hat{I}_k = \frac{V_M - E_k}{R_k} \Rightarrow \hat{I}_k = \hat{g}_k(V_M - E_k) \tag{1.19}$$

where $\hat{g}_k$ is the conductance normalized per unit of surface.

The above holds for one channel of ions. Assume now that there are N_k channels of ions. The aggregate resistance of the circuit is $N_k R_k = r_k$. Then the current through the ions' channel is written as

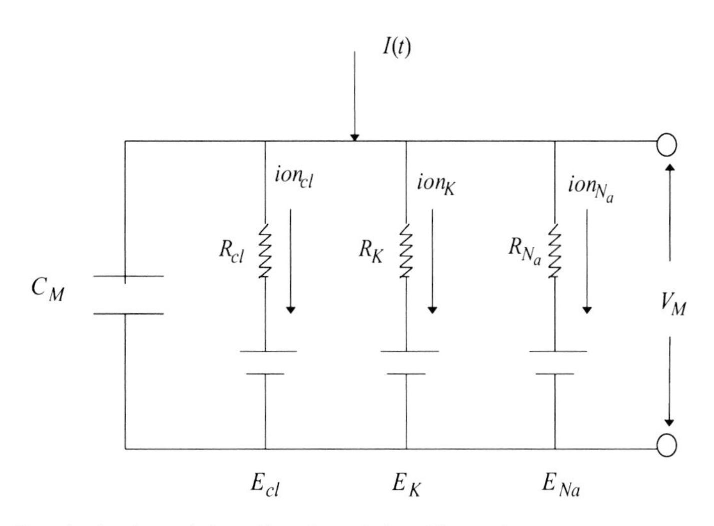

Fig. 1.4 Electric circuit consisting of ion channels in cell's membrane

$$I_k = \frac{V_M - E_k}{r_k} \tag{1.20}$$

By applying Kirchhoff's law in the loop, in the transient stage, one gets [16, 65]

$$I_{\text{capac}} + I_k \Rightarrow c_M \frac{dV_M}{dt} + \frac{(V_M - E_k)}{r_k} = 0 \tag{1.21}$$

that is

$$c_M \frac{dV_M}{dt} = -\frac{V_M - E_k}{r_k} = -g_k(V_M - E_k) \tag{1.22}$$

For example, in the case of $N = 3$ ion channels of the Cl, K and Na type it holds that the aggregate ions current is

$$I_{\text{ion}} = -\frac{V_M - E_{\text{Cl}}}{R_{\text{Cl}}} - \frac{V_M - E_{\text{K}}}{R_{\text{K}}} - \frac{V_M - E_{\text{Na}}}{R_{\text{Na}}} \tag{1.23}$$

or using the normalized conductance coefficients g_{Cl}, g_{K}, and g_{Na} and the normalized capacitance c_M one has

$$I_{\text{ion}} = -g_{\text{Cl}}(V_M - E_{\text{Cl}}) - g_{\text{K}}(V_M - E_{\text{K}}) - g_{\text{Na}}(V_M - E_{\text{Na}}) \tag{1.24}$$

Equivalently, assuming an external input current I (normalized per surface of the membrane), it holds that [16, 65]

$$c_M \frac{dV_M}{dt} = -g_{\text{Cl}}(V_M - E_{\text{Cl}}) - g_{\text{K}}(V_M - E_{\text{K}}) - g_{\text{Na}}(V_M - E_{\text{Na}}) + \frac{I(t)}{A} \tag{1.25}$$

where A denoted membrane's surface (Fig. 1.4). Using the notation

$$E_R = (g_{\text{Cl}} E_{\text{Cl}} + g_{\text{K}} E_{\text{K}} + g_{\text{Na}} E_{\text{Na}}) r_M \tag{1.26}$$

$$r_M = \frac{1}{g_{\text{Cl}} + g_{\text{K}} + g_{\text{Na}}} \tag{1.27}$$

it holds that

$$c_M \frac{dV_M}{dt} = -\frac{(V_M - E_R)}{r_M} + \frac{I(t)}{A} \tag{1.28}$$

Considering that the conductance coefficients are constants and that the currents have reached their steady state one obtains

$$V_{\text{ss}} = \frac{g_{\text{Cl}} E_{\text{Cl}} + g_{\text{K}} E_{\text{K}} + g_{\text{Na}} E_{\text{Na}} + I/A}{g_{\text{Cl}} + g_{\text{K}} + g_{\text{Na}}} \tag{1.29}$$

Therefore, in case that there is no external current, the membrane's voltage V_M in the steady state is the weighted sum of the equilibrium potentials E_{Cl}, E_{K}, and E_{Na} (according to Nernst's equation) for the three ionic channels.

1.2.2 Membrane's Time Constant

It is assumed that the external current applied to the membrane is $I(t) = I_0$ between time instant $t = 0$ and $t = T$. Assuming that the cell's shape is approximated by a sphere with surface $A = 4\pi\rho^2$, the normalized (over surface and time unit) current is

$$I_M(t) = \frac{I(t)}{4\pi\rho^2} = \begin{cases} \frac{I_0}{4\pi\rho^2} & \text{if } 0 \leq t < T \\ 0 & \text{otherwise} \end{cases} \tag{1.30}$$

Using Eq. (1.28) and considering $E_R = 0$ for the membrane's potential, the following partial differential equation holds

$$c_M \frac{dV_M}{dt} = -\frac{V_M}{r_M} + I_M(t) \tag{1.31}$$

Solving the differential equation with respect to $V_M(t)$ one obtains the change in time of the membrane's potential which is described by Eq. (1.32) and is depicted in Fig. 1.5

$$V_M(t) = \begin{cases} \frac{r_M I_0}{4\pi\rho^2}(1 - e^{-\frac{t}{\tau_m}}) & \text{for } 0 < t < T \\ V_M(T) e^{-\frac{t}{\tau_M}} & \text{for } t > T \end{cases} \tag{1.32}$$

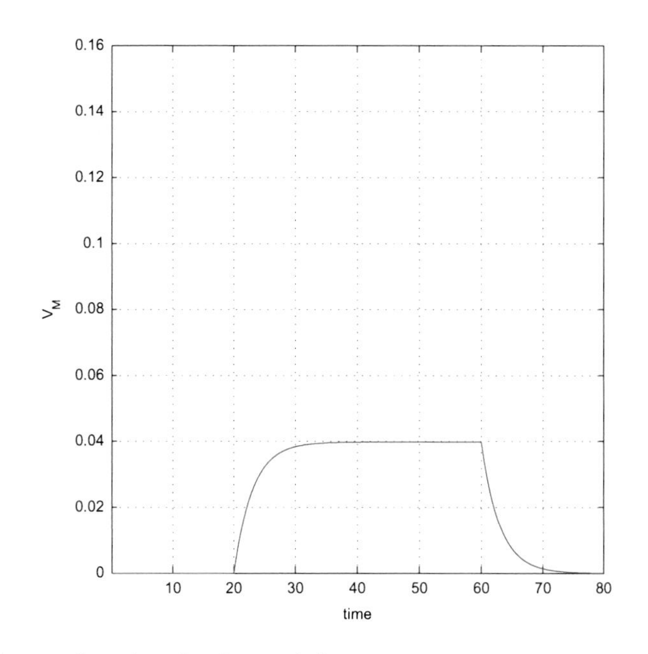

Fig. 1.5 Diagram of membrane's voltage variation

1.3 Describing Membrane's Voltage Dynamics with Cable's Equation

The previous analysis has considered that the neurons have a spherical shape. This assumption does not hold for neurons' axon and in the case of neurons' dendrites where the neurons' shape is better approximated by a cylinder. In the latter case it is considered that the neurons have the shape of a long cylinder or cable, or cable of radius a as depicted in Fig. 1.6.

The equation of the membrane's voltage according to the cable's model is a partial differential equation that describes the voltage's spatiotemporal variation $V_M(x,t)$. Voltage $V_M(x,t)$ depends also on the coaxial currents I_L that pass through the cell [21].

The aggregate resistance of the cable, for radius a and length Δx is given by

$$R_L = \frac{r_L \Delta_x}{\pi \alpha^2} \tag{1.33}$$

where r_L is a special resistance. According to Ohm's law, for the change of voltage along the cables's axis holds [16, 65]

$$V_M(x + \Delta x, t) - V_M(x,t) = -I_L(x,t)R_L = -I_L(x,t)\frac{\Delta x}{\pi \alpha^2} r_L \tag{1.34}$$

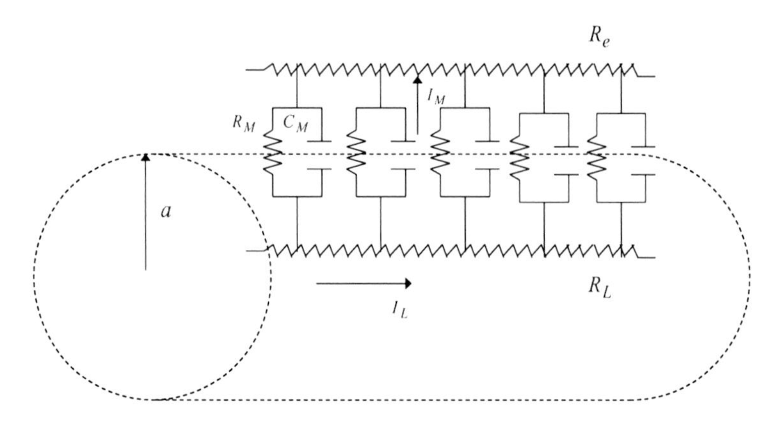

Fig. 1.6 Cable equivalent circuit of the neuron's membrane

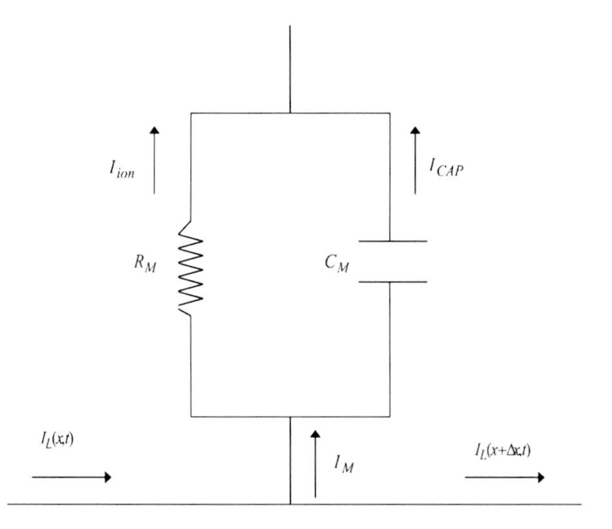

Fig. 1.7 Currents of the membrane at point x

Therefore, for $\Delta_x \to 0$ it holds

$$I_L(x,t) = -\frac{\pi \alpha^2}{r_L} \frac{\partial V_M}{\partial x}(x,t) \tag{1.35}$$

Assuming that along the cable (dendrite) there are parallel ion channels, which are noted by the resistance R_M, for the variation of the current along the neuron's axis it holds (see Fig. 1.7).

$$I_L(x + \Delta x, t) = I_L(x,t) - I_{\text{ion}}(x,t) - I_{\text{cap}}(x,t) \tag{1.36}$$

or equivalently

$$I_{cap}(x,t) + I_{ion}(x,t) = -I_L(x+\Delta x,t) + I_L(x,t) \tag{1.37}$$

out of which one obtains [16, 65]

$$(2\pi\alpha\Delta x)c_M\frac{\partial V_M}{\partial t} + (2\pi\alpha\Delta x)i_{ion} = \frac{\pi\alpha^2}{r_L}\frac{\partial V_M}{\partial x}(x+\Delta x,t) - \frac{\pi\alpha^2}{r_L}\frac{\partial V_M}{\partial x}(x,t) \tag{1.38}$$

After dividing both sides of the equation with $2\pi\alpha\Delta x$ and taking $\Delta x \Rightarrow 0$ cable's equation is produced

$$c_M\frac{\partial V_M}{\partial t} = \frac{\alpha}{2r_L}\frac{\partial^2 V_M}{\partial x^2} - i_{ion} \tag{1.39}$$

Setting $i_{ion} = \frac{V_M(x,t)}{r_M}$ one gets

$$c_M\frac{\partial V_M}{\partial t} = \frac{\alpha}{2r_L}\frac{\partial^2 V_M}{\partial x^2} - \frac{V_M}{r_M} \tag{1.40}$$

Cable's equation for the neural cell is formulated as follows [16, 65]

$$\tau_M\frac{\partial V_M}{\partial t} = \lambda^2\frac{\partial^2 V_M}{\partial x^2} - V_M \tag{1.41}$$

where $\lambda = \sqrt{\frac{\alpha r_M}{2r_L}}$ and $\tau_M = c_M r_M$.

Next, it is assumed that the neural cell has the form of a semi-infinite cable for $x > 0$, in which a step current $I_M(0) = I_0$ is injected, while the associated boundary condition is $V_M(0)$. The solution of the partial differential equation given in Eq. (1.41) provides the spatiotemporal variation of the voltage $V(x,t)$.

By examining the solution of the PDE in steady state, i.e. when $\frac{\partial V_M}{\partial t} = 0$, the partial differential equation becomes an ordinary differential equation of the form

$$\lambda^2\frac{d^2V_{ss}}{dx^2} - V_{ss} = 0 \tag{1.42}$$

From Eq. (1.35), using that $I_0 = -\frac{\pi\alpha^2}{r_L}\frac{\partial V_M}{\partial x}$ one obtains the boundary condition

$$\frac{dV_{ss}}{dx}(0) = -\frac{r_L}{\pi\alpha^2}I_0 \tag{1.43}$$

while another boundary condition is that V tends to 0 as $x \to \infty$. The solution of the steady state (spatial component) of the cable equation is finally written as

$$V_{ss}(x) = \frac{r_L I_0 \lambda}{\pi\alpha^2}e^{\frac{-x}{\lambda}} \tag{1.44}$$

Time constant λ affects the variation of voltage.

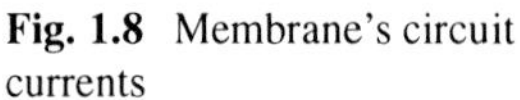

Fig. 1.8 Membrane's circuit currents

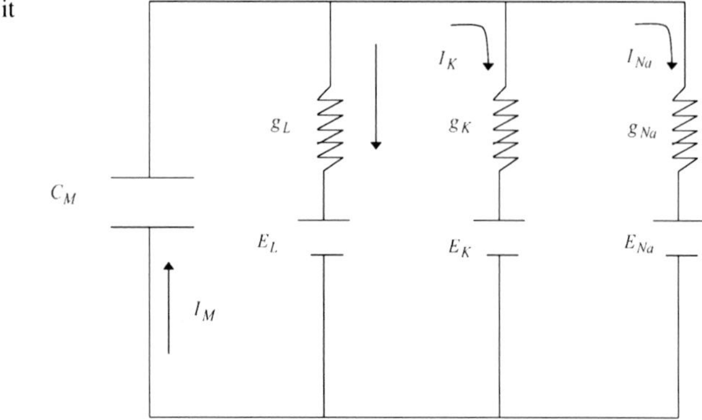

1.4 The Hodgkin–Huxley Model

1.4.1 Derivation of the Hodgkin–Huxley Model

The Hodgkin–Huxley model is obtained from a modification of the cable's PDE, which describes the change of the voltage $V(x,t)$ along dendrites' axis. It holds that [15, 39]

$$I_L = I_{\text{cap}} + I_{\text{ion}} \tag{1.45}$$

Cable's equation comprises as inputs the currents which are developed in the ions' channels

$$\frac{\alpha}{2r_L}\frac{\partial^2 V_M}{\partial^2 x^2} = c_M \frac{\partial V_M}{\partial t} + I_{\text{K}} + I_{\text{Na}} + I_L \tag{1.46}$$

which can be also written in the following form

$$c_M \frac{\partial V_M}{\partial t} = \frac{\alpha}{2r_L}\frac{\partial^2 V_M}{\partial^2 x^2} - g_{\text{K}}(V_M - E_{\text{K}}) - g_N(V_M - E_{\text{Na}}) - g_L(V_M - E_L) \tag{1.47}$$

where g_k is the conductance of the K^+ channel, g_{Na} is the conductance of the Na^+ channel, and g_L is the conductance of the leakage channel (Fig. 1.8).

Through an identification procedure, Hodgkin and Huxley derived specific relations for the conductances in the ionic channels K^+ and Na^+.

$$g_{\text{K}} = \bar{g_{\text{K}}} n^4 \quad g_{\text{Na}} = \bar{g}_{\text{Na}} m^3 h \tag{1.48}$$

where $\bar{g}_{\text{K}}$ and $\bar{g}_{\text{Na}}$ are the maximum values for the conductances and n, m, h are gating variables which take values between 0 and 1. Thus, n^4 represents the probability channel K^+ to be open. Moreover, m^3 represents the probability that the sodium activation gate is open, and h is the probability that the sodium inactivation gate is open. It is noted that independently from the model derived by Hodgkin and Huxley it is possible to obtain the values of the parameters of the neuron membrane's model through nonlinear estimation and Kalman Filtering methods [84, 99, 175].

The change in time of parameters m, h, and n in time is given by

$$\begin{aligned} \frac{dn}{dt} &= a_n(V)(1-n) - \beta_n(V)\cdot n = (n_\infty(V) - n)/\tau_n(V) \\ \frac{dm}{dt} &= a_m(V)(1-m) - \beta_m(V)\cdot m = (m_\infty(V) - m)/\tau_m(V) \\ \frac{dh}{dt} &= a_h(V)(1-h) - \beta_h(V)\cdot h = (h_\infty(V) - h)/\tau_h(V) \end{aligned} \tag{1.49}$$

For the boundary conditions of the aforementioned differential equations one has (denoting the general variable $X = n, m$ or h) that

$$X_\infty(V) = \frac{a_x(V)}{a_x(V)+\beta_x(V)} \quad \tau_X(V) = \frac{1}{a_x(V)+\beta_x(V)} \tag{1.50}$$

Moreover, the following parameter values have been experimentally used [16]

$$\begin{aligned} \bar{g}_{\mathrm{Na}} &= 120\,\mathrm{mS/cm^3} \;\; \bar{g}_{\mathrm{K}} = 36\,\mathrm{mS/cm^3} \;\; \bar{g}_L = 0.3\,\mathrm{mS/cm^3} \\ E_{\mathrm{Na}} &= 50\,\mathrm{mV} \;\; E_{\mathrm{K}} = -77\,\mathrm{mV} \;\; E_L = 54.4\,\mathrm{mV} \end{aligned}$$

$$\begin{aligned} a_n(V) &= 0.01(V+55)/(1-\exp(-(V+55)/10)) \\ \beta_n(V) &= 0.125\exp(-(V+65)/80) \\ a_m(V) &= 0.1(V+40)/(1-\exp(-(V+40)/10)) \\ \beta_m(V) &= 4\exp(-(V+65)/18) \\ a_h(V) &= 0.07\exp(-(V+65)/20) \\ \beta_h(V) &= 1/(1+\exp(-(V+35)/10)) \end{aligned} \tag{1.51}$$

1.4.2 Outline of the Hodgkin–Huxley Equations

By considering that there is no spatial variation of $V(x,t)$, i.e. that $\frac{\partial V(x,t)}{\partial t} = 0$, and moreover that the conductances $\bar{g}_k$, $\bar{g}_{\mathrm{Na}}$, and $\bar{g}_L$ vary in time, the Hodgkin–Huxley equation becomes an ordinary differential equation

$$\begin{aligned} c_M \frac{dV}{dt} &= -\bar{g}_{\mathrm{Na}} m^3 h (V - E_{\mathrm{Na}}) - \bar{g}_{\mathrm{K}} n^4 (V - E_{\mathrm{K}}) - \bar{g}_L (V - E_L) \\ \frac{dn}{dt} &= \phi[a_n(V)(1-n) - \beta_n(V)n] \\ \frac{dm}{dt} &= \phi[a_m(V)(1-m) - \beta_m(V)m] \\ \frac{dh}{dt} &= \phi[a_h(V)(1-h) - \beta_h(V)h] \end{aligned} \tag{1.52}$$

Coefficient ϕ varies with the system's temperature according to the relation

$$\phi = {Q_{10}}^{(T - T_{\mathrm{base}})/10} \text{ where } T_{\mathrm{base}} = 6.3°\mathrm{C} \text{ and } Q_{10} = 3 \tag{1.53}$$

If the currents I_L, I_K, I_{Na} which are applied to the membrane are lower than a specific threshold, then the variation of the membrane's potential returns fast to the equilibrium. If the currents I_L, I_K, I_{Na} which are applied to the membrane are greater than a specific threshold, then the variation of the membrane's potential exhibits oscillations.

1.5 The FitzHugh–Nagumo Model of Neurons

The FitzHugh–Nagumo model is a two-dimensional simplified form of the Hodgkin–Huxley model and in several cases represents efficiently the membrane's voltage dynamics in neurons [16, 55]. The model captures all the qualitative properties of the Hodgkin–Huxley dynamics, in the sense that it results in the same bifurcation diagram (loci of fixed points with respect to model's parameters). In Eq. (1.54) variable V stands for the membrane's voltage, while variable w is known as recovery variable and is associated with the ionic currents of the membrane. Finally, variable I is an external input current.

The FitzHugh–Nagumo model comprises two differential equations

$$\begin{aligned} \frac{dV}{dt} &= V(V-a)(1-V) - w + I \\ \frac{dw}{dt} &= \epsilon(V - \gamma w) \end{aligned} \tag{1.54}$$

where typically $0 < a < 1$, $\epsilon > 0$, and $\gamma \geq 0$. The FitzHugh–Nagumo model is equivalent to Van Der Pol oscillator, the latter being described by

$$C\frac{dV}{dt} = -F(V) - J \tag{1.55}$$

$$L\frac{dJ}{dt} = V \tag{1.56}$$

As shown in Fig. 1.9 the model of the FitzHugh–Nagumo neuron, for specific values of its parameters or suitable feedback control input I may exhibit sustained oscillations in its state variables and in such a case limit cycles will also appear in the associated phase diagram. The detailed stability analysis of models of biological neurons will be given in Chap. 3.

1.6 The Morris–Lecar Model of Neurons

The dynamics of a neuron (cell's membrane and the associated potential) can be also studied using not cable's equation for the Hodgkin–Huxley model but a simpler model that is called Morris–Lecar model [16,112]. The Morris–Lecar model

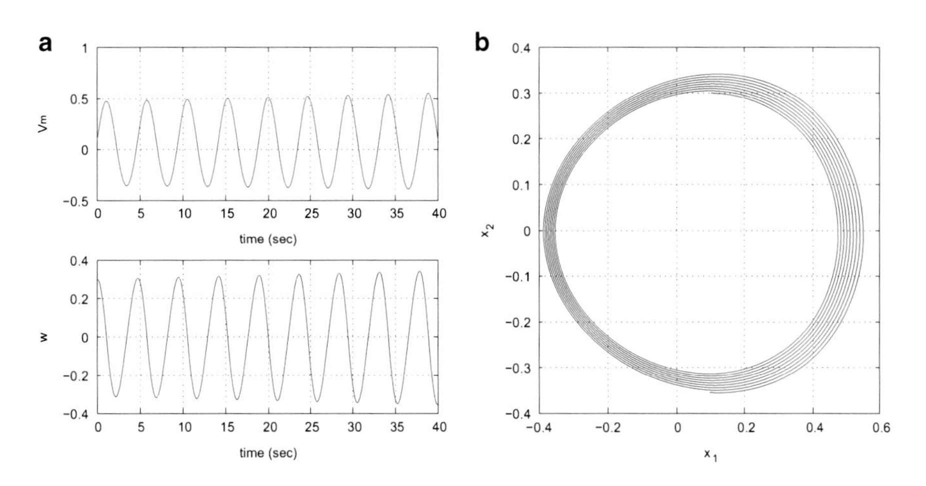

Fig. 1.9 (**a**) State variables of the FitzHugh–Nagumo neuron exhibiting sustained oscillations, (**b**) limit cycles in the phase diagram of the FitzHugh–Nagumo neuron

examines the variation of the membrane's potential in a neuron as a function of the conductances of the associated ionic channels and contains also a second differential equation which expresses the probability an ionic channel to be open. The model's equations are:

$$\begin{aligned} C_M \frac{dV}{dt} &= I_{\text{app}} - g_L(V - E_L) - g_{k\eta}(V - E_k) - g_{\text{Ca}} m_\infty(V)(V - E_{\text{Ca}}) \\ &= I_{\text{app}} - I_{\text{ion}}(V, n) \\ \frac{d\eta}{dt} &= \phi(\eta_\infty(V) - \eta)/\tau_\eta(V) \end{aligned} \tag{1.57}$$

where

$$\begin{aligned} m_\infty(V) &= \frac{1}{2}\left[1 + \tan h\left(\frac{V - V_1}{V_2}\right)\right] \\ \tau_\eta(V) &= \frac{1}{\cos h((V - V_1)/2V_4)} \\ m_\infty(V) &= \frac{1}{2}\left[1 + \tan h\left(\frac{V - V_3}{V_4}\right)\right] \end{aligned} \tag{1.58}$$

Here, parameters V_i, $i = 1, \cdots, 4$ are chosen to fit data obtained from an identification procedure. Although it is more simple comparing to the Hodgkin–Huxley model (the latter expressed either in a PDE or set of ODEs form), the Morris–Lecar model captures efficiently several properties of the biological neurons.

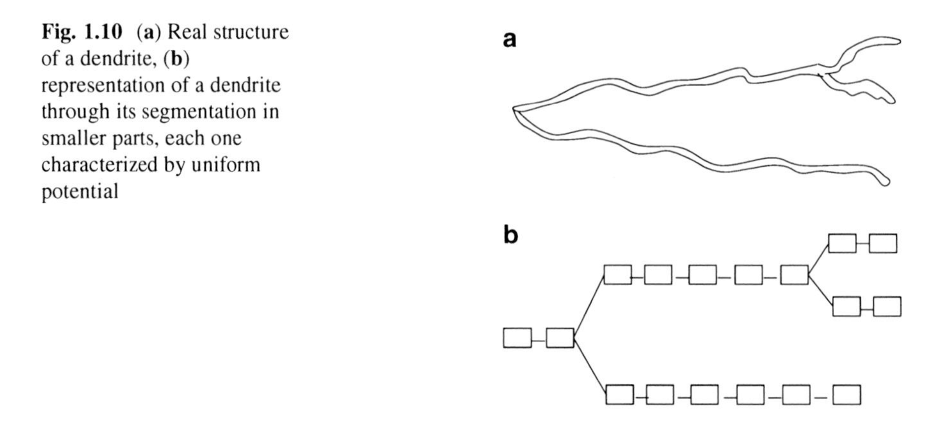

Fig. 1.10 (**a**) Real structure of a dendrite, (**b**) representation of a dendrite through its segmentation in smaller parts, each one characterized by uniform potential

1.7 Modelling Dendrites in Terms of Electrical Circuits

1.7.1 Dendrites

The greater part of the cumulative surface of the membrane in neural cells is occupied by dendrites. Dendrites permit the creation of connections between different neurons. Moreover, dendrites direct signals of a specific voltage amplitude from the neighboring neurons to the soma of a particular neuron. The dendrites, as parts of the membrane, have ionic channels which open or close according to the change of the membrane's potential [16, 65]. As a final result, dendrites define the response of the neurons to synaptic inputs (voltages $V(x,t)$; Fig. 1.10).

For the study of the voltage dynamics in dendrites, an established approach is to partition their tree-like structure in smaller parts which are characterized by spatially uniform voltage potential, i.e. $\frac{\partial V(x,t)}{\partial x}=0$. It is considered that there exist differences in the value and distribution of the voltage $V(x,t)$, in different parts of the dendrites (Fig. 1.11).

For each smaller part in the dendrite that is characterized by spatially uniform potential, the major parameters are: the cylinder's radius α_i, the length L_i, the membrane's potential V_i, the capacitance (normalized per unit of surface) c_i, and the resistance of the membrane (normalized per unit of surface) r_{L_i}. It is assumed that in each compartment there is an electrode to which external current $I_{\text{electrode}}$ is applied.

From the analysis of the circuit of the elementary part of the dendrite one has

$$I^i_{\text{cap}}+I^i_{\text{ion}}=I^i_{\text{long}}+I^i_{\text{electrode}} \tag{1.59}$$

where I^i_{cap} and I^i_{ion} are the capacitance and ionic currents per unit area of membrane for segment i, and additionally

$$I^i_{\text{cap}}=C_i\frac{dV_i}{dt},\quad I^i_{\text{ion}}=\frac{V_i}{r^i_M} \tag{1.60}$$

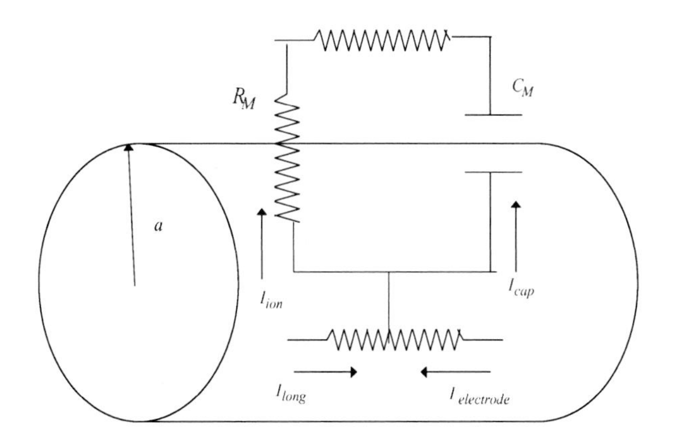

Fig. 1.11 Electric circuit of local segment of a dendrite

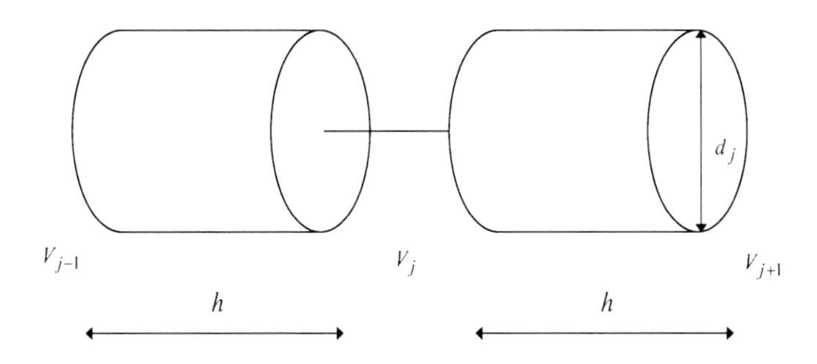

Fig. 1.12 Segments of the membrane

Next, two interconnected elementary parts of the dendrite are examined (Fig. 1.12). The associated longitudinal resistance R_{long} is equal to the sum of the resistances of two successive semi-cylinders

$$R_{\text{long}} = \frac{r_L L_1}{2\pi\alpha^2} + \frac{r_L L_2}{2\pi\alpha^2} \tag{1.61}$$

Defining the conductances of the two elementary parts

$$g_{1,2} = \frac{\alpha_1 \alpha_2^2}{r_L L_1 (a_2^2 L_1 + a_1^2 L_2)} \tag{1.62}$$

$$g_{2,1} = \frac{\alpha_2 \alpha_1^2}{r_L L_1 (a_2^2 L_1 + a_1^2 L_2)} \tag{1.63}$$

the associated currents in the two segments are (there is longitudinal voltage variation while the membrane's potential in the i-th segment is spatially unchanged)

$$\begin{aligned} I_{\text{long}}^1 &= g_{1,2}(V_2 - V_1) \\ I_{\text{long}}^2 &= g_{2,1}(V_1 - V_2) \end{aligned} \tag{1.64}$$

The current of the i-th electrode is

$$i^i_{\text{electrode}} = \frac{I_{\text{electrode}}}{A_i} \tag{1.65}$$

with $A_i = 2\pi\alpha_i L$ denoting the surface of segment i. Thus, one arrives at a set of coupled differential equations

$$\begin{aligned} c_1\frac{dV_1}{dt} + \frac{V_1}{r_{m_1}} &= g_{1,2}(V_2 - V_1) + \frac{I^1_{\text{electrode}}}{A_1} \\ c_2\frac{dV_2}{dt} + \frac{V_2}{r_{m_2}} &= g_{2,1}(V_1 - V_2) + \frac{I^1_{\text{electrode}}}{A_2} \end{aligned} \tag{1.66}$$

or equivalently

$$\begin{aligned} c_1\frac{dV_1}{dt} + \frac{V_1}{r_{M_1}} &= \frac{V_2 - V_1}{r_1} + i_1 \\ c_2\frac{dV_2}{dt} + \frac{V_2}{r_{M_2}} &= \frac{V_1 - V_2}{r_2} + i_2 \end{aligned} \tag{1.67}$$

Next, it is assumed that the current is injected in only one out of the two cylinders, that is $i_2 = 0$. It is assumed that $r_1 = r_2 = r$ and $r_{M_1} = r_{M_2} = r_M$. From Eqs. (1.66) and (1.67) it holds

$$c_1\frac{dV_1}{dt} + c_2\frac{dV_2}{dt} + \frac{V_1 + V_2}{r_M} = i_1 \tag{1.68}$$

Thus, finally about the segment's resistance one has

$$\frac{V_1}{i_1} = \frac{r_M(r + r_M)}{r + 2r_M} \tag{1.69}$$

while the variation of voltage V at the i-th segment is generalized as follows

$$C_j\frac{dV_j}{dt} = -\frac{V_j}{R_j} + \sum_{k,j}\frac{V_k - V_j}{R_{jk}} + I_j \tag{1.70}$$

where R_{kj} denotes the resistance of the connection between the k-th and the j-th segment.

1.7.2 Computing Cable's Equation in Neurons

It is assumed that the cable is defined in the interval $(0, l)$ and that its cross-section is circular with diameter $d(x)$. The cable is segmented in n-parts and the distance from the origin is defined as $x_j = j \cdot h$, where $h = \frac{l}{n}$ is the elementary length. Each part has surface equal to $A_j = \pi d_j h$ where $d_j = d(x_j)$ is the diameter and the associated surface of the cross-section is equal to $\pi\frac{d_j^2}{4}$. It is assumed that the special resistance and the capacitance of the membrane are c_M and r_M, respectively, while the longitudinal resistance is r_L. By ignoring the final points, the voltage varies according to the following relation [16,65]

$$c_M A_j \frac{dV_j}{dt} = -\frac{V_j}{r_M/A_j} + \frac{V_{j+1}-V_j}{4r_L h/(\pi d_{j+1}^2)} + \frac{V_{j-1}-V_j}{4r_L h/(\pi d_{j+1}^2)} \quad (1.71)$$

Dividing by h the two last terms (coupling terms) one obtains

$$\frac{\pi}{h}\left(d_{j+1}^2\left(\frac{V_{j+1}-V_j}{4r_L h}\right) - d_j^2\left(\frac{V_j-V_{j-1}}{4r_L h}\right)\right) \quad (1.72)$$

As $h \to 0$ one obtains the diffusion operator

$$\frac{\pi}{4r_L}\frac{\partial}{\partial x}(d^2(x)\frac{\partial V}{\partial x}) \quad (1.73)$$

Moreover, by dividing both members of Eq. (1.71) with πdx, cable's equation comes to the form

$$c_M \frac{\partial V}{\partial t} = -\frac{V}{r_M} + \frac{1}{4r_L d(x)}\frac{\partial}{\partial x}(d^2(x)\frac{\partial V}{\partial x}) \quad (1.74)$$

It is noted that the term $\frac{\pi d_j^2(V_{j-1}-V_j)}{4r_L h}$ has current dimensions and at the limit $h \to 0$ it corresponds to longitudinal current $I_L = -\frac{\pi d^2(x)}{4r_L}\frac{\partial V}{\partial x}$.

If $d(x) = d$ is the constant diameter of the segment, then it is possible to multiply the two segments with r_M, and this leads to the linear cable equation

$$\tau \frac{\partial V}{\partial t} = -V + \lambda^2 \frac{\partial^2 V}{\partial x^2} \quad (1.75)$$

where $\lambda = \sqrt{\frac{d r_M}{4r_L}}$ is the space constant and $\tau = r_M c_M$ is the time constant.

Considering that as $x \to \infty$ the voltage of the membrane approaches zero, then the steady-state solution of the cable's equation takes the form $V(x) = Ae^{-\frac{x}{\lambda}}$.

1.7.3 The Infinite Length Cable

The relation given in Eq. (1.75) is modified in case of an infinite length cable. Including an excitation current to the previous equation

$$\tau \frac{\partial V}{\partial t} + V(x,t) - \lambda^2 \frac{\partial^2 V}{\partial x^2} = r_M I(x,t) \quad (1.76)$$

where $-\infty < x < +\infty$ and $V(x,0) = V_0(x)$. The current $I(x,t)$ has units $\mu A/\text{cm}^2$. For this linear differential equation of the 1st order, the solution is [16,65]

$$\hat{V}(k,t) = e^{\frac{-(1+\lambda^2 k^2)t}{\tau}}\hat{V}_0(k) + (\frac{r_M}{\tau})\int_0^t e^{-(1+\lambda^2 k^2)\frac{(t-s)}{\tau}}\hat{I}(\kappa,s)ds \quad (1.77)$$

where

$$\hat{V}(k,t) = \int_{-\infty}^{+\infty} e^{-ikx} V(x,t)dx$$
$$\hat{V}_0 = \int_{-\infty}^{+\infty} e^{-ikx} V_o(x)dx \tag{1.78}$$
$$\hat{I}(k,t) = \int_{-\infty}^{+\infty} e^{-ikx} I(x,t)dx$$

Using the definition of the inverse Fourier transform one has

$$V(x,t) = \frac{1}{2\pi}\int_{-\infty}^{+\infty} e^{ikx}\hat{V}(k,t)dk \tag{1.79}$$

one finds that the solution $V(x,t)$ is given by

$$V(x,t) = \int_{-\infty}^{+\infty} G(x-y,t)V_o(y)dy + \frac{r_M}{\tau}\int_0^t \int_{-\infty}^{+\infty} G(x-y,t-s)I(y,s)dyds \tag{1.80}$$

where the Green function $G(x,t)$ is defined as

$$G(x,t) = \frac{1}{\sqrt{\frac{4\pi\lambda^2 t}{\tau}}} e^{\frac{-t}{\tau}} e^{-x^2/(\frac{4\lambda^2 t}{\tau})} \tag{1.81}$$

If the membrane is in rest state at time instant 0, i.e. $V_0(x) = 0$ and for $I(x,t) = I_0\delta(x)\delta(t)$, then

$$V(x,t) = \frac{r_M I_0}{\tau\lambda\sqrt{\frac{4\pi t}{\tau}}} e^{\left(-\frac{tx^2}{4\lambda^2\tau}\right)} e^{\left(-\frac{t}{\tau}\right)} \tag{1.82}$$

If $I(x,t) = I_o u(x)$, i.e. a step input is applied at position x, then

$$V(x,t) = \frac{r_M I_o \lambda}{4}\left[e^{\frac{-x}{\lambda}} erfc\left(\frac{x\sqrt{\tau}}{2\lambda\sqrt{t}} - \sqrt{\frac{t}{\tau}}\right) - e^{\frac{x}{\lambda}} erfc\left(\frac{x\sqrt{\tau}}{2\lambda\sqrt{t}} + \sqrt{\frac{t}{\tau}}\right)\right] \tag{1.83}$$

where $erfc = \frac{2}{\sqrt{\pi}}\int_x^{\infty} e^{-y^2}dy$, while in steady-state the solution becomes

$$V_{ss}(x) = \frac{r_M I_0}{4\lambda} e^{\frac{-|x|}{\lambda}} \tag{1.84}$$

1.8 Ion Channels and Their Characteristics

1.8.1 Types of Ion Channels

In the set of equations that constitute the Hodgkin–Huxley model one can distinguish the K^+ and Na^+ ion channels, whereas in the Morris–Lecar model one can distinguish the Ca^+ model. However, there may be more ion channels across the neuron's membrane which are described as follows [16, 60, 65].

According to the Hodgkin–Huxley formulation, the aggregate current that goes through an ion channel is given by a relation that has the generic form:

$$I_{\text{channel}} = m^p h^q I_{\text{drive}}(V) \tag{1.85}$$

where m and h are variables taking values between 0 and 1, p and q are non-negative integers, and V is the membrane's potential. In general, the more the membrane's potential grows, the more parameter h gets smaller (inactivation) and the more parameter m grows (activation). Sodium channels of the Hodgkin–Huxley model have both activation and inactivation.

Channel K^+ of the Hodgkin–Huxley model, and channels Ca^{2+} and K^+ of the Morris–Lecar model do not exhibit inactivation. In general, the drive current can take the following form

$$\begin{aligned} I_{\text{lin}} &= g_{\max}(V - V_{\max}) \\ I_{\text{cfe}} &= P_{\max}\frac{z^2F^2}{RT}V\left(\frac{[C]_{\text{in}} - [C]_{\text{out}}e^{\frac{-zVF}{RT}}}{1-e^{\frac{-zVF}{RT}}}\right) \end{aligned} \tag{1.86}$$

where the units of $g_{\max}$ are Siemens/cm^2 and cfe means constant field equation. In the equation about the channel's current Eq. (1.85), the variation of parameters m and h is given by differential equations of the form

$$\begin{aligned} \frac{dx}{dt} &= a_x(1-x) - b_x x \\ \frac{dx}{dt} &= (x_\infty - x)/\tau_x \end{aligned} \tag{1.87}$$

while for the voltage-dependent parameters a_x and b_x, as well as for parameters x_∞ and τ_x the functions which approximate them have the form

$$\begin{aligned} F_e(V,A,B,C) &= Ae^{(V-B)/C} \\ F_l(V,A,B,C) &= \frac{A(V-B)}{1-e^{(V-B)/C}} \\ F_h(V,A,B,C) &= A/(1+e^{-(V-B)/C}) \end{aligned} \tag{1.88}$$

whereas parameters x_∞ and τ_x are described by

$$\begin{aligned} x_\infty(V) &= \frac{1}{(1+e^{(V-V_T)/K})} \\ \tau_x(V) &= \tau_{\min} + \tau_{\text{amp}}/\cosh\left(\frac{V-V_{\max}}{\sigma}\right) \end{aligned} \tag{1.89}$$

A key parameter for the appearance and density of voltage pulses is the leak potential E_L. According to the value of this parameter, at points of the membrane, the variation of the voltage V in time may take the form of spikes or bursts.

1.8.2 Sodium Channels Na^+

Sodium channels are distinguished into (1) transient or fast sodium current channels, (2) persistent or slow sodium channel currents. As an example of persistent sodium channel currents one can consider the ones appearing in the so-called pre-Bötzinger complex, that is neurons that coordinate respiration pulses. The associated model has the form described in Eq. (1.52) of the Hodgkin–Huxley dynamics and is given by

$$\begin{gathered} C_m \frac{dV}{dt} = -g_L(V - E_L) - g_k n^4(V - E_k) \\ -g_{\mathrm{Na}} m_\infty(V)^3(1 - \eta)(V - E_{\mathrm{Na}}) - g_{\mathrm{Nap}} w_\infty(V) h(V - E_{\mathrm{Na}}) \\ \frac{d\eta}{dt} = (n_\infty(V) - \eta)/\tau_\eta(V) \\ \frac{dh}{dt} = (h_\infty(V) - h)/\tau_\eta(V) \end{gathered} \tag{1.90}$$

The leak potential E_L is a key parameter for introducing bursting in such a model.

1.8.3 Calcium Channels Ca^{2+}

The equation of the current of the associated channel is given by the constant field equation, given in Eq. (1.86). Calcium channels are distinguished in two large categories:

1. T-type calcium currents $I_{\mathrm{Ca},T}$ which exhibit inactivation with respect to parameter h in the current equation $I_{\mathrm{Channel}} = m^p h^q I_{\mathrm{drive}}(V)$.
2. L-type calcium currents $I_{\mathrm{Ca},L}$ which do not exhibit inactivation with respect to parameter h in the current equation $I_{\mathrm{Channel}} = m^p h^q I_{\mathrm{drive}}(V)$.

T-currents exhibit bursts and subthreshold oscillations. They are described by a set of equations of the form

$$\begin{gathered} C \frac{dV}{dt} = I_o g_L(V - E_L) - I_T \\ \frac{dh}{dt} = (h_\infty(V) - h)/\tau_h(V) \\ I_T = m_\infty(V)^2 h I cfe(V, [\mathrm{Ca}]_o, [\mathrm{Ca}]_i) \\ m_\infty(V) = 1/(1 + \exp(-(V + 59)/6.2)) \\ h_\infty(V) = 1/(1 + \exp((V + 83)/4)) \\ \tau_h(V) = 22.7 + 0.27/[\exp((V + 48)/4) + \exp(-(V + 407/50))] \end{gathered} \tag{1.91}$$

The response of this model depends on the activation variable m and on the deactivation variable h.

1.8.4 *Voltage-Gated Potassium Channels* K^+

These are ion channels which open or close depending on the value of the membrane's potential V.

1.8.4.1 A-Current

The transient sodium current and the delayed rectifier current are similar to those of the Hodgkin–Huxley model, although in the present case they are faster. To this model an additional potassium current is introduced.

An example about the voltage gated channel comes from a model which has the form

$$I_A = g_A a^3 b(V - E_A) \tag{1.92}$$

The activation variable a increases as current increases. The de-activation variable b decreases as current increases. Steady-state current is almost zero.

$$I_{A,ss} = g_A \alpha_\infty(V)^3 b(\infty)(V)(V - E_A) \tag{1.93}$$

The conductivity variable g_A affects the number of spikes that the membrane may exhibit and is defined by the relation $g_A + g_{\mathrm{K}} = g_{\mathrm{total}} = 67.7$, where a typical value for parameter g_{K} is $g_{\mathrm{K}} = 20$.

1.8.4.2 M-Current

These are slow K^+ currents which may affect the firing rate (spiking) of the membrane. The M current and the associated with it slow currents K^+ can inhibit neurons' firing. Apart from slowing down the neuron's firing rate the M current affects also the steady-state value of the neuron's voltage [140].

If there is no M-current (g_m very small), then the model may exhibit bifurcations on limit cycles. As g_m grows then Hopf bifurcations may appear.

With reference to the dynamic model of Eq. (1.90) in the present model one should consider to introduce the following input currents [16]

$$\begin{aligned} I_M &= \bar{g}_M p(V - E_k) \\ \tfrac{dp}{dt} &= (\mu_\infty(V) - \mu)/\tau_p(V) \\ \mu_\infty(V) &= \tfrac{1}{1+e^{-(V+35)/100}} \\ \tau_p(V) &= \tfrac{\tau_{\mathrm{mac}}}{3.5+e^{(V+35/20)}+e^{-(V+35/20)}} \end{aligned} \tag{1.94}$$

1.8.4.3 Inward Rectifier

This current appears when the neuron becomes hyperpolarized. The current's equation has the form

$$\begin{aligned} I_{\mathrm{Kin}} &= g_{\mathrm{Kin}} h(V)(V - E_k) \\ h(V) &= 1/[1 + \exp((V - V_{\mathrm{th}})/\mathrm{K})] \end{aligned} \tag{1.95}$$

Typical values of the parameters appearing in the above model are $V_{\mathrm{th}} = -85\,\mathrm{mV}$ and $\mathrm{K} = 5\,\mathrm{mV}$. Considering the existence of a leak current, the steady-state current becomes

$$I = g_L(V - E_L) + g_{\mathrm{Kin}} h(V)(V - E_{\mathrm{K}}) \tag{1.96}$$

Differentiating with respect to V gives

$$\frac{dI}{dt} = g_L + g_{\mathrm{Kin}} h(V) + g_{\mathrm{Kin}} h'(V)(V - E_{\mathrm{K}}) \tag{1.97}$$

1.8.5 *Voltage Sags*

After hyperpolarization of the neuron takes place then an inward current appears with values between −43 and 0 mV. The current dynamics is [16]

$$\begin{aligned} I_h &= g_h y(V + 43) \\ \frac{dy}{dt} &= (y_\infty(V) - y)/\tau_y(V) \\ y_\infty(V) &= 1/[1 + \exp((V - V_{\mathrm{th}})/\mathrm{K})] \\ \tau_y(V) &= t_o \mathrm{sech}((V - V_m)/b) \end{aligned} \tag{1.98}$$

1.8.6 *Currents and Concentration of Ions*

The assumption that the concentration of ions in the inner and outer part of the membrane remains constant is in general acceptable, apart from the case of Ca^{2+} ions. The concentration of Ca^{2+} ions in the inner part of the membrane is very small (of the order of 10^{-4} mM) and a small inflow of Ca^{2+} suffices to change this concentration significantly. Due to this, cascading reactions in the internal part of the cell are caused.

Similarly, the concentration of ions K^+ in the outer part of the membrane is very small and in case that several neurons fire simultaneously then this concentration may change.

1.8.7 Calcium-Dependent Channels

Modelling of channels where there is transfer of Ca^{2+} is important because the concentration of these ions affects several signaling mechanisms. The two most important channels of this type are: (1) $I_{K,Ca}$ which is the calcium dependent potassium channel, and (2) I_{can} which is the inward calcium-dependent current. The first current appears in several neurons and is responsible for low hyperpolarization (AHP) and spike-frequency adaptation. It is known as AHP current.

1. Calcium-dependent potassium channel—the afterhyperpolarization

   A typical model for the $I_{K,Ca}$ current comes from [140] and is given by

$$\begin{aligned} I_{K,Ca} &= g_{K,Ca} m^2 (V - E_k) \\ \tfrac{dm}{dt} &= (m_\infty(c) - m)/\tau_m(c) \\ m_\infty(c) &= \tfrac{c^2}{k^2 + c^2} \\ \tau_m(c) &= \max(\tau_{min}, \tau_o/(1 + (c/k)^2)) \end{aligned} \tag{1.99}$$

   Typical values for the parameters of the above model are $k = 0.0025$ mM, $\tau_{min} = 0.1$ ms, τ_o variable. The firing rate f is affected by the value of the current $I_{K,Ca}$ according to the relation

$$\frac{df}{dI} = \frac{F'(I - a\cdot g\cdot f)}{1 + a\cdot g F'(I-)a\cdot g\cdot f)} \tag{1.100}$$

   and for large F' this relation is written approximately as

$$\frac{df}{dI} = \frac{1}{a\cdot g} \tag{1.101}$$

   A solution which has been also proposed about the firing rate is $f(I) = -\kappa + \sqrt{\kappa^2 + A^2 I}$, where $\kappa = A^2 ag/2$. Moreover about the voltage spikes that appear in the calcium-dependent potassium channel it holds

$$\begin{aligned} C\tfrac{dV}{dt} &= I - I_{fast} - g\cdot z(V - E_k) \\ \tfrac{dz}{dt} &= \epsilon[q(V)(1 - z) - z] \end{aligned} \tag{1.102}$$

2. Calcium-activated nonspecific cation current

   This case is concerned with the inwards calcium current I_{CAN}. This is described by the relations

$$I_{CAN} = g_{CAN} m_{CAN}^p (V - E_{CAN}) \tag{1.103}$$

   where the gate parameter m_{CAN} varies according to the dynamic model

$$\frac{dm_{CAN}}{dt} = (g(c)(1 - m_{CAN}) - m_{CAN})/\tau_{CAN} \tag{1.104}$$

where $g(c)$ is a monotonous function of the concentration of Ca^{2+}. According to the model of Destexhe–Paré it holds

$$\begin{aligned} I_{CAN} &= g_{CAN} m_C (V + 20) \\ \frac{dm_c}{dt} &= 0.005[Ca^{2+}]^2(1 - m_c) - m_c/3000 \end{aligned} \tag{1.105}$$

The rate of the voltage spikes in a membrane characterized by calcium-activated nonspecific cation (CAN) current increases each time calcium stimuli arrives.

1.9 Conclusions

In this chapter it has been shown that the functioning of the cells membrane can be represented as an electric circuit. To this end: (1) the ion channels have been represented as resistors, (2) the gradients of the ions concentration have been represented as voltage sources, (3) the capability of the membrane for charge storage has been represented as a capacitor. An assumption made was that the neurons have the shape of a long cylinder, or of a cable with specific radius.The Hodgkin–Huxley model was derived from a modification of the cable's PDE, which describes the change of the voltage along dendrites axis. It has been shown that cables' equation comprises as inputs the currents which are developed in the ions channels. Additional models of reduced dimensionality that describe voltage variations along the neurons membrane have been introduced. These were the FitzHugh–Nagumo model and the Morris–Lecar model. It has been also demonstrated that cable's equation is also shown to be suitable for describing voltage variations along dendrites. Finally, the various types of ionic channels and ion currents across the neurons membrane have been analyzed.

Chapter 2
Systems Theory for the Analysis of Biological Neuron Dynamics

Abstract The chapter analyzes the basics of systems theory which can be used in the modelling of biological neurons dynamics. To understand the oscillatory behavior of biological neurons benchmark examples of oscillators are given. Moreover, using as an example, the model of biological neurons the following properties are analyzed: phase diagram, isoclines, attractors, local stability, bifurcations of fixed points, and chaotic dynamics.

2.1 Characteristics of the Dynamics of Nonlinear Systems

Main features characterizing the stability of nonlinear dynamical systems are defined as follows [92, 209]:

1. *Finite escape time*: It is the finite time within which the state-vector of the nonlinear system converges to infinity.
2. *Multiple isolated equilibria*: A linear system can have only one equilibrium to which converges the state vector of the system in steady-state. A nonlinear system can have more than one isolated equilibria (fixed points). Depending on the initial state of the system, in steady-state the state vector of the system can converge to one of these equilibria.
3. *Limit cycles*: For a linear system to exhibit oscillations it must have eigenvalues on the imaginary axis. The amplitude of the oscillations depends on initial conditions. In nonlinear systems one may have oscillations of constant amplitude and frequency, which do not depend on initial conditions. This type of oscillations is known as *limit cycles*.
4. *Sub-harmonic, harmonic, and almost periodic oscillations*: A stable linear system under periodic input produces an output of the same frequency. A nonlinear system, under periodic excitation can generate oscillations with frequencies which are several times smaller (subharmonic) or multiples of the frequency of the input (harmonic). It may also generate almost periodic oscillations with

G.G. Rigatos, *Advanced Models of Neural Networks*,
DOI 10.1007/978-3-662-43764-3_2,

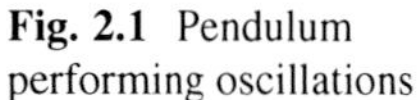

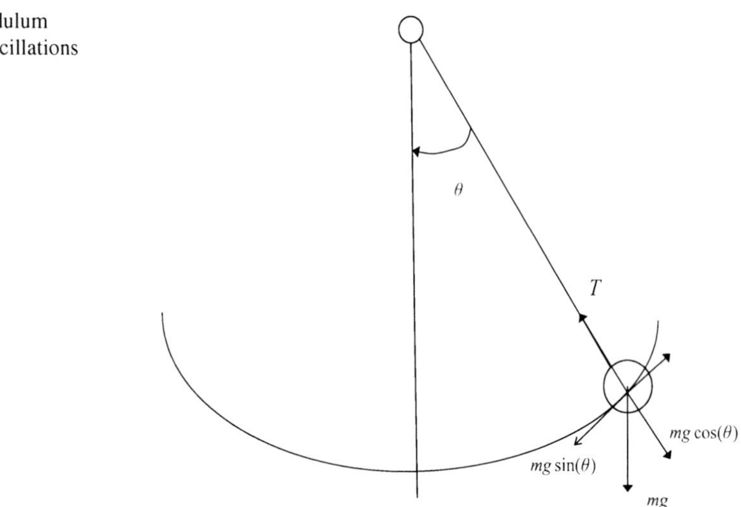

Fig. 2.1 Pendulum performing oscillations

frequencies which are not necessarily multiples of a basis frequency (almost periodic oscillations).

5. *Chaos*: A nonlinear system in steady-state can exhibit a behavior which is not characterized as equilibrium, periodic oscillation, or almost periodic oscillation. This behavior is characterized as chaos. As time advances the behavior of the system changes in a random-like manner, and this depends on the initial conditions. Although the dynamic system is deterministic, it exhibits randomness in the way it evolves in time.
6. *Multiple modes of behavior*: It is possible the same dynamical system to exhibit simultaneously more than one of the aforementioned characteristics (1)–(5). Thus, a system without external excitation may exhibit simultaneously more than one limit cycles. A system receiving a periodic external input may exhibit harmonic or subharmonic oscillations, or an even more complex behavior in steady state which depends on the amplitude and frequency of the excitation.

Example 1. Oscillations of a pendulum (Fig. 2.1).

The equation of the rotational motion of the pendulum under friction is given by

$$\begin{aligned} ml^2\ddot{\theta} &= mg\sin(\theta)l - klv \Rightarrow \\ ml^2\ddot{\theta} &= mg\sin(\theta)l - kl^2\dot{\theta} \Rightarrow \\ ml\ddot{\theta} &= mg\sin(\theta) - kl\dot{\theta} \end{aligned} \tag{2.1}$$

By defining the state variables $x_1 = \theta$ and $x_2 = \dot{\theta}$ one has the state-space description

$$\begin{aligned} \dot{x}_1 &= x_2 \\ \dot{x}_2 &= \tfrac{g}{l}\sin(x_1) - \tfrac{k}{m}x_2 \end{aligned} \tag{2.2}$$

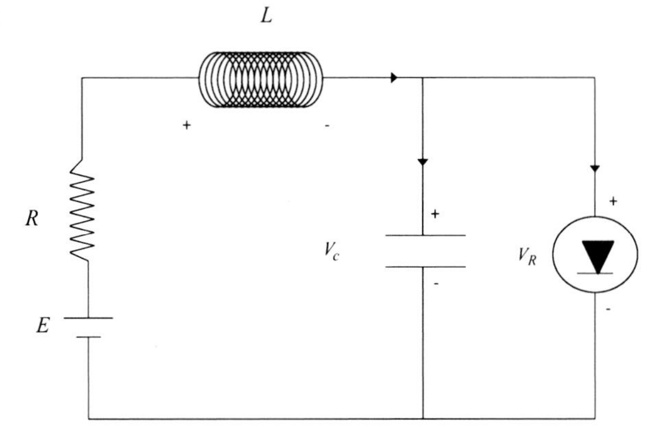

Fig. 2.2 Tunnel diode circuit

To compute the equilibrium one has

$$\begin{array}{c}\dot{x}_1 = \dot{x}_2 = 0 \Rightarrow \\ x_2 = 0 \text{ and } x_1 = n\pi \;\; n = 0, \pm 1, \pm 2, \cdots \end{array} \tag{2.3}$$

Indicative forms of oscillations for the pendulum are defined as follows:

(a) The effect of friction is neglected, therefore one has

$$\begin{array}{c}\dot{x}_1 = x_2 \\ \dot{x}_2 = \frac{g}{l}\sin(x_1)\end{array} \tag{2.4}$$

(b) An external torque T is applied to the pendulum

$$\begin{array}{c}\dot{x}_1 = x_2 \\ \dot{x}_2 = \frac{g}{l}\sin(x_1) - \frac{k}{m}x_2 + \frac{1}{ml^2}T\end{array} \tag{2.5}$$

Example 2. Tunnel diode circuit (Fig. 2.2).

The following equations hold

$$V_c = \frac{1}{sC}I_c(s) \Rightarrow I_c(s) = sC\,V_c(s) \Rightarrow I_c(t) = C\frac{dV_c(t)}{dt} \tag{2.6}$$

$$V_L(s) = sLI_L(s) \Rightarrow V_L(t) = L\frac{dI_L(t)}{dt} \tag{2.7}$$

$$I_L(s) = I_C + I_R \tag{2.8}$$

$$I_R(s) = h(V_R) \tag{2.9}$$

The following state variables are defined $x_1 = V_C$ and $x_2 = i_L$. From Eq. (2.8) it holds

$$I_L - I_C - I_R = 0 \Rightarrow I_C = I_L - I_R \Rightarrow I_C = x_2 - h(x_1) \tag{2.10}$$

Moreover, it holds

$$\begin{array}{c} I_L R - E + V_L + V_C = 0 \Rightarrow E = V_L + V_C + i_L R \Rightarrow E = L\frac{dI_L}{dt} + V_C + I_L R \Rightarrow \\ E = L\dot{x}_2 + V_C + x_2 R \Rightarrow \dot{x}_2 = \frac{1}{L}[-V_C - Rx_2 + E] \Rightarrow \dot{x}_2 = \frac{1}{L}[-x_1 - Rx_2 + u] \end{array} \tag{2.11}$$

where $u = E$. Additionally, from Eq. (2.6) it holds

$$\frac{dV_c(t)}{dt} = \frac{1}{C} I_c(t) \tag{2.12}$$

By replacing Eqs. (2.10) into (2.12) one obtains

$$\frac{dV_c(t)}{dt} = \frac{1}{C}[x_2 - h(x_1)] \tag{2.13}$$

From Eqs. (2.11) and (2.13) one has the description of the system is state-space form

$$\begin{array}{c} \dot{x}_1 = \frac{1}{C}[x_2 - h(x_1)] \\ \dot{x}_2 = \frac{1}{L}[-x_1 - Rx_2 + u] \end{array} \tag{2.14}$$

The associated equilibrium is computed from the condition $\dot{x}_1 = 0$ and $\dot{x}_2 = 0$ which gives

$$\begin{array}{c} -h(x_1) + x_2 = 0 \\ -x_1 - Rx_2 + u = 0 \end{array} \tag{2.15}$$

which gives

$$\begin{array}{c} -h(x_1) + \frac{E - x_1}{R} = 0 \\ x_2 = \frac{E - x_1}{R} \end{array} \tag{2.16}$$

Therefore, finally the equilibrium point is computed from the solution of the relation

$$h(x_1) = \frac{E}{R} - \frac{1}{R}x_1 \tag{2.17}$$

Example 3. Spring-mass system (Fig. 2.3).

It holds that

$$\begin{array}{c} m\ddot{x} = F - F_f - F_{\text{sp}} \Rightarrow \\ m\ddot{x} = F - b\dot{x} - kx \end{array} \tag{2.18}$$

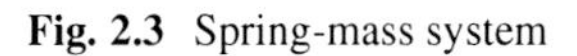

Fig. 2.3 Spring-mass system

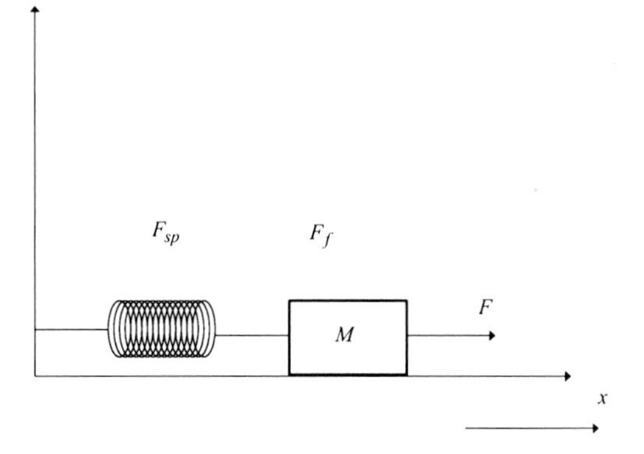

One can also consider a model with nonlinear spring dynamics given by

$$\begin{aligned} g(x) &= k(1-a^2x^2)x \quad |ax|<1 \quad \text{model of softening spring} \\ g(x) &= k(1+a^2x^2)x \quad x>x_{\text{thres}} \quad \text{model of hardening spring} \end{aligned} \tag{2.19}$$

The combination of a hardening spring, a linear viscous damping term and of a periodic external force $F = A\cos(\omega t)$ results into the Duffing oscillator

$$m\ddot{x} + c\dot{x} + kx + ka^2x^3 = A\cos(\omega t) \tag{2.20}$$

The combination of a linear spring, a linear viscous damping term, a dry friction term and of a zero external force generates the following oscillator model

$$m\ddot{x} + kx + c\dot{x} + \eta(x,\dot{x}) = 0 \tag{2.21}$$

where

$$\eta(x,\dot{x}) = \begin{cases} \mu_k mg{\cdot}\text{sign}(\dot{x}) \text{ if } |\dot{x} > 0| \\ -kx \text{ if } |\dot{x}| = 0 \text{ and } |x| \leq \mu_s mg/k \\ -\mu_s mg\text{sign}(x) \text{ if } \dot{x} = 0 \text{ and } |x| > \mu_s mg/k \end{cases} \tag{2.22}$$

By defining the state variables $x_1 = x$ and $x_2 = \dot{x}$ one has

$$\begin{aligned} \dot{x}_1 &= x_2 \\ \dot{x}_2 &= -\tfrac{k}{m}x_1 - \tfrac{c}{m}x_2 - \tfrac{1}{m}\eta(x,\dot{x}) \end{aligned} \tag{2.23}$$

For $x_2 = \dot{x} > 0$ one obtains the state-space description

$$\begin{aligned} \dot{x}_1 &= x_2 \\ \dot{x}_2 &= -\tfrac{k}{m}x_1 - \tfrac{c}{m}x_2 - \mu_k g \end{aligned} \tag{2.24}$$

For $x_2 = \dot{x} < 0$ one obtains the state-space description

$$\begin{aligned} \dot{x}_1 &= x_2 \\ \dot{x}_2 &= -\tfrac{k}{m}x_1 - \tfrac{c}{m}x_2 + \mu_k g \end{aligned} \tag{2.25}$$

2.2 Computation of Isoclines

An autonomous second order system is described by two differential equations of the form

$$\begin{aligned} \dot{x}_1 &= f_1(x_1, x_2) \\ \dot{x}_2 &= f_2(x_1, x_2) \end{aligned} \tag{2.26}$$

The method of the isoclines consists of computing the slope (ratio) between f_2 and f_1 for every point of the trajectory of the state vector (x_1, x_2).

$$s(x) = \tfrac{f_2(x_1, x_2)}{f_1(x_1, x_2)} \tag{2.27}$$

The case $s(x) = c$ describes a curve in the $x_1 - x_2$ plane along which the trajectories $\dot{x}_1 = f_1(x_1, x_2)$ and $\dot{x}_2 = f_2(x_1, x_2)$ have a constant slope.

The curve $s(x) = c$ is drawn in the $x_1 - x_2$ plane and along this curve one also draws small linear segments of length c. The curve $s(x) = c$ is known as isocline. The direction of these small linear segments is according to the sign of the ratio $f_2(x_1, x_2)/f_1(x_1, x_2)$.

Example 1. The following simplified pendulum equation is considered

$$\begin{aligned} \dot{x}_1 &= x_2 \\ \dot{x}_2 &= -\sin(x_1) \end{aligned} \tag{2.28}$$

The slope $s(x)$ is given by the relation

$$s(x) = \tfrac{f_2(x_1, x_2)}{f_1(x_1, x_2)} \Rightarrow s(x) = -\tfrac{\sin(x_2)}{x_2} \tag{2.29}$$

Setting $s(x) = c$ it holds that the isoclines are given by the relation

$$x_2 = -\tfrac{1}{c}\sin(x_1) \tag{2.30}$$

For different values of c one has the following isoclines diagram depicted in Fig. 2.4.

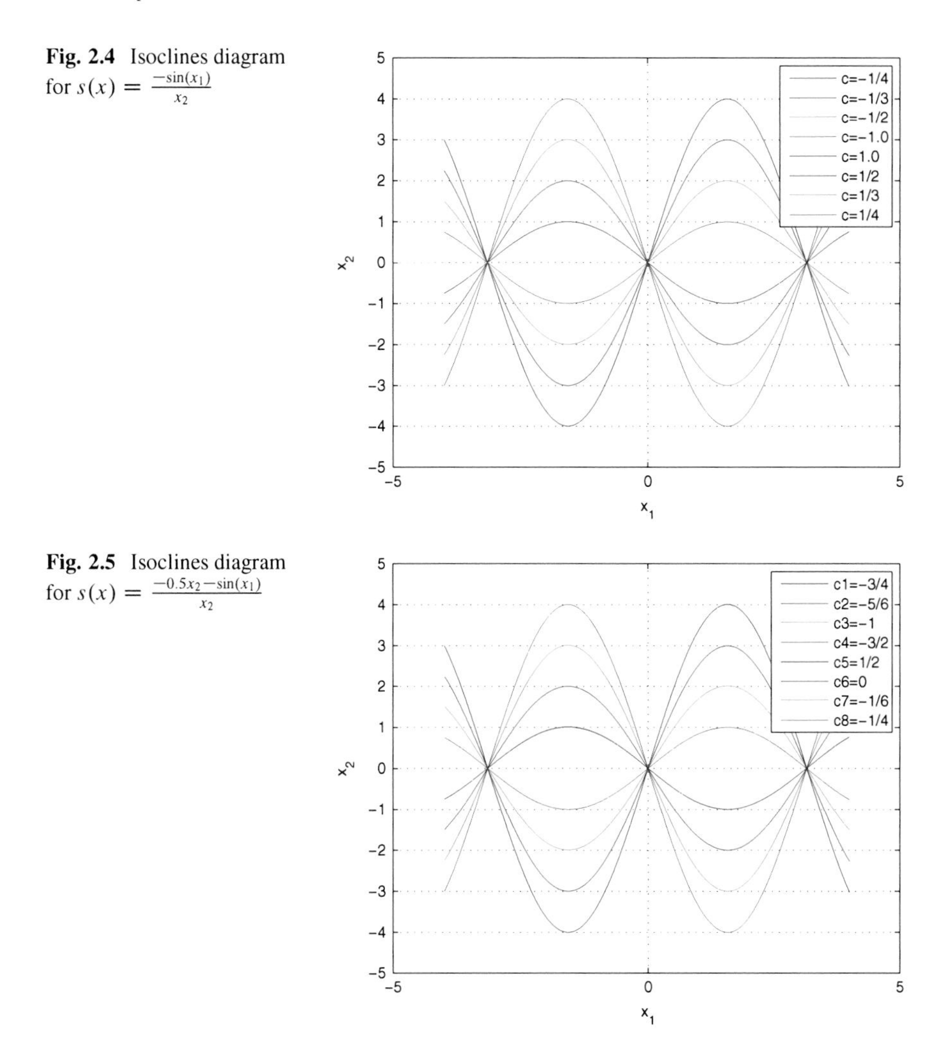

Fig. 2.4 Isoclines diagram for $s(x) = \frac{-\sin(x_1)}{x_2}$

Fig. 2.5 Isoclines diagram for $s(x) = \frac{-0.5x_2 - \sin(x_1)}{x_2}$

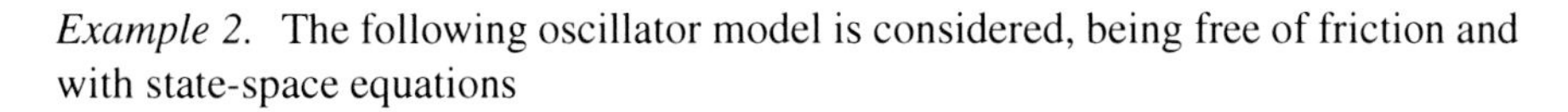

Example 2. The following oscillator model is considered, being free of friction and with state-space equations

$$\begin{aligned}\dot{x}_1 &= x_2 \\ \dot{x}_2 &= -0.5x_2 - \sin(x_1)\end{aligned} \tag{2.31}$$

To compute isoclines one has

$$\begin{aligned}s(x) = \tfrac{-0.5x_2 - \sin(x_1)}{x_2} = c \Rightarrow -0.5x_2 - \sin(x_1) = cx_2 \Rightarrow \\ (0.5 + c)x_2 = \sin(x_1) \Rightarrow x_2 = -\tfrac{1}{0.5+c}\sin(x_1)\end{aligned} \tag{2.32}$$

For different values of parameter c the isoclines are depicted in Fig. 2.5.

2.3 Systems Theory and Neurodynamics

Basic features that are important in the study of neurodynamics are (1) equilibria (fixed points), (2) limit cycles, (3) phase diagrams, (4) periodic orbits, and (5) bifurcations of fixed points [92, 209]. The definition of these features will be given through examples in the case neuron models.

2.3.1 *The Phase Diagram*

One can consider the Morris–Lecar model with the two state variables, V and η. The dynamics of the Morris–Lecar model can be written as

$$\begin{aligned}\frac{dV}{dt} &= f(V,t)\\ \frac{d\eta}{dt} &= g(V,\eta)\end{aligned} \tag{2.33}$$

The phase diagram consists of the points on the trajectories of the solution of the associated differential equation, i.e. $(V(t_k),\eta(t_k))$.

At a fixed point or equilibrium it holds $f(V_R,\eta_R) = 0$ and $g(V_R,\eta_R) = 0$. The closed trajectories are associated with periodic solutions. If there are closed trajectories, then $\exists T > 0$ such that $(V(t_k),\eta(t_k)) = (V(t_k+T),\eta(t_k+T))$.

Another useful parameter is the nullclines. The V-nullcline is characterized by the relation $\dot{V} = f(V,\eta) = 0$. The η-nullcline is characterized by the relation $\dot{\eta} = g(V,\eta) = 0$. The fixed points (or equilibria) are found on the intersection of nullclines.

2.3.2 *Stability Analysis of Nonlinear Systems*

2.3.2.1 Local Linearization

A manner to examine stability in nonlinear dynamical systems is to perform local linearization around an equilibrium. Assume the nonlinear system $\dot{x} = f(x,u)$ and $f(x_0,u) = 0$ that is x_0 is the local equilibrium. The linearization of $f(x,u)$ with respect to u round x_0 (Taylor series expansion) gives the equivalent description

$$\dot{x} = Ax + Bu \tag{2.34}$$

where

$$A = \nabla_x f = \begin{pmatrix} \frac{\partial f_1}{\partial x_1} & \frac{\partial f_1}{\partial x_2} & \cdots & \frac{\partial f_1}{\partial x_n} \\ \frac{\partial f_2}{\partial x_1} & \frac{\partial f_2}{\partial x_2} & \cdots & \frac{\partial f_2}{\partial x_n} \\ \cdots & \cdots & \cdots & \cdots \\ \frac{\partial f_n}{\partial x_1} & \frac{\partial f_n}{\partial x_2} & \cdots & \frac{\partial f_n}{\partial x_n} \end{pmatrix} |_{x=x_0} \tag{2.35}$$

and

$$B = \nabla_u f = \begin{pmatrix} \frac{\partial f_1}{\partial u_1} & \frac{\partial f_1}{\partial u_2} & \cdots & \frac{\partial f_1}{\partial u_n} \\ \frac{\partial f_2}{\partial u_1} & \frac{\partial f_2}{\partial u_2} & \cdots & \frac{\partial f_2}{\partial u_n} \\ \cdots & \cdots & \cdots & \cdots \\ \frac{\partial f_n}{\partial u_1} & \frac{\partial f_n}{\partial u_2} & \cdots & \frac{\partial f_n}{\partial u_n} \end{pmatrix} |_{x=x_0} \tag{2.36}$$

The eigenvalues of matrix A define the local stability features of the system:

Example 1. Assume the nonlinear system

$$\frac{d^2x}{dt^2} + 2x\frac{dx}{dt} + 2x^2 - 4x = 0 \tag{2.37}$$

By defining the state variables $x_1 = x$, $x_2 = \dot{x}$ the system can be written in the following form

$$\begin{aligned} \dot{x}_1 &= x_2 \\ \dot{x}_2 &= -2x_1x_2 - 2x_1^2 + 4x_1 \end{aligned} \tag{2.38}$$

It holds that $\dot{x} = 0$ if $f(x) = 0$ that is $(x_1^{*,1}, x_2^{*,1}) = (0,0)$ and $(x_1^{*,2}, x_2^{*,2}) = (2,0)$. Round the first equilibrium $(x_1^{*,1}, x_2^{*,1}) = (0,0)$ the system's dynamics is written as

$$A = \nabla_x f = \begin{pmatrix} 0 & 1 \\ 4 & 0 \end{pmatrix} \text{ or } \begin{pmatrix} \dot{x}_1 \\ \dot{x}_2 \end{pmatrix} = \begin{pmatrix} 0 & 1 \\ 4 & 0 \end{pmatrix} \begin{pmatrix} x_1 \\ x_2 \end{pmatrix} \tag{2.39}$$

The eigenvalues of the system are $\lambda_1 = 2$ and $\lambda_2 = -2$. This means that the fixed point $(x_1^{*,1}, x_2^{*,1}) = (0,0)$ is an unstable one.

Next, the fixed point $(x_1^{*,2}, x_2^{*,2}) = (2,0)$ is analyzed. The associated Jacobian matrix is computed again. It holds that

$$A = \nabla_x f = \begin{pmatrix} 0 & 1 \\ -4 & -4 \end{pmatrix} \tag{2.40}$$

The eigenvalues of the system are $\lambda_1 = -2$ and $\lambda_2 = -2$. Consequently, the fixed point $(x_1^{*,2}, x_2^{*,2}) = (2,0)$ is a stable one.

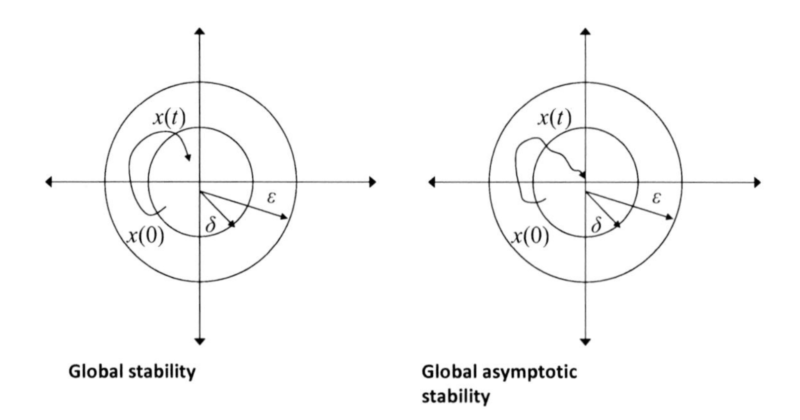

Fig. 2.6 Global stability and global asymptotic stability

2.3.2.2 Lyapunov Stability Approach

The Lyapunov method analyzes the stability of a dynamical system without the need to compute explicitly the trajectories of the state vector $x = [x_1, x_2, \cdots, x_n]^T$.

Theorem. *The system described by the relation $\dot{x} = f(x)$ is asymptotically stable in the vicinity of the equilibrium $x_0 = 0$ if there is a function $V(x)$ such that*

(i) $V(x)$ to be continuous and to have a continuous first order derivative at x_0
(ii) $V(x) > 0$ if $x \neq 0$ and $V(0) = 0$
(iii) $\dot{V}(x) < 0$, $\forall x \neq 0$.

The Lyapunov function is usually chosen to be a quadratic (and thus positive) energy function of the system; however, there in no systematic method to define it.

Assume now, that $\dot{x} = f(x)$ and $x_0 = 0$ is the equilibrium. Then the system is globally asymptotically stable if for every $\epsilon > 0$, $\exists \delta(\epsilon) > 0$, such that if $||x(0)|| < \delta$ then $||x(t)|| < \epsilon$, $\forall t \geq 0$.

This means that if the state vector of the system starts in a disc of radius δ then as time advances it will remain in the same disc, as shown in Fig. 2.6. Moreover, if $\lim_{t\to\infty}||x(t)|| = x_0 = 0$, then the system is globally asymptotically stable.

Example 1. Consider the system

$$\begin{aligned} \dot{x}_1 &= x_2 \\ \dot{x}_2 &= -x_1 - x_3^2 \end{aligned} \tag{2.41}$$

The following Lyapunov function is defined

$$V(x) = x_1^2 + x_2^2 \tag{2.42}$$

The equilibrium point is $(x_1 = 0, x_2 = 0)$. It holds that $V(x) > 0 \; \forall \; (x_1, x_2) \neq (0,0)$ and $V(x) = 0$ for $(x_1, x_2) = (0,0)$. Moreover, it holds

$$\begin{aligned} \dot{V}(x) = 2x_1\dot{x}_1 + 2x_2\dot{x}_2 = 2x_1x_2 + 2x_2(-x_1 - x_2^3) \Rightarrow \\ \dot{V}(x) = -2x_2^4 < 0 \; \forall \; (x_1, x_2) \neq (0,0) \end{aligned} \tag{2.43}$$

Therefore, the system is asymptotically stable and $\lim_{t \to \infty}(x_1, x_2) = (0,0)$.

Example 2. Consider the system

$$\begin{aligned} \dot{x}_1 &= -x_1(1 + 2x_1x_2^2) \\ \dot{x}_2 &= x_1^3 x_2 \end{aligned} \tag{2.44}$$

The following Lyapunov function is considered

$$V(x) = \tfrac{1}{2}x_1^2 + x_2^2 \tag{2.45}$$

The equilibrium point is $x_1 = 0, x_2 = 0$. It holds that

$$\begin{aligned} \dot{V}(x) = x_1\dot{x}_1 + 2x_2\dot{x}_2 = -x_1^2(1 + 2x_1x_2^2 + 2x_2(x_1^3x_2)) \Rightarrow \\ \dot{V}(x) = -x_1^2 < 0 \; \forall \; (x_1, x_2) \neq (0,0) \end{aligned} \tag{2.46}$$

Therefore, the system is asymptotically stable and $\lim_{t \to \infty}(x_1, x_2) = (0,0)$.

2.3.3 *Stability Analysis of the Morris–Lecar Nonlinear Model*

Local stability of the Morris–Lecar model can be studied round the associated equilibria. Local linearization can be performed round equilibria. Using the set of differential equations that describe the Morris–Lecar neuron that was given in Eq. (1.57) and performing Taylor series expansion, that is $\dot{x} = h(x) \Rightarrow \dot{x} = h(x_0)|_{x_0} + \nabla_x h(x - x_0) + \cdots$ one obtains the system's local linear dynamics.

The Morris–Lecar neuron model has the generic form

$$\begin{pmatrix} \dot{x}_1 \\ \dot{x}_2 \end{pmatrix} = \begin{pmatrix} f(x_1, x_2) \\ g(x_1, x_2) \end{pmatrix} \tag{2.47}$$

where $f(x_1, x_2) = I_{app} - g_L(V - E_L) - g_{k\eta}(V - E_k) - g_{Ca}m_{\infty}(V)(V - E_{Ca}) = I_{app} - I_{ion}(V, n)$ and $g(x_1, x_2) = \phi(\eta_{\infty}(V) - \eta)/\tau_{\eta}(V)$. The fixed points of this model are computed from the condition $\dot{x}_1 = 0$ and $\dot{x}_2 = 0$. The Jacobian matrix $\nabla_x h = M$ is given by

$$M = \begin{pmatrix} \frac{\partial f}{\partial V}(V_R, \eta_R) & \frac{\partial f}{\partial \eta}(V_R, \eta_R) \\ \frac{\partial g}{\partial V}(V_R, \eta_R) & \frac{\partial g}{\partial \eta}(V_R, \eta_R) \end{pmatrix} \tag{2.48}$$

which results into the matrix

$$J = \begin{pmatrix} \frac{\partial I_{\rm ion}}{\partial V}(V_R, \eta_R)/C_M & -g_k(V_R - E_k)/C_M \\ \phi \eta'_\infty/\tau_\eta(V_R) & -\phi/\tau_\eta(V_R) \end{pmatrix} \tag{2.49}$$

The eigenvalues of matrix M define stability round fixed points (stable or unstable fixed point). To this end, one has to find the roots of the associated characteristic polynomial that is given by $\det(\lambda I - J) = 0$ where I is the identity matrix.

2.4 Phase Diagrams and Equilibria of Neuronal Models

2.4.1 *Phase Diagrams for Linear Dynamical Systems*

The following autonomous linear system is considered

$$\dot{x} = Ax \tag{2.50}$$

The eigenvalues of matrix A define the system dynamics. Some terminology associated with fixed points is as follows.

A fixed point for the system of Eq. (2.50) is called hyperbolic if none of the eigenvalues of matrix A has zero real part. A hyperbolic fixed point is called a saddle if some of the eigenvalues of matrix A have real parts greater than zero and the rest of the eigenvalues have real parts less than zero. If all of the eigenvalues have negative real parts, then the hyperbolic fixed point is called a stable node or sink. If all of the eigenvalues have positive real parts, then the hyperbolic fixed point is called an unstable node or source. If the eigenvalues are purely imaginary, then one has an elliptic fixed point which is said to be a center.

Case 1: Both eigenvalues of matrix A are real and unequal, that is $\lambda_1 \neq \lambda_1 \neq 0$.

For $\lambda_1 < 0$ and $\lambda_2 < 0$ the phase diagram for z_1 and z_2 is shown in Fig. 2.7:
In case that λ_2 is smaller than λ_1, the term $e^{\lambda_2 t}$ decays faster than $e^{\lambda_1 t}$.
For $\lambda_1 > 0 > \lambda_2$ the phase diagram of Fig. 2.8 is obtained:
In the latter case there are stable trajectories along eigenvector v_1 and unstable trajectories along eigenvector v_2 of matrix A. The stability point $(0, 0)$ is said to be a saddle point.
When $\lambda_1 > \lambda_2 > 0$ then one has the phase diagrams of Fig. 2.9:

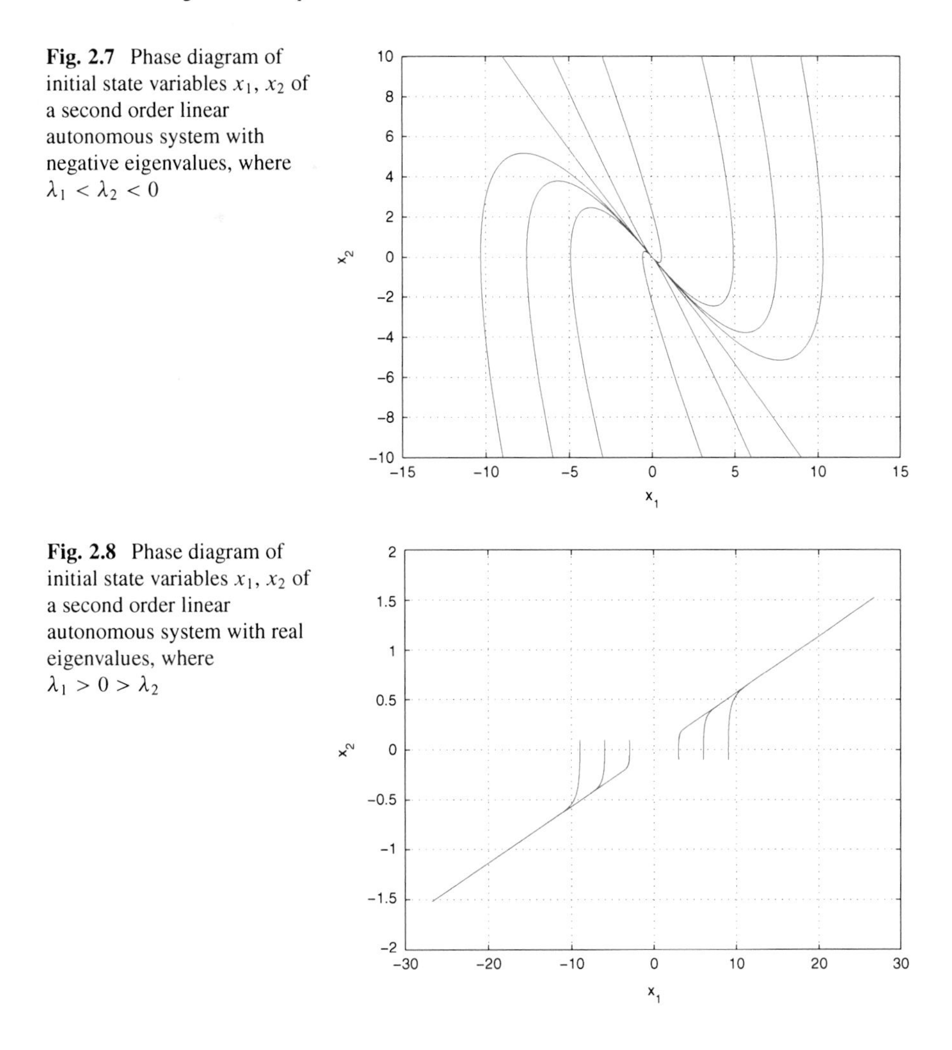

Fig. 2.7 Phase diagram of initial state variables x_1, x_2 of a second order linear autonomous system with negative eigenvalues, where $\lambda_1 < \lambda_2 < 0$

Fig. 2.8 Phase diagram of initial state variables x_1, x_2 of a second order linear autonomous system with real eigenvalues, where $\lambda_1 > 0 > \lambda_2$

Case 2: Complex eigenvalues:

Typical phase diagrams in the case of stable complex eigenvalues are given in Fig. 2.10.

Typical phase diagrams in the case of unstable complex eigenvalues are given in Fig. 2.11.

Typical phase diagrams in the case of imaginary eigenvalues are given in Fig. 2.12.

Case 3: Matrix A has non-zero eigenvalues which are equal to each other. The associated phase diagram is given in Fig. 2.13.

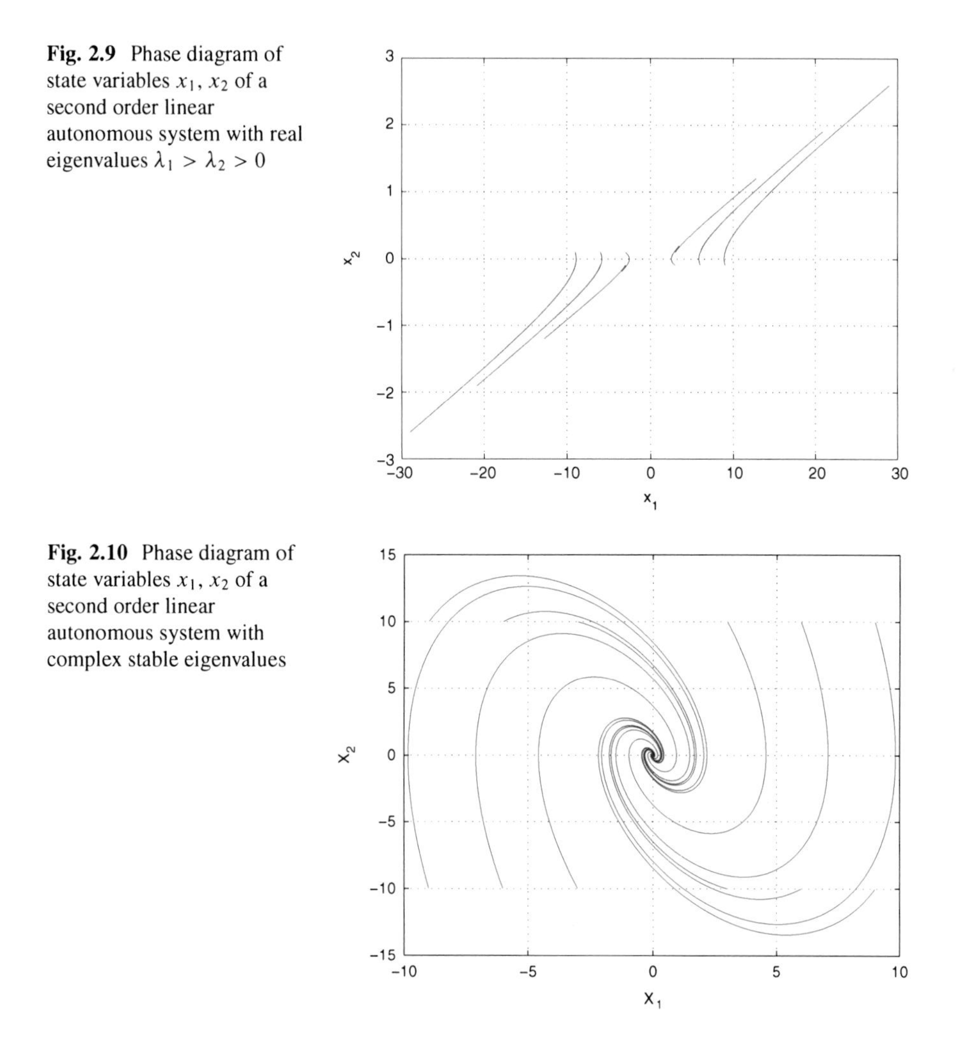

Fig. 2.9 Phase diagram of state variables x_1, x_2 of a second order linear autonomous system with real eigenvalues $\lambda_1 > \lambda_2 > 0$

Fig. 2.10 Phase diagram of state variables x_1, x_2 of a second order linear autonomous system with complex stable eigenvalues

2.4.2 Multiple Equilibria for Nonlinear Dynamical Systems

A nonlinear system can have multiple equilibria as shown in the following example. Consider, for instance, the model of a pendulum under friction

$$\begin{aligned} \dot{x}_1 &= x_2 \\ \dot{x}_2 &= -\tfrac{g}{l}\sin(x_1) - \tfrac{K}{m}x_2 \end{aligned} \tag{2.51}$$

The associated phase diagram is designed for different initial conditions and is given in Fig. 2.14.

For the previous model of the nonlinear oscillator, local linearization round equilibria with the use of Taylor series expansion enables analysis of the local

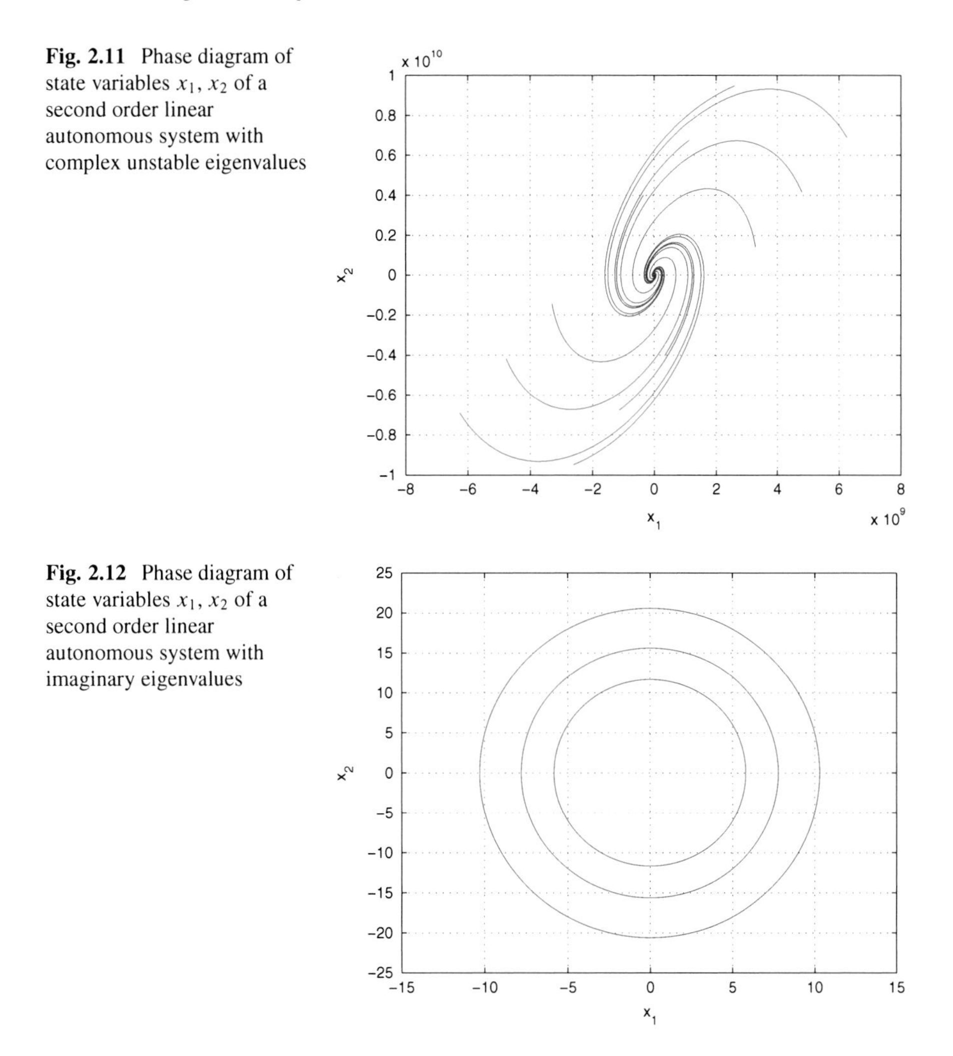

Fig. 2.11 Phase diagram of state variables x_1, x_2 of a second order linear autonomous system with complex unstable eigenvalues

Fig. 2.12 Phase diagram of state variables x_1, x_2 of a second order linear autonomous system with imaginary eigenvalues

stability properties of nonlinear dynamical systems

$$\begin{aligned} \dot{x}_1 &= f_1(x_1, x_2) \\ \dot{x}_2 &= f_2(x_1, x_2) \end{aligned} \tag{2.52}$$

which gives

$$\begin{aligned} \dot{x}_1 &= f_1(p_1, p_2) + \alpha_{11}(x_1 - p_1) + \alpha_{12}(x_2 - p_2) + \text{h.o.t} \\ \dot{x}_2 &= f_2(p_1, p_2) + \alpha_{21}(x_1 - p_1) + \alpha_{22}(x_2 - p_2) + \text{h.o.t} \end{aligned} \tag{2.53}$$

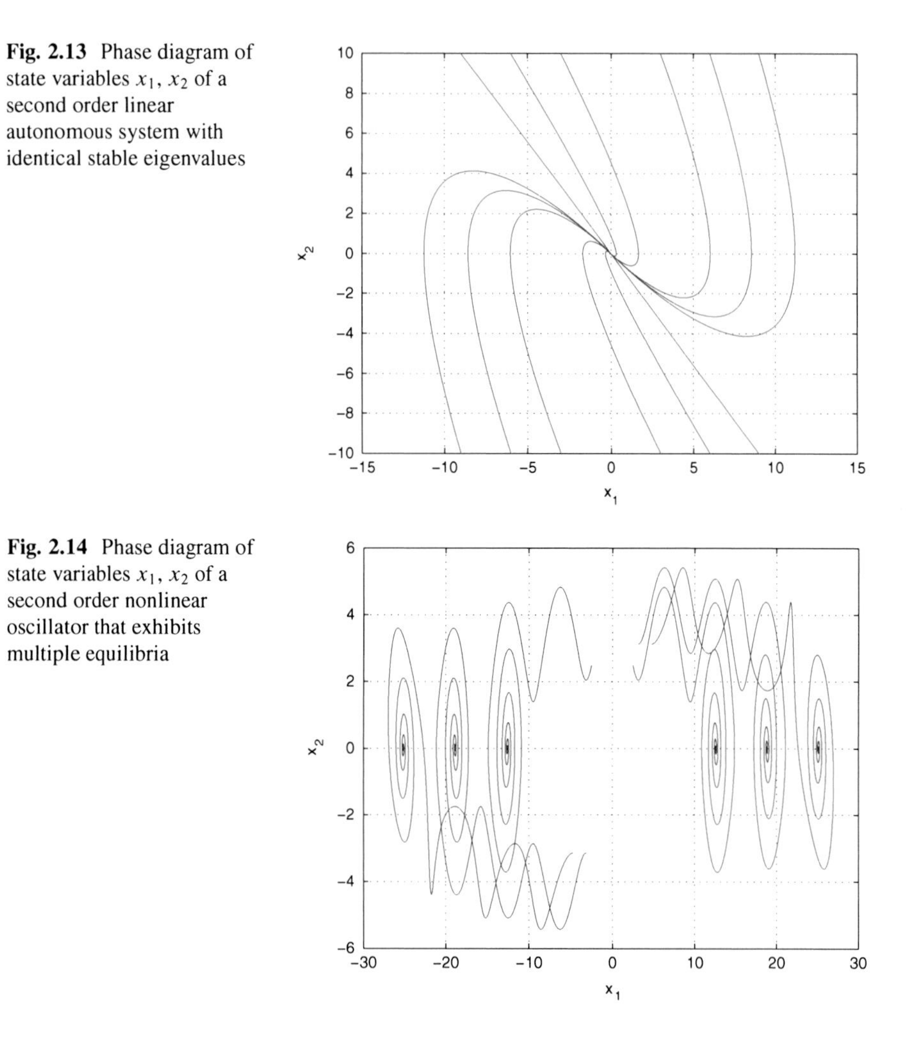

Fig. 2.13 Phase diagram of state variables x_1, x_2 of a second order linear autonomous system with identical stable eigenvalues

Fig. 2.14 Phase diagram of state variables x_1, x_2 of a second order nonlinear oscillator that exhibits multiple equilibria

where

$$\begin{aligned} \alpha_{11} &= \frac{\partial f_1(x_1,x_2)}{\partial x_1}\Big|_{x_1=p_1,x_2=p_2} & \alpha_{12} &= \frac{\partial f_1(x_1,x_2)}{\partial x_2}\Big|_{x_1=p_1,x_2=p_2} \\ \alpha_{21} &= \frac{\partial f_2(x_1,x_2)}{\partial x_1}\Big|_{x_1=p_1,x_2=p_2} & \alpha_{22} &= \frac{\partial f_2(x_1,x_2)}{\partial x_2}\Big|_{x_1=p_1,x_2=p_2} \end{aligned} \tag{2.54}$$

For the equilibrium it holds:

$$\begin{aligned} f_1(p_1, p_2) &= 0 \\ f_2(p_1, p_2) &= 0 \end{aligned} \tag{2.55}$$

Next, by defining the new variables $y_1 = x_1 - p_1$ and $y_2 = x_2 - p_2$ one can rewrite the state-space equation

$$\begin{aligned}\dot{y}_1 &= \dot{x}_1 = \alpha_{11}y_1 + \alpha_{12}y_2 + \text{h.o.t.}\\ \dot{y}_2 &= \dot{x}_2 = \alpha_{21}y_1 + \alpha_{22}y_2 + \text{h.o.t.}\end{aligned} \tag{2.56}$$

By omitting the higher order terms one can approximate the initial nonlinear system with its linearized equivalent

$$\begin{aligned}\dot{y}_1 &= \alpha_{11}y_1 + \alpha_{12}y_2\\ \dot{y}_2 &= \alpha_{21}y_1 + \alpha_{22}y_2\end{aligned} \tag{2.57}$$

which in matrix form is written as

$$\dot{y} = Ay \tag{2.58}$$

where

$$A = \begin{pmatrix}\alpha_{11} & \alpha_{12}\\ \alpha_{21} & \alpha_{22}\end{pmatrix} = \begin{pmatrix}\frac{\partial f_1}{\partial x_1} & \frac{\partial f_1}{\partial x_2}\\ \frac{\partial f_2}{\partial x_1} & \frac{\partial f_2}{\partial x_2}\end{pmatrix}|_{x=p} = \frac{\partial f}{\partial x}|_{x=p} \tag{2.59}$$

Matrix $A = \frac{\partial f}{\partial x}$ is the Jacobian matrix of the system that is computed at point $(x_1, x_2) = (p_1, p_2)$. It is anticipated that the trajectories of the phase diagram of the linearized system in the vicinity of the equilibrium point will be also close to the trajectories of the phase diagram of the nonlinear system.

Therefore, if the origin (equilibrium) in the phase diagram of the linearized system is (1) a stable node (matrix A has stable linear eigenvalues), (2) a stable focus (matrix A has stable complex eignevalues), (3) a saddle point (matrix A has some eigenvalues with negative real part while the rest of the eigenvalues have positive real part), then one concludes that the same properties hold for phase diagram of the nonlinear system.

Example. A nonlinear oscillator of the following form is considered:

$$\begin{aligned}\dot{x} = f(x) \Rightarrow\\ \begin{pmatrix}\dot{x}_1\\ \dot{x}_2\end{pmatrix} = \begin{pmatrix}x_2\\ -\sin(x_1) - 0.5x_2\end{pmatrix}\end{aligned} \tag{2.60}$$

The associated Jacobian matrix is

$$\frac{\partial f}{\partial x} = \begin{pmatrix}0 & 1\\ -\cos(x_1) & -0.5\end{pmatrix} \tag{2.61}$$

There are two equilibrium points $(0,0)$ and $(\pi, 0)$. The linearization round the equilibria gives

$$A_1 = \begin{pmatrix} 0 & 1 \\ -1 & -0.5 \end{pmatrix} \qquad A_2 = \begin{pmatrix} 0 & 1 \\ 1 & -0.5 \end{pmatrix}$$

$$\text{with eigenvalues} \qquad \text{with eigenvalues} \tag{2.62}$$

$$\lambda_{1,2} = -0.25 \pm j0.97 \qquad \lambda_1 = -1.28 \; \lambda_2 = 0.78$$

Consequently, the equilibrium $(0, 0)$ is a stable focus (matrix A_1 has stable complex eignevalues) and the equilibrium $(\pi, 0)$ is a saddle point (matrix A_2 has an unstable eigenvalue).

2.4.3 *Limit Cycles*

A dynamical system is considered to exhibit limit cycles when it admits the nontrivial periodic solution

$$x(t + T) = x(t) \; \forall \; t \geq 0 \tag{2.63}$$

for some $T > 0$ (the trivial periodic solution is the one associated with $x(t) =$ constant). An example about the existence of limit cycles is examined in the case of the Van der Pol oscillator (actually this is a model that approximates the functioning of neural oscillators such as Central Pattern Generators). The state equations of the oscillator are

$$\begin{aligned} \dot{x}_1 &= x_2 \\ \dot{x}_2 &= -x_1 + \epsilon(1 - x_1^2)x_2 \end{aligned} \tag{2.64}$$

Next the phase diagram of the Van der Pol oscillator is designed for three different values of parameter ϵ, namely $\epsilon = 0.2$, $\epsilon = 1$, and $\epsilon = 5.0$. In Figs. 2.15 and 2.16, it can be observed that in all cases there is a closed trajectory to which converge all curves of the phase diagram that start from points far from it.

To study conditions under which dynamical systems exhibit limit cycles, a second order autonomous nonlinear system is considered next, given by

$$\begin{pmatrix} \dot{x}_1 \\ \dot{x}_2 \end{pmatrix} = \begin{pmatrix} f_1(x_1, x_2) \\ f_2(x_1, x_2) \end{pmatrix} \tag{2.65}$$

Next, the following theorem defines the appearance of limit cycles in the phase diagram of dynamical systems [92, 209].

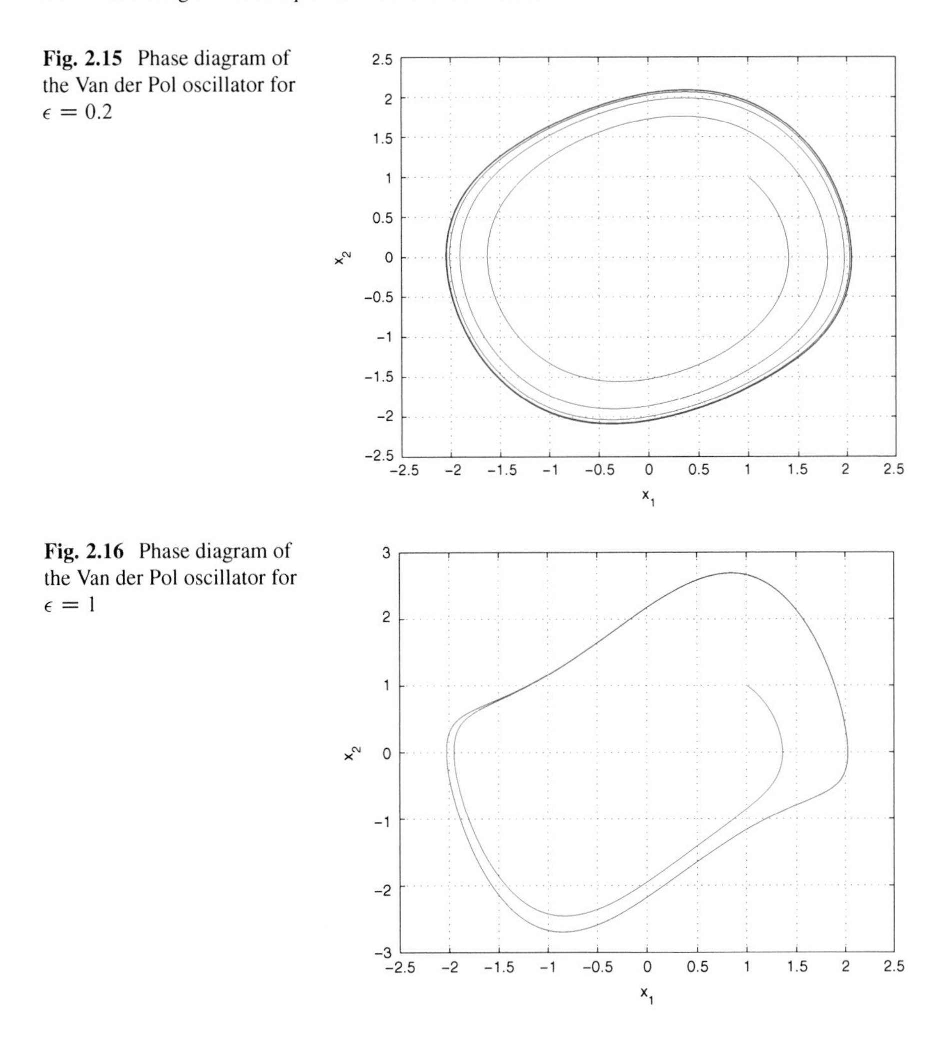

Fig. 2.15 Phase diagram of the Van der Pol oscillator for $\epsilon = 0.2$

Fig. 2.16 Phase diagram of the Van der Pol oscillator for $\epsilon = 1$

Theorem 1 (Poincaré–Bendixson). *If a trajectory of the nonlinear system of Eq. (2.65) remains in a finite region Ω, then one of the following is true: (i) the trajectory goes to an equilibrium point, (ii) the trajectory tends to a limit cycle, (iii) the trajectory is itself a limit cycle.*

Moreover, the following theorem provides a sufficient condition for the nonexistence of limit cycles:

Theorem 2 (Bendixson). *For the nonlinear system of Eq. (2.65), no limit cycle can exist in a region Ω of the phase plane in which $\frac{\partial f_1}{\partial x_1} + \frac{\partial f_2}{\partial x_2}$ does not vanish and does not change sign.*

2.5 Bifurcations in Neuronal Dynamics

2.5.1 *Bifurcations of Fixed Points of Biological Neuron Models*

As the parameters of the nonlinear model of the neuron are changed, the stability of the equilibrium point can also change and also the number of equilibria may vary. Values of these parameters at which the locus of the equilibrium (as a function of the parameter) changes and different branches appear are known as critical or bifurcation values. The phenomenon of bifurcation has to do with quantitative changes of parameters leading to qualitative changes of the system's properties [36,54,76,97,113].

Another issue in bifurcation analysis has to do with the study of the segments of the bifurcation branches in which the fixed points are no longer stable but either become unstable or are associated with limit cycles. The latter case is called Hopf bifurcation and the system's Jacobian matrix has a pair of complex imaginary eigenvalues.

2.5.2 *Saddle-Node Bifurcations of Fixed Points in a One-Dimensional System*

The considered dynamical system is given by $\dot{x} = \mu - x^2$. The fixed points of the system result from the condition $\dot{x} = 0$ which for $\mu > 0$ gives $x^* = \pm\sqrt{\mu}$. The first fixed point $x = \sqrt{\mu}$ is a stable one whereas the second fixed point $x = -\sqrt{\mu}$ is an unstable one. The phase diagram of the system is given in Fig. 2.17. Since there is one stable and one unstable fixed point the associated bifurcation (locus of the fixed points in the phase plane) will be a saddle-node one.

The bifurcations diagram is given next. The diagram shows how the fixed points of the dynamical system vary with respect to the values of parameter μ. In the above case it represents a parabola in the $\mu - x$ plane as shown in Fig. 2.18.

For $\mu > 0$ the dynamical system has two fixed points located at $\pm\sqrt{\mu}$. The one fixed point is stable and is associated with the upper branch of the parabola. The other fixed point is unstable and is associated with the lower branch of the parabola. The value $\mu = 0$ is considered to be a bifurcation value and the point $(x, \mu) = (0, 0)$ is a bifurcation point. This particular type of bifurcation where one branch is associated with fixed points and the other branch is not associated with any fixed points is known as saddle-node bifurcation.

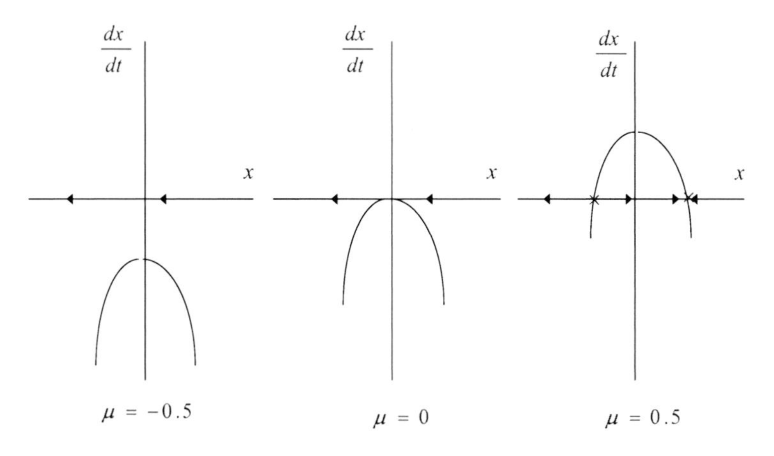

Fig. 2.17 Phase diagram and fixed points of the system $\dot{x} = \mu - x^2$. *Converging arrows* denote a stable fixed point

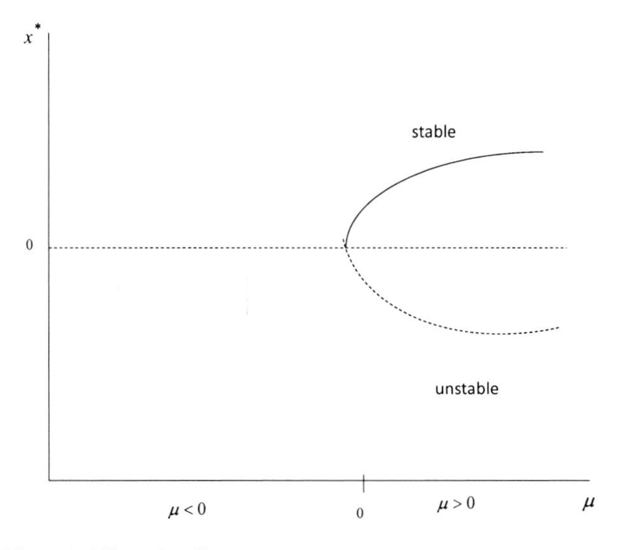

Fig. 2.18 Saddle-node bifurcation diagram

2.5.3 *Pitchfork Bifurcation of Fixed Points*

In pitchfork bifurcations the number of fixed points varies with respect to the values of the bifurcation parameter. The dynamical system $\dot{x} = x(\mu - x^2)$ is considered. The associated fixed points are found by the condition $\dot{x} = 0$. For $\mu < 0$ there is one fixed point at zero which is stable. For $\mu = 0$ there is still one fixed point at zero which is still stable. For $\mu > 0$ there are three fixed points, one at $x = 0$,

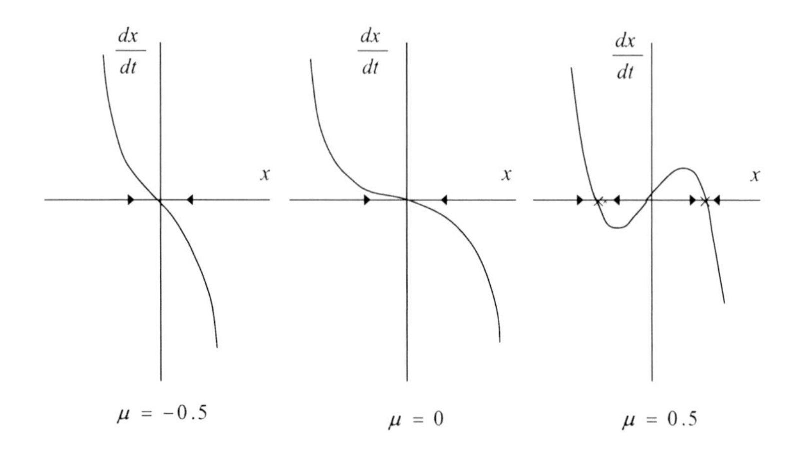

Fig. 2.19 Phase diagram and fixed points of a system exhibiting pitchfork bifurcations. *Converging arrows* denote a stable fixed point

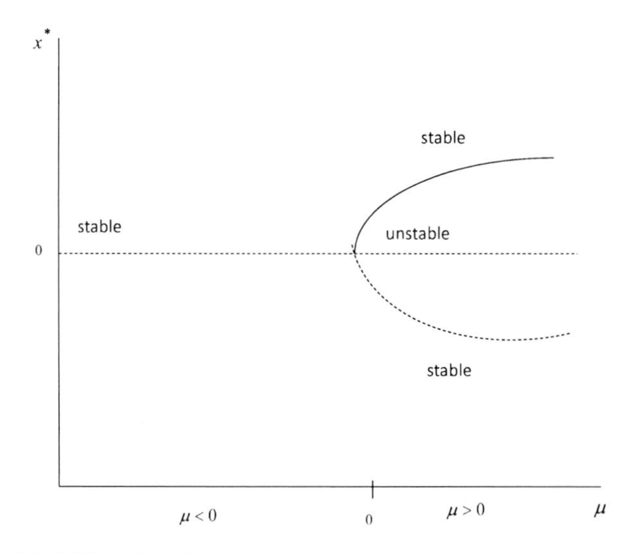

Fig. 2.20 Pitchfork bifurcation diagram

one at $x = +\sqrt{\mu}$ which is stable, and one at $x = -\sqrt{\mu}$ which is also stable. The associated phase diagrams and fixed points are presented in Fig. 2.19.

The bifurcations diagram is given next. The diagram shows how the fixed points of the dynamical system vary with respect to the values of parameter μ. In the above case it represents a parabola in the $\mu - x$ plane as shown in Fig. 2.20.

2.5.4 The Hopf Bifurcation

Bifurcation of the equilibrium point means that the locus of the equilibrium on the phase plane changes due to variation of the system's parameters. Equilibrium x^* is a hyperbolic equilibrium point if the real parts of all its eigenvalues are non-zero. A Hopf bifurcation appears when the hyperbolicity of the equilibrium point is lost due to variation of the system parameters and the eigenvalues become purely imaginary. By changing the values of the parameters at a Hopf bifurcation, an oscillatory solution appears [131].

The stages for finding Hopf bifurcations in nonlinear dynamical systems are described next. The following autonomous differential equation is considered:

$$\tfrac{dx}{dt} = f_1(x,\lambda) \tag{2.66}$$

where x is the state vector $x \in R^n$ and $\lambda \in R^m$ is the vector of the system parameters. In Eq. (3.15) a point x^* satisfying $f_1(x^*) = 0$ is an equilibrium point. Therefore from the condition $f_1(x^*) = 0$ one obtains a set of equations which provide the equilibrium point as function of the bifurcating parameter. The stability of the equilibrium point can be evaluated by linearizing the system's dynamic model around the equilibrium point and by computing eigenvalues of the Jacobian matrix. The Jacobian matrix at the equilibrium point can be written as

$$J_{f_1(x^*)} = \tfrac{\partial f_1(x)}{\partial x}|_{x=x^*} \tag{2.67}$$

and the determinant of the Jacobian matrix provides the characteristic equation which is given by

$$\det(\mu_i I_n - J_{f_1(x^*)}) = \lambda_1^n + \alpha_1 \lambda_i^{n-1} + \cdots + \alpha_{n-1}\lambda_i + \alpha_n = 0 \tag{2.68}$$

where I_n is the $n \times n$ identity matrix, μ_i with $i = 1, 2, \cdots, n$ denotes the eigenvalues of the Jacobian matrix $Df_1(x^*)$. From the requirement the eigenvalues of the system to be purely imaginary one obtains a condition, i.e. values that the bifurcating parameter should take, for the appearance of Hopf bifurcations.

As example, the following nonlinear system is considered:

$$\dot{x}_1 = x_2 \dot{x}_2 = -x_1 + (m - x_1^2)x_1 \tag{2.69}$$

Setting $\dot{x}_1 = 0$ and $\dot{x}_2 = 0$ one obtains the system's fixed points. For $m \leq 1$ the system has the fixed point $(x_1^*, x_2^*) = (0,0)$. For $m > 1$ the system has the fixed points $(x_1^*, x_2^*) = (0,0)$, $(x_1^*, x_2^*) = (\sqrt{m-1}, 0)$, $(x_1^*, x_2^*) = (-\sqrt{m-1}, 0)$. The system's Jacobian is

$$J = \begin{pmatrix} 0 & 1 \\ -1 + m - 3x_1^2 & 0 \end{pmatrix} \tag{2.70}$$

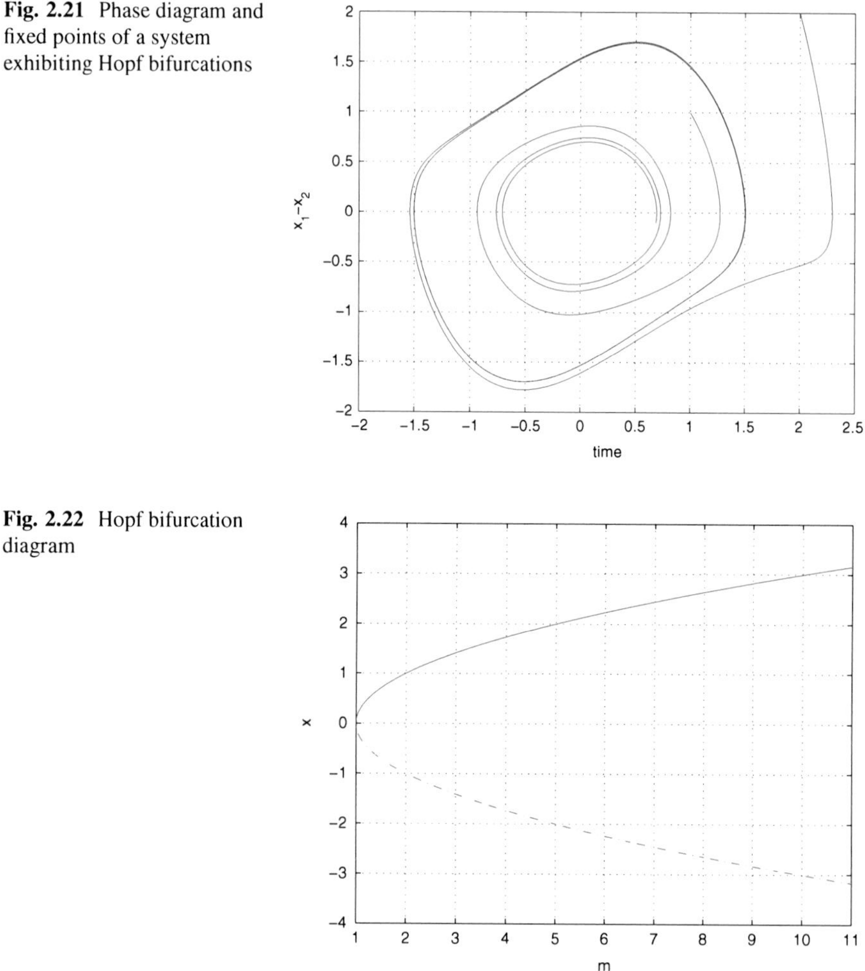

Fig. 2.21 Phase diagram and fixed points of a system exhibiting Hopf bifurcations

Fig. 2.22 Hopf bifurcation diagram

If $m \leq 1$, the eigenvalues of the Jacobian at the fixed point $(x_1^*, x_2^*) = (0, 0)$ will be imaginary. Thus the system exhibits Hopf bifurcations, as shown in Fig. 2.21. If $m > 1$, then the fixed point $(x_1^*, x_2^*) = (0, 0)$ is an unstable one. On the other hand, if $m > 1$, the eigenvalues of the Jacobian at the fixed points $(x_1^*, x_2^*) = (\sqrt{m-1}, 0)$ and $(x_1^*, x_2^*) = (-\sqrt{m-1}, 0)$ will be imaginary and the system exhibits again Hopf bifurcations.

The bifurcations diagram is given next. The diagram shows how the fixed points of the dynamical system vary with respect to the values of parameter μ. In the above case it represents a parabola in the $\mu - x$ plane as shown in Fig. 2.22.

2.6 Chaos in Neurons

2.6.1 *Chaotic Dynamics in Neurons*

Nonlinear dynamical systems, such as neurons, can exhibit a phenomenon called chaos, by which it is meant that the system's output becomes extremely sensitive to initial conditions. The essential feature of chaos is the unpredictability of the system's output. Even if the model of the chaotic system is known, the system's response in the long run cannot be predicted. Chaos is distinguished from random motion. In the latter case, the system's model or input contain uncertainty and as a result, the time variation of the output cannot be predicted exactly (only statistical measures can be computed). In chaotic systems, the involved dynamical model is deterministic and there is little uncertainty about the system's model, input, and initial conditions. In such systems, by slightly varying initial conditions or parameters values, a completely different phase diagram can be produced [214,216].

Some known chaotic dynamical systems are the Van der Pol oscillator which as explained in Sect. 1.5. is equivalent to the model of the FitzHugh–Nagumo neuron [207, 224]. Other chaotic models are the Duffing oscillator which is considered to be equivalent to variants of the FitzHugh–Nagumo neuron and which has been proposed to model the dynamics of neuronal groups recorded by the Electroengephalogram (EEG).

2.6.2 *Chaotic Dynamics in Associative Memories*

The variations of the parameters of associative memories (elements of the weight matrix) within specific ranges can result in the appearance of chaotic attractors or in the appearance of limitcycles. The change in the branches of the locus where these attractors reside, which is due to changes in the values of the dynamic model's parameters, is also noted as bifurcation. An example about chaotic dynamics emerging in Hopfield associated memories is shown in Fig. 2.23.

2.7 Conclusions

In this chapter, the main elements of systems theory are overviewed, thus providing the basis for modelling of biological neurons dynamics. To explain oscillatory phenomena and consequently the behavior of biological neurons benchmark examples

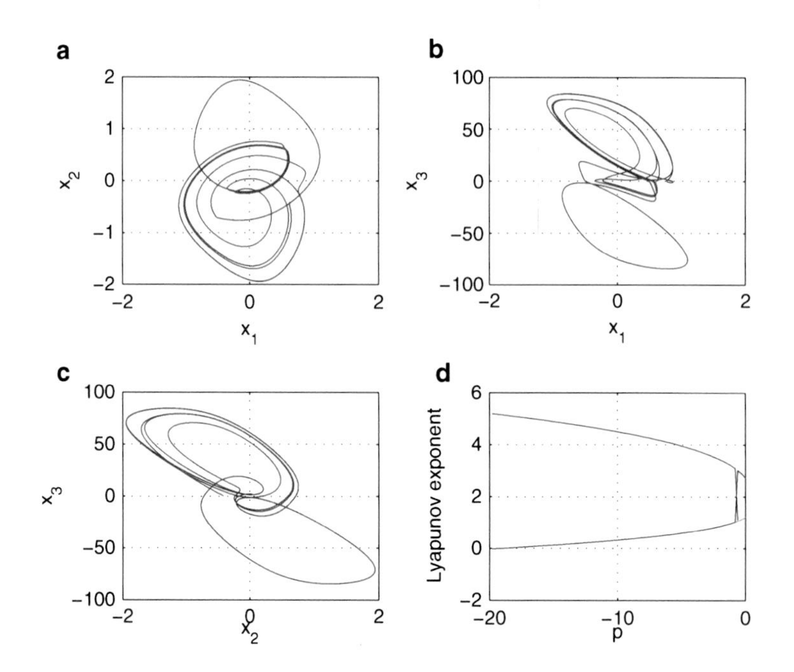

Fig. 2.23 (**a**)–(**c**) Phase diagrams showing limit cycles of the Hopfield model, (**d**) Lyapunov exponents vs varying parameters

of oscillators have been given. Moreover, using mathematical models of oscillators and biological neurons the following properties have been analyzed: phase diagram, isoclines, attractors, local stability, bifurcations of fixed points, and chaos properties.

Chapter 3
Bifurcations and Limit Cycles in Models of Biological Systems

Abstract The chapter proposes a systematic method for fixed point bifurcation analysis in circadian cells and similar biological models using interval polynomials theory. The stages for performing fixed point bifurcation analysis in such biological systems comprise (i) the computation of fixed points as functions of the bifurcation parameter and (ii) the evaluation of the type of stability for each fixed point through the computation of the eigenvalues of the Jacobian matrix that is associated with the system's nonlinear dynamics model. Stage (ii) requires the computation of the roots of the characteristic polynomial of the Jacobian matrix. This problem is nontrivial since the coefficients of the characteristic polynomial are functions of the bifurcation parameter and the latter varies within intervals. To obtain a clear view about the values of the roots of the characteristic polynomial and about the stability features they provide to the system, the use of interval polynomials theory and particularly of Kharitonov's stability theorem is proposed. In this approach the study of the stability of a characteristic polynomial with coefficients that vary in intervals is equivalent to the study of the stability of four polynomials with crisp coefficients computed from the boundaries of the aforementioned intervals. The efficiency of the proposed approach for the analysis of fixed points bifurcations in nonlinear models of biological neurons is tested through numerical and simulation experiments.

3.1 Outline

The chapter analyzes the use of interval polynomial theory for the study of fixed point bifurcations and the associated stability characteristics in circadian cells and similar biological models. The bifurcation parameter is usually one of the coefficients of the biological system or a feedback control gain that is used to modify the system's dynamics. The bifurcation parameter varies within intervals; therefore, assessment of the stability features for all fixed points of the system (each fixed point depends on a different value of the bifurcation parameter) requires extended computations. Therefore, it is significant to develop methods that enable to draw

G.G. Rigatos, *Advanced Models of Neural Networks*,
DOI 10.1007/978-3-662-43764-3_3,

a conclusion about the stability features of larger sets of fixed points where each set is associated with an interval from which the bifurcation parameter takes its values. A solution to this problem can be obtained from interval polynomials theory and Kharitonov's theorem [162, 192].

Aiming at understanding how the dynamics of neurons is modified due to external stimuli and parametric variations the topic of fixed point bifurcations in such biological models has been extensively studied during the last years [116, 204, 214]. Similar results and methods have been extended to artificial neural networks, such as Hopfield associative memories [76, 178]. Additionally, studies on bifurcations appearing in biological models such as circadian cells show how the dynamics and the stability of such models can be affected by variations of external inputs and internal parameters [128, 193]. A subject of particular interest has been also the control of bifurcations appearing in biological systems which is implemented with state feedback and which finally succeeds to modify the biological oscillator's dynamics and its convergence to attractors [49, 50, 54, 131, 132, 199, 205].

In this chapter, as a case study, bifurcations of fixed points in two particular types of biological models will be examined. The first model is the FitzHugh–Nagumo neuron. The voltage of the neurons membrane exhibits oscillatory variations after receiving suitable external excitation either when the neuron is independent from neighboring neural cells or when the neuron is coupled to neighboring neural cells through synapses or gap junctions, as shown in Chap. 1. The FitzHugh–Nagumo model of biological neurons is a simplified model (second order) of the Hodgkin–Huxley model (fourth order), where the latter is considered to reproduce efficiently in several cases the neuron's dynamics. The FitzHugh–Nagumo model describes the variations of the voltage of the neuron's membrane as a function of the ionic currents that get through the membrane and of an external current that is applied as input to the neuron [5, 16, 161, 220]. The second model is that of Neurospora circadian cells. Circadian cells perform protein synthesis within a feedback loop. The concentration of the produced proteins exhibits periodic variations in time. The circadian cells regulate the levels of specific proteins' concentration through an RNA transcription and translation procedure [102, 104, 189]. Circadian neurons in the hypothalamus are also responsible for synchronization and periodic functioning of several organs in the human body.

Typically, the stages for analyzing the stability features of the fixed points that lie on bifurcation branches comprise (i) the computation of fixed points as functions of the bifurcation parameter and (ii) the evaluation of the type of stability for each fixed point through the computation of the eigenvalues of the Jacobian matrix that is associated with the system's nonlinear dynamics model. Stage (ii) requires the computation of the roots of the characteristic polynomial of the Jacobian matrix. This problem is nontrivial since the coefficients of the characteristic polynomial are functions of the bifurcation parameter and the latter varies within intervals [36, 113, 129]. To obtain a clear view about the values of the roots of the characteristic polynomial and about the stability features they provide to the system the use of interval polynomials theory and particularly of Kharitonov's stability theorem is proposed [162, 192]. In this approach the study of the stability of a characterized polynomial with coefficients that vary in intervals is equivalent to the study of the stability of four polynomials with crisp coefficients computed from the boundaries of the aforementioned intervals.

3.2 Generalization of the Routh–Hurwitz Criterion with Kharitonov's Theorem

3.2.1 *Application of the Routh Criterion to Systems with Parametric Uncertainty*

The following characteristic polynomial is considered

$$p(\lambda) = c_n\lambda^n + c_{n-1}\lambda^{n-1} + \cdots + c_1\lambda + c_0 \tag{3.1}$$

The polynomial's coefficients vary within intervals, that is

$$c_i^{\min} \leq c_i \leq c_i{}^{\max} \; i = 0, 1, 2, \cdots, n \tag{3.2}$$

The study of the stability of the above characteristic polynomial is now transferred to the study of the stability of four Kharitonov polynomials, where each polynomial is written as

$$p_i(\lambda) = c_n^i\lambda^n + c_{n-1}^i\lambda^{n-1} + \cdots + c_1^i\lambda + c_0^i \; i = 1, \cdots, 4 \tag{3.3}$$

The associated coefficients are written as

$$c_{2k}^i, \; c_{2k+1}^i \; k = m, m-1, \cdots, 0, 1 \tag{3.4}$$

where $m = n/2$ if n is an even number and $m = (n-1)/2$ if n is an odd number.

For the first Kharitonov polynomial $p_1(s) = c_n^1 s^n + c_{n-1}^1 s^{n-1} + \cdots + c_1^1 s + c_0^1$ the coefficients are

$$\begin{aligned} c_{2k}^1 &= c_{2k}^{\max} \text{ if } k \text{ is an even number, } c_{2k+1}^1 = c_{2k+1}^{\max} \text{ if } k \text{ is an even number} \\ c_{2k}^1 &= c_{2k}^{\min} \text{ if } k \text{ is an odd number, } c_{2k+1}^1 = c_{2k+1}^{\min} \text{ if } k \text{ is an odd number} \end{aligned} \tag{3.5}$$

For the second Kharitonov polynomial $p_2(s) = c_n^2 s^n + c_{n-1}^2 s^{n-1} + \cdots + c_1^2 s + c_0^2$ the coefficients are

$$\begin{aligned} c_{2k}^2 &= c_{2k}^{\min} \text{ if } k \text{ is an even number, } c_{2k+1}^2 = c_{2k+1}^{\min} \text{ if } k \text{ is an even number} \\ c_{2k}^2 &= c_{2k}^{\max} \text{ if } k \text{ is an odd number, } c_{2k+1}^2 = c_{2k+1}^{\max} \text{ if } k \text{ is an odd number} \end{aligned} \tag{3.6}$$

For the third Kharitonov polynomial $p_3(s) = c_n^3 s^n + c_{n-1}^3 s^{n-1} + \cdots + c_1^3 s + c_0^3$ the coefficients are

$$\begin{array}{ll} c_{2k}^3 = c_{2k}^{\min} \text{ if } k \text{ is an even number}, & c_{2k+1}^3 = c_{2k+1}^{\max} \text{ if } k \text{ is an even number} \\ c_{2k}^3 = c_{2k}^{\max} \text{ if } k \text{ is an odd number}, & c_{2k+1}^3 = c_{2k+1}^{\min} \text{ if } k \text{ is an odd number} \end{array} \quad (3.7)$$

For the fourth Kharitonov polynomial $p4(s) = c_n^4 s^n + c_{n-1}^4 s^{n-1} + \cdots + c_1^4 s + c_0^4$ the coefficients are

$$\begin{array}{ll} c_{2k}^4 = c_{2k}^{\max} \text{ if } k \text{ is an even number}, & c_{2k+1}^4 = c_{2k+1}^{\min} \text{ if } k \text{ is an even number} \\ c_{2k}^4 = c_{2k}^{\min} \text{ if } k \text{ is an odd number}, & c_{2k+1}^4 = c_{2k+1}^{\max} \text{ if } k \text{ is an odd number} \end{array} \quad (3.8)$$

3.2.2 An Application Example

As an example the following characteristic polynomial is considered $p(\lambda) = \lambda^3 + (a+b)\lambda^2 + ab\lambda + K$ with parametric uncertainty given by $K \in [8, 12]$, $a \in [2.5, 3.5]$, $b \in [3.8, 4.2]$. The coefficients of the characteristic polynomial have the following variation ranges: $c_3 = 1$, $c_2 = (a+b) \in [6.3, 7.7]$, $c1 = ab \in [9.5, 14.7]$ and $c_0 = K \in [8, 12]$. The associated Kharitonov polynomials are

$$\begin{aligned} p_1(\lambda) &= 12 + 14.7\lambda + 6.3\lambda^2 + \lambda^3 \\ p_2(\lambda) &= 8 + 14.7\lambda + 7.7\lambda^2 + \lambda^3 \\ p_3(\lambda) &= 8 + 9.5\lambda + 7.7\lambda^2 + \lambda^3 \\ p_4(\lambda) &= 12 + 9.5\lambda + 6.3\lambda^2 + \lambda^3 \end{aligned} \quad (3.9)$$

Next, Routh–Hurwitz criterion is applied for each one of the four Kharitonov polynomials. For polynomial $p_1(\lambda)$

$$\begin{array}{l|ll} \lambda^3 & 1 & 14.7 \\ \lambda^2 & 6.3 & 12 \\ \lambda^1 & 12.795 & 0 \\ \lambda^0 & 12 & \end{array} \quad (3.10)$$

Since there is no change of sign in the first column of the Routh matrix it can be concluded that polynomial $p_1(\lambda)$ has stable roots. For polynomial $p_2(\lambda)$

$$\begin{array}{l|ll} \lambda^3 & 1 & 14.7 \\ \lambda^2 & 7.7 & 8 \\ \lambda^1 & 13.6610 & 0 \\ \lambda^0 & 8 & \end{array} \quad (3.11)$$

Since there is no change of sign in the first column of the Routh matrix it can be concluded that polynomial $p_2(\lambda)$ has stable roots. For polynomial $p_3(\lambda)$

$$\begin{array}{l|ll} \lambda^3 & 1 & 9.5 \\ \lambda^2 & 7.7 & 8 \\ \lambda^1 & 8.4610 & 0 \\ \lambda^0 & 8 & \end{array} \tag{3.12}$$

Since there is no change of sign in the first column of the Routh matrix it can be concluded that polynomial $p_3(\lambda)$ has stable roots. For polynomial $p_4(\lambda)$

$$\begin{array}{l|ll} \lambda^3 & 1 & 9.5 \\ \lambda^2 & 6.3 & 12 \\ \lambda^1 & 7.5952 & 0 \\ \lambda^0 & 12 & \end{array} \tag{3.13}$$

Since there is no change of sign in the first column of the Routh matrix it can be concluded that polynomial $p_4(\lambda)$ has stable roots.

Using that all four Kharitonov characteristic polynomials $p^i(\lambda)$ have stable roots it can be concluded that the initial characteristic polynomial $p(\lambda)$ has stable roots when its parameters vary in the previously defined ranges. Equivalently, the computation of the four Kharitonov characteristic polynomials can be performed as follows:

$$\begin{array}{l} p_1(\lambda) = c_0^{\max} + c_1^{\max}\lambda + c_2^{\min}\lambda^2 + c_3^{\min}\lambda^3 + c_4^{\max}\lambda^4 + c_5^{\max}\lambda^5 + c_6^{\min}\lambda^6 + \cdots \\ p_2(\lambda) = c_0^{\min} + c_1^{\min}\lambda + c_2^{\max}\lambda^2 + c_3^{\max}\lambda^3 + c_4^{\min}\lambda^4 + c_5^{\min}\lambda^5 + c_6^{\max}\lambda^6 + \cdots \\ p_3(\lambda) = c_0^{\min} + c_1^{\max}\lambda + c_2^{\max}\lambda^2 + c_3^{\min}\lambda^3 + c_4^{\min}\lambda^4 + \lambda_5^{\max}\lambda^5 + c_6^{\max}\lambda^6 + \cdots \\ p_4(\lambda) = c_0^{\max} + c_1^{\min}\lambda + c_2^{\min}\lambda^2 + c_3^{\max}\lambda^3 + c_4^{\max}\lambda^4 + c_5^{\min}\lambda^5 + c_6^{\min}\lambda^6 + \cdots \end{array} \tag{3.14}$$

while an equivalent statement of Kharitonov's theorem is as follows [162, 192]:

Theorem. *Each characteristic polynomial in the interval polynomials family* $p(\lambda)$ *is stable if and only if all four Kharitonov polynomials* $p_i(\lambda)$, $i = 1, \cdots, 4$ *are stable.*

3.3 Stages in Bifurcations Analysis

The classification of fixed points according to their stability features has been given in Sect. 2.4. Here the classification of fixed points' bifurcations is overviewed. Bifurcation of the equilibrium point means that the locus of the equilibrium on the plane (having as horizontal axis the bifurcation parameter and as vertical axis the fixed point variable) changes due to variation of the system's parameters. Equilibrium

x^* is a hyperbolic equilibrium point if the real parts of all its eigenvalues are non-zero. One can have stable or unstable bifurcations of hyperbolic equilibria. A saddle-node bifurcation occurs when there are eigenvalues of the Jacobian matrix with negative real part and other eigenvalues with non-negative real part. A Hopf bifurcation appears when the hyperbolicity of the equilibrium point is lost due to variation of the system parameters and the eigenvalues become purely imaginary. By changing the values of the parameters at a Hopf bifurcation, an oscillatory solution appears.

The stages for finding bifurcations in nonlinear dynamical systems are described next. The following autonomous differential equation is considered:

$$\frac{dx}{dt} = f_1(x, q) \tag{3.15}$$

where x is the state vector $x \in R^n$ and $q \in R^m$ is the vector of the system parameters. In Eq. (3.15) a point x^* satisfying $f_1(x^*) = 0$ is an equilibrium point. Therefore from the condition $f_1(x^*) = 0$ one obtains a set of equations which provide the equilibrium point as function of the bifurcating parameter. The stability of the equilibrium point can be evaluated by linearizing the system's dynamic model around the equilibrium point and by computing eigenvalues of the Jacobian matrix. The Jacobian matrix at the equilibrium point can be written as

$$J_{f_1(x^*)} = \frac{\partial f_1(x)}{\partial x}|_{x=x^*} \tag{3.16}$$

and the determinant of the Jacobian matrix provides the characteristic equation which is given by

$$\det(\lambda_i I_n - J_{f_1(x^*)}) = c_n \lambda_1^n + c_{n-1} \lambda_i^{n-1} + \cdots + c_1 \lambda_i + c_0 = 0, \tag{3.17}$$

where I_n is the $n \times n$ identity matrix, μ_i with $i = 1, 2, \cdots, n$ denotes the eigenvalues of the Jacobian matrix $J_{f_1(x^*)}$. Using the above characteristic polynomial one can formulate conditions for the system to have stable or unstable fixed points or saddle-node fixed points. From the requirement the eigenvalues of the system to be purely imaginary one obtains a condition, i.e. values that the bifurcating parameter should take, so as Hopf bifurcations to appear.

3.4 Bifurcation Analysis of the FitzHugh–Nagumo Neuron

3.4.1 Equilibria and Stability of the FitzHugh–Nagumo Neuron

The model of the dynamics of the FitzHugh–Nagumo neuron is considered

$$\begin{aligned} \frac{dV}{dt} &= V(V-a)(1-V) - w + I \\ \frac{dw}{dt} &= \epsilon(V - \gamma w) \end{aligned} \tag{3.18}$$

By defining the state variables $x_1 = w$, $x_2 = V$ and setting current I as the external control input one has

$$\begin{gathered}\frac{dx_1}{dt} = -\epsilon\gamma x_1 + \epsilon x_2 \\ \frac{dx_2}{dt} = x_2(x_2 - a)(1 - x_2) - x_1 + u \\ \\ y = h(x) = x_1\end{gathered} \tag{3.19}$$

Equivalently one has

$$\begin{gathered}\begin{pmatrix}\dot{x}_1 \\ \dot{x}_2\end{pmatrix} = \begin{pmatrix}-\epsilon\gamma x_1 + \epsilon x_2 \\ x_2(x_2 - a)(1 - x_2) - x_1\end{pmatrix} + \begin{pmatrix}0 \\ 1\end{pmatrix} u \\ \\ y = h(x) = x_1\end{gathered} \tag{3.20}$$

Next, the fixed points of the dynamic model of the FitzHugh–Nagumo neuron are computed. It holds that

$$\begin{gathered}\dot{x}_1 = -\epsilon\gamma x_1 + \epsilon x_2 \\ \dot{x}_2 = x_2(x_2 - \alpha)(1 - x_2) - x_1 + I\end{gathered} \tag{3.21}$$

Setting $I = 0$, and setting $\dot{x}_1 = 0$ and $\dot{x}_2 = 0$ (nullclines), the associated fixed points are found

$$\begin{gathered}x_2 = -\gamma x_1 \\ x_1(-\gamma^3 x_1^2 + (a\gamma^2 + \gamma^2)x_1 - \alpha\gamma - 1) = 0\end{gathered} \tag{3.22}$$

The numerical values of the parameters are taken to be $\gamma = 0.2$ and $a = 7$. Then by substituting the relation for x_2 from the first row of Eq. (3.22) into the second row one obtains

$$x_1[\gamma^3 x_1^2 + \gamma^2(1 + a)x_1 + (a\gamma - 1)] = 0 \tag{3.23}$$

From the above conditions one has the first fixed point

$$x_1^{*,1} = 0 \; x_2^{*,1} = 0 \tag{3.24}$$

whereas, since the determinant $\Delta = \gamma^4(1 + a) - 2a\gamma^4 - 4\gamma^3) < 0$ no other real valued solutions for (x_1, x_2) can be found and consequently no other fixed points can be considered.

Next, the type of stability that is associated with the fixed point $x_1^{*,1} = 0$, $x_2^{*,1} = 0$ is examined. The Jacobian matrices of the right-hand side of Eq. (3.21) are computed and the associated characteristic polynomials are found

$$J = \begin{pmatrix} \frac{\partial f_1}{\partial x_1} & \frac{\partial f_1}{\partial x_2} \\ \frac{\partial f_2}{\partial x_1} & \frac{\partial f_2}{\partial x_2} \end{pmatrix} = \begin{pmatrix} -\epsilon\gamma & \epsilon \\ -1 & -3x_2^2 + 2(1+a)x_2 - a \end{pmatrix} \tag{3.25}$$

Stability at the fixed point $x_1 = 0$ and $x_2 = 0$: Substituting the values of x_1 and x_2 in Eq. (3.27) the obtained characteristic polynomial is $p(\lambda) = \lambda^2 + (a + \epsilon\gamma)\lambda + (\epsilon\alpha\gamma + \epsilon)$.

Next, it is considered that $a \in [a_{\min}, a_{\max}] = [6.8, 7.2]$, whereas $\epsilon = 1.5$ and $\gamma = 0.2$ are crisp numerical values. Then for the coefficients of the characteristic polynomial $p(s) = \lambda^2 + k_1\lambda + k_2$ one has $k_1 = (a + \epsilon\gamma) \in [7.1, 7.5]$ and $k_2 = (\epsilon a\gamma) + \epsilon \in [3.54, 3.66]$.

The application of the Routh–Hurwitz criterion gives

$$\begin{array}{l|ll} \lambda^2 & 1 & k_2 \\ \lambda^1 & k_1 & 0 \\ \lambda^0 & k_2 & \end{array} \tag{3.26}$$

Since $k_1 > 0$ and $k_2 > 0$ there is no change of sign in the first column of the Routh matrix and the roots of the characteristic polynomial $p(\lambda)$ are stable. Therefore, the equilibrium $(0, 0)$ is a stable one.

3.4.2 *Condition for the Appearance of Limit Cycles*

Finally, it is noted that the appearance of the limit cycles in the FitzHugh–Nagumo neuron model is possible. For instance, if state feedback is implemented in the form $I = -qx_2$, then $(x_1^*, x_2^*) = (0, 0)$ is again a fixed point for the model while the Jacobian matrix J at (x_1^*, x_2^*) becomes

$$J = \begin{pmatrix} -\epsilon\gamma & \epsilon \\ -1 & -3x_2^2 + 2(1+a)x_2 - a - q \end{pmatrix} \tag{3.27}$$

The characteristic polynomial of the above Jacobian matrix is $p(\lambda) = \lambda^2 + (a + q + \epsilon\gamma\lambda + (\epsilon\gamma q + \epsilon))$. Setting the feedback gain $q = -a - \epsilon\gamma$ the characteristic polynomial becomes $p(\lambda) = \lambda^2 + (\epsilon^2\gamma^2 + \alpha\epsilon\gamma - \epsilon)$. Then, the condition $(\alpha\epsilon\gamma - \epsilon) > 0$ with $a > 0$, $\epsilon > 0$ and $\gamma > 0$ suffices for the appearance of imaginary eigenvalues. In such a case the neuron model exhibits sustained oscillations (limit cycles).

3.5 Bifurcation Analysis of Circadian Oscillators

3.5.1 *Fixed Points Bifurcation Analysis Using Kharitonov's Theory*

The model of the circadian cells is given by

$$\dot{x}_1 = v_s \frac{K_i^n}{K_i^n + x_3^n} - v_m \frac{x_1}{K_m + x_1} \tag{3.28}$$

$$\dot{x}_2 = -v_d \frac{x_2}{K_d + x_2} - K_1 x_2 + K_2 x_3 + x_1 K_s \tag{3.29}$$

$$\dot{x}_3 = K_1 x_2 - K_2 x_3 \tag{3.30}$$

The model's parameters are defined as follows: v_s denotes the rate of transcription of the *frq* gene into FRQ protein inside the nucleus. K_i represents the threshold constant beyond which nuclear FRQ protein inhibits transcription of the *frq* gene. n is the Hill coefficient showing the degree of cooperativity of the inhibition process, v_m is the maximum rate of *frq* mRNA degradation, K_M is the Michaelis constant related to the transcription process. Parameter K_s defines the rate of FRQ synthesis and stands for the control input to the model. Parameter K_1 denotes the rate of transfer of FRQ protein from the cytoplasm to the nucleus and K_2 denotes the rate of transfer of the FRQ protein from the nucleus into the cytoplasm. Parameter v_d denotes the maximum rate of degradation of the FRQ protein and K_d is the Michaelis constant associated with this process.

Fixed points are computed from the nullclines $\dot{x}_1 = 0$, $\dot{x}_2 = 0$, and $\dot{x}_3 = 0$. From Eq. (3.28) one has

$$x_3 = \tfrac{K_1}{K_2} x_2 \tag{3.31}$$

By substituting Eq. (3.31) and after intermediate operations one obtains x_1 also as a function of x_2 that is

$$x_1 = \tfrac{K_2(v_d + K_1 K_d)x_2 + K_1 K_2 K_d x_2^2 - (K_2 K_d + K_2 x_2) K_1 x_2}{K_2(K_s K_d + K_s x_2)} \tag{3.32}$$

By substituting Eqs. (3.31) and (3.32) into Eq. (3.28) and after intermediate operations one obtains the equation

$$\begin{aligned}-v_m k_1^2(k_d - 1)x_2^3 + [v_m K_i^n K_1 K_2 K_d - v_m K_i^n K_1 K_2 + v_m K_1 v_d - v_s K_i^n K_1 K_2 K_d \\ + v_s K_i^n K_1 K_2]x_2^2 + [v_m K_i^n K_2 v_d - v_s K_i^n K_m K_s K_2 - v_s K_1^n K_2(v_d + K_1 K_d) + v_s \\ + K_i^n K_1 K_2 K_d]x_2 - v_s K_i^n K_m K_2 K_s K_d\end{aligned} \tag{3.33}$$

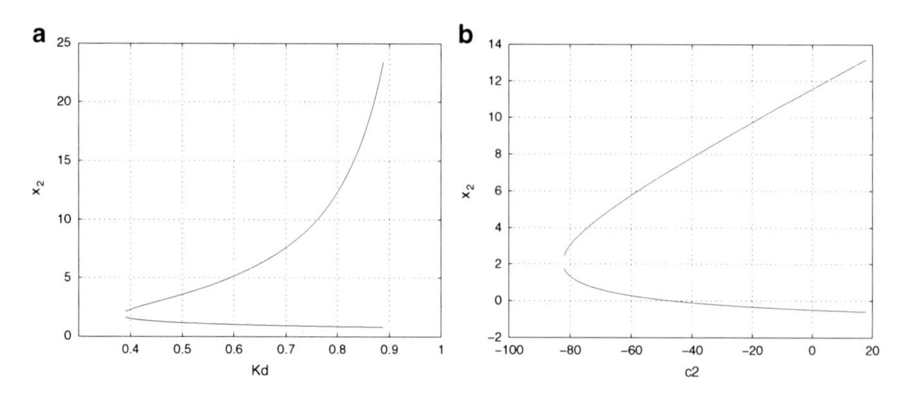

Fig. 3.1 (**a**) Bifurcation of fixed point $x_2^{*,2}$ (*blue line*) and $x_2^{*,3}$ (*red line*) of the circadian oscillators under variation of the Michaelis parameter K_d, (**b**) Bifurcation of fixed point $x_2^{*,2}$ (*blue line*) and $x_2^{*,3}$ (*red line*) of the circadian oscillators under state feedback with gain c_2

The control input of the circadian oscillator is taken to be zero, that is $K_s = 0$. It is assumed that the bifurcating parameter is K_d (Michaelis constant). By substituting the numerical values of the parameters of the model into the characteristic polynomial one obtains

$$x_2\{0.3750(k_d - 1)x_2^2 + (0.2913K_d + 0.7587)x_2 + (0.0039K_d - 0.8196)\} = 0 \tag{3.34}$$

A first fixed point is located at

$$(x_1^{*,1}, x_2^{*,1}, x_3^{*,1}) = (0, 0, 0) \tag{3.35}$$

For the binomial appearing in the above equation the determinant is $\Delta = 0.0791K_d^2 + 1.6773K_d - 0.6536$ which is positive for $K_d \geq 0.3891$. In such a case there are two more fixed points at

$$x_2 = \frac{-(0.2913K_d + 0.7587) + \sqrt{0.0791K_d^2 + 1.6773K_d - 0.6538}}{0.75(K_d - 1)} \tag{3.36}$$

$$x_2 = \frac{-(0.2913K_d + 0.7587) - \sqrt{0.0791K_d^2 + 1.6773K_d - 0.6538}}{0.75(K_d - 1)} \tag{3.37}$$

The bifurcation diagram of the fixed points considering as bifurcation parameter the Michaelis constant K_d is given in Fig. 3.1a. Next, the Jacobian matrix of the dynamic model of the system is computed, which is given by

$$J = \begin{pmatrix} -v_m \frac{K_M}{(K_M+x_1)^2} & 0 & -\frac{K_i^n n x_3^{n-1}}{(K_i^n+x_3^n)^2} \\ 0 & -\frac{v_d K_d}{(K_d+x_2)^2} - K_1 & K_2 \\ 0 & K_1 & -K_2 \end{pmatrix} \tag{3.38}$$

The associated characteristic polynomial is

$$p(\lambda) = \lambda^3 + [\tfrac{v_d}{K_d} + K_1 + K_2 + \tfrac{v_m}{K_M}]\lambda^2 + [\tfrac{K_2 v_d}{K_d} + \tfrac{v_M}{K_M}(\tfrac{v_d}{K_d}) + K_1 + K_2]\lambda + \tfrac{v_M}{K_M}\tfrac{K_2 v_d}{K_d}. \tag{3.39}$$

The variation ranges for parameter K_d are taken to be $K_d \in [K_d^{\min}, K_d^{\max}] = [0.4, 0.9]$. Then the variation ranges of the rest of the parameters of the characteristic polynomial are: $c_2 = [\frac{v_d}{K_d} + K_1 + K_2 + \frac{v_M}{K_M}] \in [3.59, 5.54]$, $c_1 = \frac{K_2 v_d}{K_d} + \frac{v_M}{K_M}(\frac{v_d}{K_d}) + K_1 + K_2 \in [3.42, 6.41]$ and $c_0 = \frac{v_M}{K_M}\frac{K_2 v_d}{K_d} \in [0.88, 1.97]$.

Next, the four Kharitonov polynomials are computed as follows:

$$\begin{aligned} p_1(\lambda) &= c_0^{\max} + c_1^{\max}\lambda + c_2^{\min}\lambda^2 + c_3^{\min}\lambda^3 \\ p_2(\lambda) &= c_0^{\min} + c_1^{\min}\lambda + c_2^{\max}\lambda^2 + c_3^{\max}\lambda^3 \\ p_3(\lambda) &= c_0^{\min} + c_1^{\max}\lambda + c_2^{\max}\lambda^2 + c_3^{\min}\lambda^3 \\ p_4(\lambda) &= c_0^{\max} + c_1^{\min}\lambda + c_2^{\min}\lambda^2 + c_3^{\max}\lambda^3 \end{aligned} \tag{3.40}$$

which take the values

$$\begin{aligned} p_1(\lambda) &= 1.97 + 6.41\lambda + 3.59\lambda^2 + \lambda^3 \\ p_2(\lambda) &= 0.88 + 3.42\lambda + 5.54\lambda^2 + \lambda^3 \\ p_3(\lambda) &= 0.88 + 6.41\lambda + 5.54\lambda^2 + \lambda^3 \\ p_4(\lambda) &= 1.97 + 3.42\lambda + 3.59\lambda^2 + \lambda^3 \end{aligned} \tag{3.41}$$

The Routh criterion is applied for each characteristic polynomial $p_i(\lambda)$ $i = 1, \cdots, 4$. For polynomial $p_1(\lambda)$ it holds

$$\begin{array}{l|ll} \lambda^3 & 1 & 6.41 \\ \lambda^2 & 3.59 & 1.97 \\ \lambda^1 & 5.86 & 0 \\ \lambda^0 & 1.97 & \end{array} \tag{3.42}$$

Since there is no change of sign in the first column of the Routh matrix it can be concluded that the characteristic polynomial is a stable one. For polynomial $p_2(\lambda)$ it holds

$$\begin{array}{l|ll} \lambda^3 & 1 & 3.42 \\ \lambda^2 & 5.54 & 0.88 \\ \lambda^1 & 3.26 & 0 \\ \lambda^0 & 0.88 & \end{array} \tag{3.43}$$

Since there is no change of sign in the first column of the Routh matrix it can be concluded that the characteristic polynomial is a stable one. For polynomial $p_3(\lambda)$ it holds

$$\begin{array}{l|ll} \lambda^3 & 1 & 6.41 \\ \lambda^2 & 5.54 & 0.88 \\ \lambda^1 & 6.25 & 0 \\ \lambda^0 & 0.88 & \end{array} \tag{3.44}$$

Since there is no change of sign in the first column of the Routh matrix it can be concluded that the characteristic polynomial is a stable one. For polynomial $p_4(\lambda)$ it holds

$$\begin{array}{l|ll} \lambda^3 & 1 & 3.42 \\ \lambda^2 & 3.59 & 1.97 \\ \lambda^1 & 2.87 & 0 \\ \lambda^0 & 1.97 & \end{array} \tag{3.45}$$

Since there is no change of sign in the first column of the Routh matrix it can be concluded that the characteristic polynomial is a stable one. Consequently, the fixed point $(x_1^{*,1}, x_2^{*,1}, x_3^{*,1}) = (0, 0, 0)$ is a stable one (sink) for the considered variations of K_p.

3.5.2 *Method to Detect Hopf Bifurcations in the Circadian Cells*

Using the characteristic polynomial of the linearized equivalent of the biological model, it is possible to formulate conditions for the emergence of Hopf bifurcations. The whole procedure makes again use of the Routh–Hurwitz criterion.

Consider the characteristic polynomial computed at a fixed point

$$P(\lambda) = c_n\lambda^n + c_{n-1}\lambda^{n-1} + c_{n-2}\lambda^{n-2} + \cdots + c_1\lambda + c_0 \tag{3.46}$$

The coefficients $c_i, i = 0\cdots, n$ are functions of the bifurcating parameter, which is denoted as μ.

The following matrix is introduced

$$\Delta_j = \begin{pmatrix} c_{n-1} & c_n & 0 & 0 & \cdots & 0 & 0 \\ c_{n-3} & c_{n-2} & c_{n-1} & 0 & \cdots & 0 & 0 \\ c_{n-5} & c_{n-4} & c_{n-3} & c_{n-2} & \cdots & 0 & 0 \\ \cdots & \cdots & \cdots & \cdots & \cdots & \cdots & \cdots \\ c_{n-(2j-1)} & c_{n-(2j-2)} & c_{n-(2j-3)} & c_{n-(2j-4)} & \cdots & c_{n-(2j-n+1)} & c_{n-(2j-n)} \end{pmatrix} \tag{3.47}$$

where coefficients $c_{n-(2j-m)}$ with $j = 1,2,\cdots,n$ and $m = 1,2,\cdots,n$ become zero if $n-(2j-m)<0$ or if $n-(2j-m)>n$.

The Routh–Hurwitz determinants show the existence of eigenvalues which are conjugate imaginary numbers. A series of matrices containing the coefficients c_j are defined as follows:

$$\Delta_1 = c_{n-1}$$
$$\Delta_2 = \begin{pmatrix} c_{n-1} & c_n \\ c_{n-3} & c_{n-2} \end{pmatrix} \tag{3.48}$$
$$\Delta_3 = \begin{pmatrix} c_{n-1} & c_n & 0 \\ c_{n-3} & c_{n-2} & c_{n-1} \\ c_{n-5} & c_{n-4} & c_{n-3} \end{pmatrix}$$

and so on. Each matrix Δ_j, $j = 1,\cdots,n-1$ is square and the first column contains every other coefficient of the characteristic polynomial $c_{n-1}, c_{n-3}, \cdots$. The roots of the characteristic polynomial $p(\lambda)$ have negative real parts if and only if $\det(\Delta_j) > 0$ for $j = 1,\cdots,n$. The following two remarks can be made:

(i) Since $\det(\Delta_n) = c_0\det(\Delta_{n-1})$ this means that a necessary condition for the roots of $p(\lambda)$ to have negative real part is $c_0 > 0$. If $c_0 = 0$, then there is a zero eigenvalue.
(ii) If $\det(\Delta_j) > 0$ for all $j = 1,2,\cdots,n-2$ and $\det(\Delta_{n-1}) = 0$, then the characteristic polynomial contains imaginary roots. Provided that the rest of the roots of the characteristic polynomial have negative real part then Hopf bifurcations appear. An additional condition for the appearance of Hopf bifurcations is that the derivative of the determinant $\det(\Delta_{n-1}) = 0$ with respect to the bifurcating parameter is not zero. That is $\frac{d}{d\mu}\det(\Delta_{n-1}) \neq 0$.

According to the above criteria (i) and (ii) one can determine where possible saddle-node bifurcations (eigenvalue 0) and Hopf bifurcations (imaginary eigenvalues) appear. Therefore, one can formulate conditions about the values of the bifurcating parameter that result in a particular type of bifurcations.

Equivalently, conditions about the appearance of Hopf bifurcations can be obtained from the Routh table according to Sect. 3.2. The zeroing of certain rows of the Routh table gives an indication that the characteristic polynomial contains imaginary conjugate eigenvalues. By requiring the elements of these rows of the Routh matrix to be set equal to zero one finds the values of the bifurcating parameter that result in the appearance of Hopf bifurcations.

For the previous characteristic polynomial of the circadian cells one has

$$p(\lambda) = \lambda^3 + [\tfrac{v_d}{K_d} + K_1 + K_2 + \tfrac{v_M}{K_M}]\lambda^2 + [\tfrac{K_2 v_d}{K_d} + \tfrac{v_M}{K_M}(\tfrac{v_d}{K_d} + K_1 + K_2)]\lambda + \tfrac{v_M}{K_M}\tfrac{K_2 v_d}{K_d} \tag{3.49}$$

which after the substitution of numerical values for its parameters becomes

$$p(\lambda) = \lambda^3 + [\tfrac{1.4}{K_d} + 2.0375]\lambda^2 + [\tfrac{2.1525}{K_d} + 1.0313]\lambda + \tfrac{0.7875}{K_d} \tag{3.50}$$

Using the Routh–Hurwitz determinants one has

$$\Delta_1 = c_1 = \tfrac{1.4}{K_d} + 2.0375 \tag{3.51}$$

For $\Delta_1 > 0$ it should hold $K_d < -0.6871$ which is not possible since K_d is a positive parameter.

For the determinant of Δ_2 to be zero, where

$$\Delta_2 = \begin{pmatrix} c_1 & c_0 \\ c_3 & c_2 \end{pmatrix} = \begin{pmatrix} \tfrac{1.4}{K_d} + 2.0375 & 1 \\ \tfrac{0.7875}{K_d} & \tfrac{2.1525}{K_d} + 1.0313 \end{pmatrix} \tag{3.52}$$

one obtains the relation $\frac{3.0135}{K_d^2} + \frac{5.0420}{K_d} + 2.1013 = 0$ which in turn gives $K_d = -1.2722$, $K_d = -1.1273$. Again, these values are not acceptable since $K_d > 0$. Therefore, no Hopf bifurcations can appear at the fixed point $(x_1^{*,1}, x_2^{*,1}, x_3^{*,1}) = (0,0,0)$, under the existing set of parameters' values.

At the same conclusion one arrives if the Routh–Hurwitz matrix is used, that is

$$\begin{array}{l|ll} \lambda^3 & 1 & \frac{2.1525}{K_d} \\ \lambda^2 & \frac{1.4}{K_d} + 2.0375 & \frac{0.7875}{K_d} \\ \lambda^1 & \frac{3.0135}{K_d^2} - \frac{5.0420}{K_d} - 2.1013 & 0 \\ \lambda^0 & \frac{0.7875}{K_d} & \end{array} \tag{3.53}$$

For the first coefficient of the second row of the matrix it holds $\frac{1.4}{K_d} + 2.0375 > 0$. To obtain zero coefficients at the line associated with λ_1 (so as imaginary eigenvalues to appear in the characteristic polynomial of the preceding line) one has again the relation $\frac{3.0135}{K_d^2} + \frac{5.0420}{K_d} + 2.1013 = 0$ which in turn gives $K_d = -1.2722$, $K_d = -1.1273$. As mentioned, these values are not acceptable since $K_d > 0$. Therefore, no Hopf bifurcations can appear at the fixed point $(x_1^{*,1}, x_2^{*,1}, x_3^{*,1}) = (0,0,0)$, under the existing set of parameters' values.

3.6 Feedback Control of Bifurcations

Next, it is considered that input K_s is a feedback control term of the form $K_s = c_2 x_2$. Then the dynamics of the circadian oscillator becomes

$$\begin{aligned} \dot{x}_1 &= v_s \frac{K_i^n}{K_i^n + x_3^n} - v_m \frac{x_1}{K_m + x_1} \\ \dot{x}_2 &= -v_d \frac{x_2}{K_d + x_2} - K_1 x_2 + K_2 x_3 + c_2 x_1 x_3 \\ \dot{x}_3 &= K_1 x_2 - K_2 x_3 \end{aligned} \tag{3.54}$$

Feedback can change the oscillation characteristics of the circadian model. It can change the type of stability characterizing the system's fixed points. Additionally fixed points bifurcations can be altered through state feedback control. The previously analyzed Eq. (3.33) is considered again for computing fixed points

$$\begin{array}{c}-v_m k_1^2(K_d-1)x_2^3+[v_m K_i^n K_1 K_2 K_d - v_m K_i^n K_1 K_2 + v_m K_1 v_d - v_s K_i^n K_1 K_2 K_d \\ +v_s K_i^n K_1 K_2]x_2^2 + [v_m K_i^n K_2 v_d - v_s K_i^n K_m K_s K_2 - v_s K_1^n K_2(v_d + K_1 K_d) \\ +v_s K_i^n K_1 K_2 K_d]x_2 - v_s K_i^n K_m K_2 K_s K_d\end{array} \tag{3.55}$$

Next, the nominal value of K_d will be considered known and will be substituted in the characteristic polynomial, while the input K_s will be considered equal to the feedback law of the form $K_s = c_2 x_2$. In such a case the equation for computing x_2 becomes

$$\begin{array}{c}-v_m k_1^2(K_d-1)x_2^3 \\ +[v_m K_i^n K_1 K_2 K_d - v_m K_i^n K_1 K_2 + v_m K_1 v_d - v_s K_i^n K_1 K_2 K_d + v_s K_i^n K_1 K_2 \\ -v_s K_i^n K_m c_2 K_2]x_2^2 + [v_m K_i^n K_2 v_d - v_s K_1^n K_2(v_d + K_1 K_d) + v_s K_i^n K_1 K_2 K_d \\ -v_s K_i^n K_m K_2 c_2 K_d]x_2\end{array} \tag{3.56}$$

Substituting numerical values for the model's parameters and taking $K_d = 1.4$ the equation for computing x_2 becomes

$$x_2\{-0.15x_2^2 + (1.1665 + 0.0126c_2)x_2 + (0.8157 + 0.0176c_2)\} = 0 \tag{3.57}$$

The fixed points for the model of the circadian cells are found to be:

First fixed point: $(x_1^{*,1}, x_2^{*,1}, x_3^{*,1}) = (0,0,0)$ whereas from the computation of the associated determinant one obtains $\Delta = ((1.6665 + 0.0126c_2)^2 + 0.6(0.8157 + 0.0176c_2))$ which is positive if c_2 takes values in the intervals $c_2 < -251.15$ and $c_2 > -82.39$. Thus, for $\Delta > 0$ one has the fixed points expressed as functions of the feedback gain c_2

Second fixed point:

$$\begin{array}{c} x_1^{*,2} = \frac{K_1(K_d-1)x_2{}^{*,2}+v_d}{c_2(k_d+1)} \\ x_2^{*,2} = \frac{-(1.665+0.0126c_2)+\sqrt{1.58\cdot 10^{-4}c_2^2+0.0527c_2+3.2694}}{-0.3} \\ x_3^{*,2} = \frac{K_1}{K_2}x_2{}^{*,2}\end{array} \tag{3.58}$$

Third fixed point:

$$x_1^{*,3} = \frac{K_1(K_d-1){x_2}^{*,3}+v_d}{c_2(k_d+1)}$$
$$x_2^{*,3} = \frac{-(1.665+0.0126c_2)-\sqrt{1.58\cdot 10^{-4}c_2^2+0.0527c_2+3.2694}}{-0.3} \tag{3.59}$$
$$x_3^{*,3} = \frac{K_1}{K_2}{x_2}^{*,3}$$

The bifurcation diagram of the fixed points considering as bifurcating parameter the feedback control gain c_2 is given in Fig. 3.1b. Next, from the right part of Eq. (3.54) the system's Jacobian is computed

$$J = \begin{pmatrix} -v_m\frac{K_m}{(K_m+x_1)^2} & 0 & v_s\frac{-K_i^n n x_3^{n-1}}{(K_i^n+x_3^n)^2} \\ c_2x_2 & -v_d\frac{K_d}{(K_d+x_2)^2} - K_1 + c_2x_1 & K_2 \\ 0 & K_1 & -K_2 \end{pmatrix} \tag{3.60}$$

From the system's Jacobian a generic form of the characteristic polynomial is computed at the fixed point (x_1^*, x_2^*, x_2^*). After intermediate operations one obtains

$$\begin{aligned} \det(\lambda I - J) = \lambda^3 + \left\{\left[k_2 + \frac{v_d k_d}{K_d+{x_2}^2}K_1 - c_2x_1\right] + \left[v_m + \frac{K_m}{K_m+{x_1}^2}\right]\right\}\lambda^2 \\ +\left\{\left[\frac{v_d k_d}{(K_d+x_2)^2}K_1 - K_1K_2\right] + \left[v_m + \frac{K_m}{(K_m+x_1)^2}\right]\left[K_2 + \left(\frac{v_d K_d}{(K_d+x_2)^2}K_1 - c_2x_1\right)\right] - \frac{v_s K_i^n n x_3^{n-1}}{(K_i^n+x_3^n)^2}\right\}\lambda \\ +\left\{\left[v_m + \frac{K_m}{K_m+{x_1}^2}\right]\left[\left(\frac{v_d K_d}{(K_d+x_2)^2}K_1 - c_2x_1\right) - K_1K_2\right] - \frac{v_s K_i^n n x_3^{n-1}}{(K_i^n+x_3^n)^2}\left(\frac{v_d K_d}{K_d+{x_2}^2} + K_1 - c_2x_2\right)\right\} \end{aligned} \tag{3.61}$$

or equivalently

$$\det(\lambda I - J) = \lambda^3 + c_{f_2}\lambda^2 + c_{f_1}\lambda + c_{f_0} \tag{3.62}$$

where

$$\begin{aligned} c_{f_0} = \left\{\left[v_m + \frac{K_m}{K_m+{x_1}^2}\right]\left[\left(\frac{v_d K_d}{(K_d+x_2)^2}K_1 - c_2x_1\right) - K_1K_2\right] \right. \\ \left. - \frac{v_s K_i^n n x_3^{n-1}}{(K_i^n+x_3^n)^2}\left(\frac{v_d K_d}{K_d+{x_2}^2} + K_1 - c_2x_2\right)\right\} \end{aligned} \tag{3.63}$$

$$\begin{aligned} c_{f_1} = \left\{\left[\frac{v_d k_d}{(K_d+x_2)^2}K_1 - K_1K_2\right] + \left[v_m + \frac{K_m}{(K_m+x_1)^2}\right] \right. \\ \left. \left[K_2 + \left(\frac{v_d K_d}{(K_d+x_2)^2}K_1 - c_2x_1\right)\right] - \frac{v_s K_i^n n x_3^{n-1}}{(K_i^n+x_3^n)^2}\right\} \end{aligned} \tag{3.64}$$

$$c_{f_2} = \left\{\left[k_2 + \frac{v_d k_d}{K_d + x_2{}^2} K_1 - c_2 x_1\right] + \left[v_m + \frac{K_m}{K_m + x_1{}^2}\right]\right\} \tag{3.65}$$

The associated Routh matrix is

$$\begin{array}{l|cc} \lambda^3 & 1 & c_{f_1} \\ \lambda^2 & c_{f_2} & c_{f_0} \\ \lambda^1 & c_{f_0} - c_{f_1}c_{f_2} & 0 \\ \lambda^0 & c_{f_0} & \end{array} \tag{3.66}$$

The condition for obtaining Hopf bifurcation is $c_{f_2} > 0$ and $c_{f_0} - c_{f_1}c_{f_2} = 0$. It can be confirmed that for the fixed point $(x_1^{*,2}, x_2^{*,2}, x_3^{*,2})$ and for the previously used set of values of the parameters of the circadian oscillator model, solutions c_2 obtained from the condition $c_{f_0} - c_{f_1}c_{f_2} = 0$ result in $c_{f_2} < 0$, therefore Hopf bifurcations do not appear.

The bifurcation diagram of Fig. 3.1b shows the limit values of c_2 beyond which $x_2^{*,2}$ remains positive. Such a value is $c_2 < -45.2$. It is examined next if for specific range of variations of parameter c_2 the fixed point $(x_1^{*,2}, x_2^{*,2}, x_3^{*,2})$ is a stable or an unstable one. To this end Kharitonov's theorem will be used again. For $c_2 \in [-60.2, -50.2]$ it holds that $c_{f_2} \in [31.31, 31.91]$, while $c_{f_1} \in [68.33, 70.39]$ and $c_{f_0} \in [23.27, 23.67]$. The associated four Kharitonov polynomials take the values

$$\begin{aligned} p_1(\lambda) &= 23.67 + 70.39\lambda + 31.31\lambda^2 + \lambda^3 \\ p_2(\lambda) &= 23.27 + 68.33\lambda + 31.91\lambda^2 + \lambda^3 \\ p_3(\lambda) &= 23.27 + 70.39\lambda + 31.91\lambda^2 + \lambda^3 \\ p_4(\lambda) &= 23.67 + 68.33\lambda + 31.31\lambda^2 + \lambda^3 \end{aligned} \tag{3.67}$$

The Routh table for Kharitonov polynomial $p_1(\lambda)$ is

$$\begin{array}{l|ll} \lambda^3 & 1 & 70.39 \\ \lambda^2 & 31.31 & 23.67 \\ \lambda^1 & 69.63 & 0 \\ \lambda^0 & 23.67 & \end{array} \tag{3.68}$$

Since there is no change of sign of the coefficients of the first column of the Routh table it can be concluded that the characteristic polynomial $p_1(\lambda)$ is a stable one. The associated Routh table for the second Kharitonov polynomial $p_2(\lambda)$ is

$$\begin{array}{l|ll} \lambda^3 & 1 & 68.33 \\ \lambda^2 & 31.91 & 23.27 \\ \lambda^1 & 67.59 & 0 \\ \lambda^0 & 23.27 & \end{array} \tag{3.69}$$

Since there are no changes of sign in the coefficients of the first column of the Routh table it can be concluded that polynomial $p_2(\lambda)$ is also stable. The associated Routh table for Kharitonov polynomials $p_3(\lambda)$ and $p_4(\lambda)$ are

$$\begin{array}{l|ll} \lambda^3 & 1 & 70.39 \\ \lambda^2 & 31.91 & 23.27 \\ \lambda^1 & 69.66 & 0 \\ \lambda^0 & 23.27 & \end{array} \tag{3.70}$$

Since there is no change of sign of the coefficients of the first column of the Routh table it can be concluded that the characteristic polynomial $p_3(\lambda)$ is a stable one. Finally, the associated Routh table for the fourth Kharitonov polynomial $p_4(\lambda)$ is

$$\begin{array}{l|ll} \lambda^3 & 1 & 68.33 \\ \lambda^2 & 31.31 & 23.67 \\ \lambda^1 & 67.57 & 0 \\ \lambda^0 & 23.67 & \end{array} \tag{3.71}$$

Since there are no changes of sign in the coefficients of the first column of the Routh tables it can be concluded that polynomial $p_4(\lambda)$ is stable. From the previous analysis it can be seen that all Kharitonov polynomials have stable roots therefore the fixed point $(x_1^{*,2}, x_2^{*,2}, x_3^{*,2})$ of the circadian oscillator under feedback gain $c_2 \in [-60.2, -50.2]$ is a stable one.

Next, the case that the feedback gain $c_2 \in [-270.2, -260.2]$ is examined. In such a case the variation ranges for the coefficients of the characteristic polynomial are $c_{f_2} \in [-58.52, -55.12]$, $c_{f_1} \in [-175.03, -163.82]$ and $c_{f_0} \in [-88.17, -84.13]$. The four Kharitonov polynomials of the system are:

$$\begin{aligned} p_1(\lambda) &= -84.13 - 163.82\lambda - 58.52\lambda^2 + \lambda^3 \\ p_2(\lambda) &= -88.17 - 175.03\lambda - 55.12\lambda^2 + \lambda^3 \\ p_3(\lambda) &= -88.17 - 163.82\lambda - 55.12\lambda^2 + \lambda^3 \\ p_4(\lambda) &= -84.13 - 175.03\lambda - 58.12\lambda^2 + \lambda^3 \end{aligned} \tag{3.72}$$

The associated Routh table for the first Kharitonov polynomial $p_1(\lambda)$ is

$$\begin{array}{l|ll} \lambda^3 & 1 & -163.82 \\ \lambda^2 & -58.52 & -84.14 \\ \lambda^1 & -165.25 & 0 \\ \lambda^0 & -84.14 & \end{array} \tag{3.73}$$

Since there is change of sign of the coefficients of the first column of the Routh table it can be concluded that the characteristic polynomial $p_1(\lambda)$ is an unstable one. The associated Routh table for Kharitonov polynomial $p_1(\lambda)$ and $p_2(\lambda)$ are

$$\begin{array}{l|ll} \lambda^3 & 1 & -175.03 \\ \lambda^2 & -55.12 & -88.17 \\ \lambda^1 & -176.63 & 0 \\ \lambda^0 & -88.17 & \end{array} \tag{3.74}$$

From the changes of sign of the coefficients of the first column of the Routh tables it can be concluded that polynomial $p_2(\lambda)$ is unstable. The associated Routh table for the third Kharitonov polynomial $p_3(\lambda)$ is

$$\begin{array}{l|ll} \lambda^3 & 1 & -163.82 \\ \lambda^2 & -55.12 & -88.17 \\ \lambda^1 & -165.42 & 0 \\ \lambda^0 & -88.17 & \end{array} \tag{3.75}$$

Since there is change of sign of the coefficients of the first column of the Routh table it can be concluded that the characteristic polynomial $p_3(\lambda)$ is an unstable one. Finally, the associated Routh table for the fourth Kharitonov polynomial $p_4(\lambda)$ is

$$\begin{array}{l|ll} \lambda^3 & 1 & -175.03 \\ \lambda^2 & -58.12 & -84.13 \\ \lambda^1 & -176.48 & 0 \\ \lambda^0 & -84.13 & \end{array} \tag{3.76}$$

From the changes of sign of the coefficients of the first column of the Routh tables it can be concluded that the characteristic polynomial $p_4(\lambda)$ is unstable. From the previous analysis it can be seen that all Kharitonov polynomials have unstable roots; therefore, the fixed point $(x_1^{*,2}, x_2^{*,2}, x_3^{*,2})$ of the circadian oscillator under feedback gain $c_2 \in [-270.2, -260.2]$ is an unstable one.

3.7 Simulation Tests

The following example is concerned with bifurcations of the fixed point of the FitzHugh–Nagumo neuron. Assuming zero external current input and known parameters of the model it has been found the associated fixed point is $(x_1^{*,1}, x_2^{*,1}, x_2^{*,1}) = (0, 0, 0)$. As confirmed by the results given in Fig. 3.2 this fixed point is a stable one. The nominal values of the model's parameters were $\epsilon = 1.5$, $\gamma = 0.1$, and $a = 7$.

Next, examples on the stability of fixed points in the circadian oscillator model are given under variation of specific parameters of the dynamical model. The nominal values of the model's parameters were: $v_s = 0.02$, $K_i = 0.9$, $v_m = 1.5$, $K_m = 1.6$, $K_s = 1$, $v_d = 1.4$, $K_1 = 0.5$, $K_2 = 0.6$, $n = 4$, while K_d

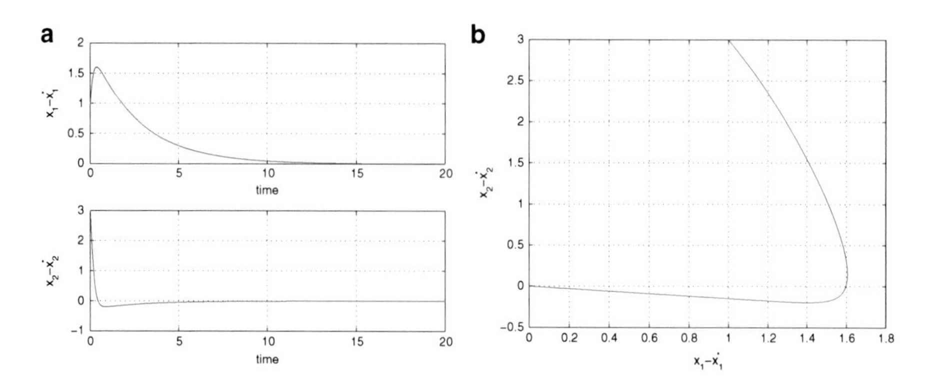

Fig. 3.2 The fixed point of the FitzHugh–Nagumo neuron is a stable one: (**a**) diagrams of the state variables, (**b**) phase diagrams

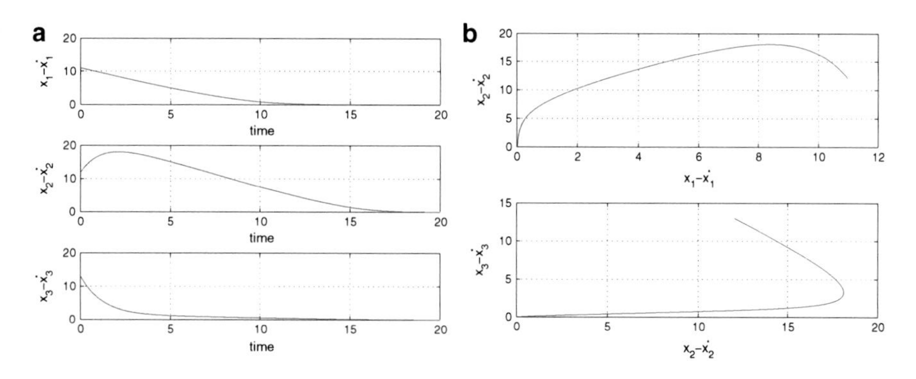

Fig. 3.3 The first fixed point of the circadian oscillator becomes a stable one for parameter $K_d \in [0.3, 0.9]$: (**a**) diagrams of the state variables, (**b**) phase diagrams

was an uncertain parameter. It was assumed that the Michaelis constant parameter varies within intervals, that was $K_d \in [0.4, 0.9]$ and based on this interval the ranges of variation of the Jacobian matrix of the model computed at the fixed point $(x_1^{*,1}, x_2^{*,1}, x_2^{*,1}) = (0, 0, 0)$ were found. By applying Kharitonov's theorem it was possible to show that conditions about the stability of the fixed point were satisfied. The associated results are depicted in Fig. 3.3.

Moreover, the case of state feedback was considered in the form $K_s = c_2 x_2$ and using parameter c_2 as feedback gain. The variation range of c_2 was taken to be the interval $[-60.2, -50.2]$ and based on this the ranges of variation of the coefficient of the characteristic polynomial of the system's Jacobian matrix were computed. Stability analysis followed, using Kharitonov's theorem. First, it was shown that by choosing $-60.2 < c_2 < -50.2$ e.g. $c_2 = -55$ then the fixed point $(x_1^{*,2}, x_2^{*,2}, x_3^{*,2})$ remained stable.

Next, it was shown that by setting $-270.2 < c_2 < -260.2$ e.g. $c_2 = -265$ the fixed point became unstable. Indeed in the first case, that is when $c_2 = -55$

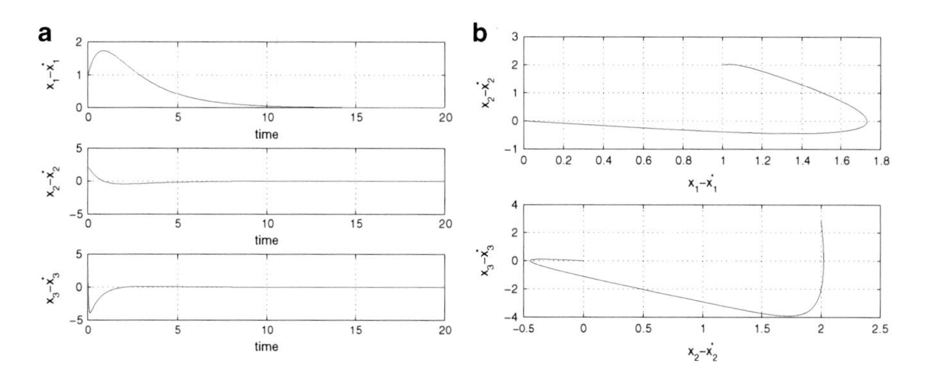

Fig. 3.4 The second fixed point of the circadian oscillator is stable for feedback gain value $c_2 \in [-60.2, -50.2]$: (**a**) diagrams of the state variables, (**b**) phase diagrams

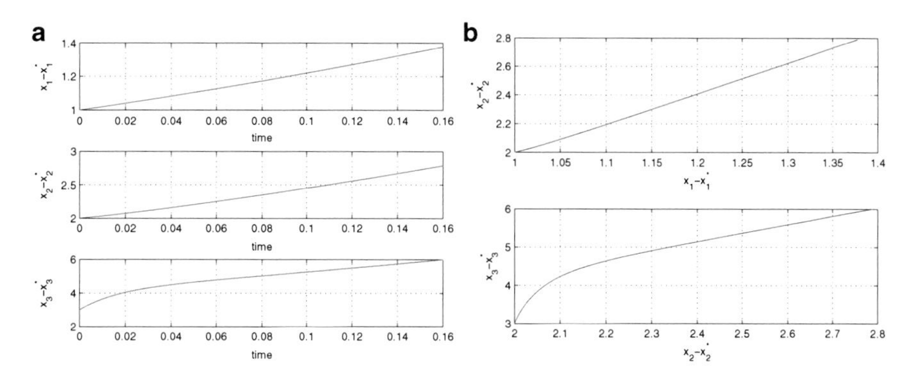

Fig. 3.5 The second fixed point of the circadian oscillator is unstable one for feedback gain value $c_2 \in [-270.2, -260.2]$: (**a**) diagrams of the state variables, (**b**) phase diagrams

the roots of the characteristic polynomial that is associated with the Jacobian of the system are $\lambda_1 = -29.49$, $\lambda_2 = -1.52$, and $\lambda_3 = -0.42$, which means that the fixed point is a stable one. In the second case, that is for $c_2 = -265$ the roots of the characteristics polynomial became $\lambda_1 = 59.5$, $\lambda_2 = -2.21$, and $\lambda_3 = -0.65$ which means that the fixed point is an unstable one. The associated results are depicted in Figs. 3.4 and 3.5.

3.8 Conclusions

The chapter has studied bifurcations in biological models, such as circadian cells. It has been shown that the stages for analyzing the stability features of the fixed points that lie on bifurcation branches comprise (i) the computation of fixed points as functions of the bifurcation parameter and (ii) the evaluation of the type of stability for each fixed point through the computation of the eigenvalues of the Jacobian matrix that is associated with the system's nonlinear dynamics model. Stage (ii) requires the computation of the roots of the characteristic polynomial of the Jacobian matrix. This problem is nontrivial since the coefficients of the characteristic polynomial are functions of the bifurcation parameter and the latter varies within intervals. To obtain a clear view about the values of the roots of the characteristic polynomial and about the stability features they provide to the system, the use of interval polynomials theory and particularly of Kharitonov's stability theorem has been proposed. In this approach the study of the stability of a characteristic polynomial with coefficients that vary in intervals is equivalent to the study of the stability of four polynomials with crisp coefficients computed from the boundaries of the aforementioned intervals. The method for fixed points stability analysis was tested on two different biological models: (1) the FitzHugh–Nagumo neuron model, (2) a model of circadian cells which produce proteins out of an RNA transcription and translation procedure. The efficiency of the proposed stability analysis method has been confirmed through numerical and simulation tests.

Chapter 4
Oscillatory Dynamics in Biological Neurons

Abstract The voltage of the neurons membrane exhibits oscillatory variations after receiving suitable external excitation either when the neuron is independent from neighboring neural cells or when the neuron is coupled to neighboring neural cells through synapses or gap junctions. In the latter case it is significant to analyze conditions under which synchronization between coupled neural oscillators takes place, which means that the neurons generate the same voltage variation pattern possibly subject to a phase difference. The loss of synchronism between neurons can cause several neurodegenerative disorders. Moreover, it can affect several basic functions of the body such as gait, respiration, and heart's rhythm. For this reason synchronization of coupled neural oscillators has become a topic of significant research during the last years. The associated results have been also used in several engineering applications, such as biomedical engineering and robotics. For example, synchronization between neural cells can result in a rhythm generator that controls joints motion in quadruped, multi-legged, and biped robots.

4.1 Neural Oscillators

Neurons exhibit complicated activation functions, denoted as spiking and bursting. This activation mechanism is characterized by a silent stage (resting) and a stage with intensive spike-like oscillations. Bursting activity appears in specific thalamic cells (this affects, for example, the phases of sleep in people with parkinsonian tremor and in such a case increased bursting activity appears in neurons which are found in basal ganglia). Moreover, bursting activity appears in cells which coordinate breath's rhythm, inside the so-called pre-Botzinger complex.

There is need for two biophysical mechanisms to generate bursting: (1) a mechanism for the generation of the bursting oscillations, (2) a mechanism for the switching between the silent phase and the bursting phase.

Spikes are voltage oscillations which are generated by the interaction of an incoming Na^+ current and an outgoing K^+ current. The slow phase is due to an ionic

G.G. Rigatos, *Advanced Models of Neural Networks*,

DOI 10.1007/978-3-662-43764-3_4,

current which is activated or deactivated at a slower pace than the rest of the currents. A slow current which forms the basis for the slow modulation phase is the calcium-dependent potassium current I_{KCa}. Ions of Ca^{2+} enter the cell's membrane during the active spiking phase and this results in raise (activation) of the current I_{KCa} (outward current). When the outward current becomes sufficiently large the cell can no longer produce the spiking activity and the active phase is terminated. During the silent phase, there is an outflow of Ca^{2+} from the cell's membrane and the calcium channels close.

A generic mathematical model of the bursting activity is described by a set of differential equations which evolve at multiple scales

$$\begin{aligned} \frac{dx}{dt} &= f(x, y) \\ \frac{dy}{dt} &= \epsilon g(x, y) \end{aligned} \tag{4.1}$$

where $\epsilon > 0$ takes a small positive value. In the above set of equations $x \in R^n$ are fast variables which are responsible for spikes generation, and $y \in R^m$ are slow variables which are responsible for the slow modulation of the silent and active phases.

There are different types of bursting which are different in the amplitude and frequency of oscillations that the neuron's membrane performs during the active phase. These are [3, 16, 49, 65]:

1. square-wave bursting: This form of bursting was first observed in pancreatic β-cells. In square-wave bursting the amplitude of the membrane's voltage variation remains practically unchanged, whereas the frequency of spiking slows down during the active phase.
2. elliptic bursting: In elliptic bursting one can notice small amplitude oscillations during the silent phase, and progressively the amplitude of the oscillations increases and remains unchanged for a long time interval. During the active phase the frequency of spikes initially increases and subsequently drops. Elliptic bursting appears in models of thalamic neurons, and particularly in neurons of the basal ganglia.
3. parabolic bursting: In parabolic bursting there are no oscillations during the silent phase and there is an abrupt transition to the active phase which consists of oscillations of large amplitude. The values of the parameters of the dynamic model of the neuron are the ones which define which type of bursting will finally appear.

Indicative diagrams of biological neurons under spiking activity are given in Fig. 4.1. Moreover, representative diagrams of neurons under bursting activity are shown in Fig. 4.2.

It is noted that by changing the parameters of the dynamic model, the neuron bursting can exhibit chaotic characteristics which have to do with the number of spikes during the active phase, the frequency under which bursts appear and the transition from bursting to spiking [88].

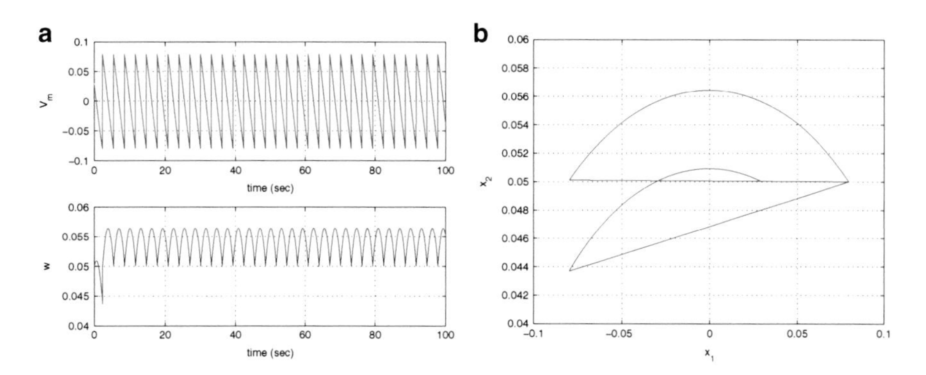

Fig. 4.1 (**a**) State variables of the FitzHugh–Nagumo neuron model under spiking activity, (**b**) The associated phase diagram

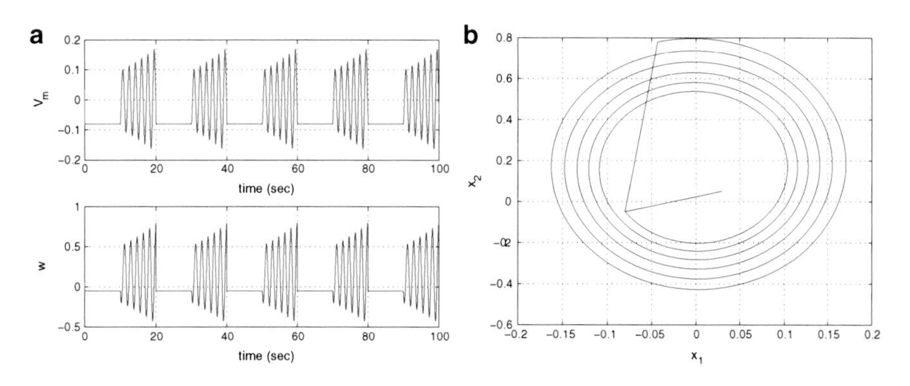

Fig. 4.2 (**a**) State variables of the FitzHugh–Nagumo neuron model under bursting activity, (**b**) The associated phase diagram

4.2 Synaptic Channels

In the isolated neurons studied up to now, ion-gated and voltage-gated channels have been studied, i.e. channels which allow the passage of specific ions (K^+, Na^+, and Ca^{2+}) and which open or close if the value of the membrane's potential exceeds a threshold or if the concentration of ions also exceeds a certain threshold. There are also other channels, among which the most important are the synaptic channels [16, 26, 65].

The events which lead to opening of synaptic channels include several stages. The wave-type spatiotemporal variation of the membrane's potential $V(x,t)$ travels along the membrane of the axon and arrives at pre-synaptic areas which are called synaptic terminals. These terminals contain ions, such as Ca^{2+}. When the terminals get depolarized Ca^{2+} ions are released. A protein (vesicle) binds the Ca^{2+} ions, transfers them through the synapse, and releases them at a post-synaptic site. Actually, the neurotransmitter is diffused through the synapse and once reaching

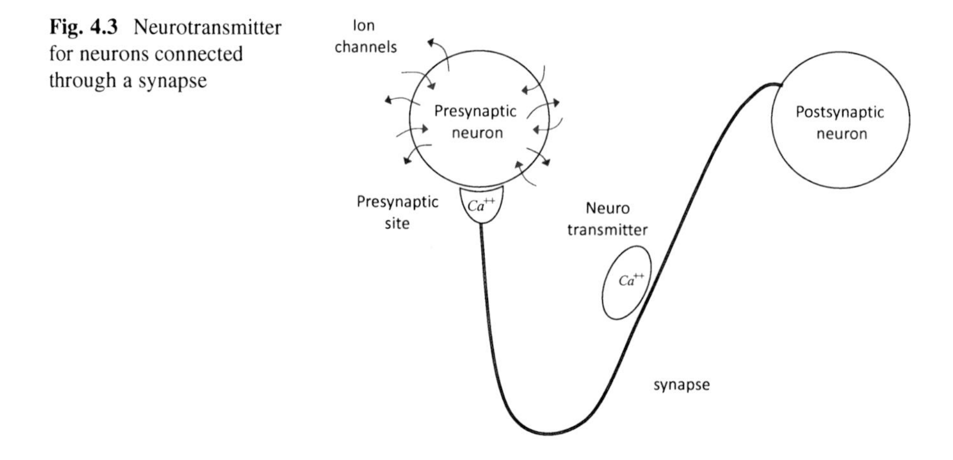

Fig. 4.3 Neurotransmitter for neurons connected through a synapse

the post-synaptic neuron it is attached to various receptors (usually on the so-called protuberances, which are part of dendrites known as spines).

In their turn, these receptors open channels which either hyperpolarize or depolarize the neuron according to the type of the neurotransmitter. The presence of other chemical substances in the presynaptic channel may facilitate or inhibit the binding of ions by the vesicle. Moreover, the binding or release of ions by the vesicle can be performed in a stochastic manner. Additionally, the presynaptic excitation can lead to activation of more proteins-transmitters which means that a larger quantity of ions will be transferred and released at the post-synaptic neuron. This increase of ion transfer is called potentiation or facilitation. There is also the inverse phenomenon, in which after several presynaptic spikes, the transfer and release of ions at the post-synaptic neuron decreases. This is known as depression (Fig. 4.3).

4.3 Dynamics of the Synapses

The following cases of synapses dynamics are examined: (1) the glutamate: it stimulates the post-synaptic cell and open channels [16, 65], (2) the γ aminobutyric acid, also known as GABA: it inhibits the postsynaptic cell and closes channels.

The synaptic currents which are developed in such a case depend on a conductivity parameter and on the difference between the membrane's voltage and voltage threshold value.

$$I_{\text{syn}} = g(t)(V_{\text{post}} - V_{\text{rev}}) \tag{4.2}$$

A common assumption made is that conductivity is a sum of constant functions which depend only on the time instances in which the presynaptic cell has given a spike.

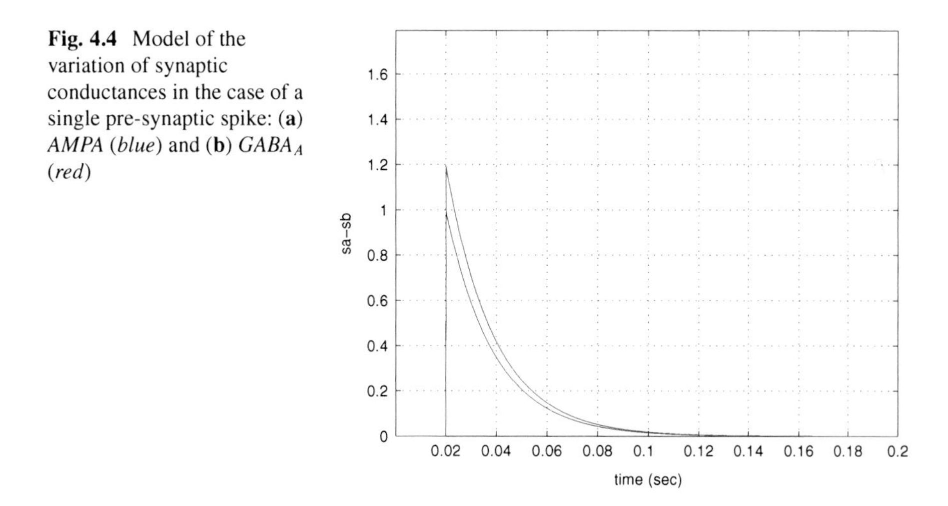

Fig. 4.4 Model of the variation of synaptic conductances in the case of a single pre-synaptic spike: (**a**) *AMPA* (*blue*) and (**b**) $GABA_A$ (*red*)

$$g(t) = \bar{g}\sum_k a(t - t_k) = \bar{g}z(t) \tag{4.3}$$

where

$$a(t) = \frac{a_d a_r}{a_r - a_d}(e^{-a_d t} - e^{-a_r t}) \tag{4.4}$$

or $a(t) = a_d^2 e^{-a_d t}$. As far as parameter $z(t)$ is concerned its variation in time is given by the differential equation

$$z'' + (a_r + a_d)z' + a_r a_d z = 0 \tag{4.5}$$

Another manner for modelling conductivity in the synapse is to use

$$g(t) = \bar{g}s(t) \tag{4.6}$$

where $s(t)$ denotes the fraction of the open channel which satisfies the relation

$$\frac{ds}{dt} = a_r[T](1 - s) - a_d s \tag{4.7}$$

where $[T]$ is the concentration of the transmitter coming from the pre-synaptic neuron and released at the post-synaptic neuron (Fig. 4.4).

The time is set at $t = t_0$. Then the concentration $[T]$ becomes equal to $T_{\max}$ and at time instant t_1 it returns to zero. Then

$$s(t - t_0) = s_\infty + (s(t_0) - s_\infty)e^{(t - t_0)/t_0} \tag{4.8}$$

for $t_0 < t < t_1$ where

$$s_\infty = \frac{a_r T_{\max}}{a_r T_{\max} + a_d} \quad (4.9)$$

and

$$\tau_s = \frac{1}{a_r T_{\max} + a_d} \quad (4.10)$$

After the pulse of the transmitter has gone, $s(t)$ decays as

$$s(t) = s(t_1)e^{-a_d(t-t_1)} \quad (4.11)$$

A model about the concentration of the transmitter at the pre-synaptic neuron is

$$[T](V_{\text{pre}}) = \frac{T_{\max}}{1+\exp(-(V_{\text{pre}} - V_T/\text{K}+p))} \quad (4.12)$$

4.4 Study of the Glutamate Neurotransmitter

Glutamate neurotransmitter activates two different types of receptors: (1) AMPA/Kainate which is vary fast, (2) NMDA which is associated with memory. Both receptors lead to activation of the post-synaptic neuron [16, 65].

(a) AMPA/Kainate
The current associated with the neuro-transmitter is

$$I_{\text{AMPA}} = \bar{g}_{\text{AMPA}}(V - V_{\text{AMPA}}) \quad (4.13)$$

The responses of AMPA synapses can be very fast, for example in the case of acoustic neurons they can have rise and decay times of the order of milliseconds. Similarly, in the case of cortical neurons the rise and decay time can be of the order of 0.4–0.8 ms. AMPA synapses exhibit intensive depression in case of subsequent spikes. Each time that a spike is generated the amplitude of the AMPA current at the post-synaptic neuron drops.

(b) NMDA
The NMDA receptor is also sensitive to glutamate with a response that lasts longer than the one of AMPA. Under normal conditions the NMDA receptor is partially found blocked by Mg^{2+} ions. One of the basic types of ions that is transferred by the NMDA current is Ca^{2+} which is responsible for many long-term changes in neurons. This ion is considered to affect significantly the mechanism of short-term memory.

The NMDA current is modelled as

$$I_{\mathrm{NMDA}} = \bar{g}_{\mathrm{NMDA}} \cdot s \cdot B(V)(V - V_{\mathrm{NMDA}}) \tag{4.14}$$

where s is given by the previous relation

$$\frac{ds}{dt} = a_r[T](1 - s) - a_d s \tag{4.15}$$

and

$$B(V) = \frac{1}{1 + e^{-0.062V}[\mathrm{Mg}^{2+}]/3.57} \tag{4.16}$$

4.5 Study of the GABA Neurotransmitter

GABA is the main inhibitory neurotransmitter in the cortex and can be distinguished into two types, i.e. $GABA_A$ and $GABA_B$ [16,65].

(a) $GABA_A$
The neurotransmitter $GABA_A$ is responsible for inhibiting the stimulation of the post-synaptic neuron and as in the case of AMPA and NMDA this neurotransmitter is activated by a spike.
The associated current equation is

$$I_{\mathrm{GABA}_A} = \bar{G}_{\mathrm{GABA}_A} s(V - V_{\mathrm{GABA}_A}) \tag{4.17}$$

(b) $GABA_B$
At the postsynaptic neuron a receptor protein binds the neurotransmitter and activates an intracellular circuit (G-protein). Next a K^+ channel is activated which polarizes the membrane of the post-synaptic neuron. The equations of current associated with the $GABA_B$ neurotransmitter are as follows:

$$\begin{aligned} I_{\mathrm{GABA}_B} &= \bar{g}_{\mathrm{GABA}_B} \frac{s^n}{Kd + s^n}(V - E_k) \\ \frac{dr}{dt} &= a_r[T](1 - r) - b_r r \\ \frac{ds}{dt} &= k_3 r - k_d s \end{aligned} \tag{4.18}$$

(c) Permanent connections between neurons (Gap or electrical junctions)
There are connections between neurons which do not depend on a neurotransmitter (vesicle). The associated current is

$$I_{\mathrm{gap}} = \bar{g}_{\mathrm{gap}}(V_{\mathrm{post}} - V_{\mathrm{pre}}) \tag{4.19}$$

where g_{gap} denotes conductivity. This model is usually met in neurons of the cerebral cortex.

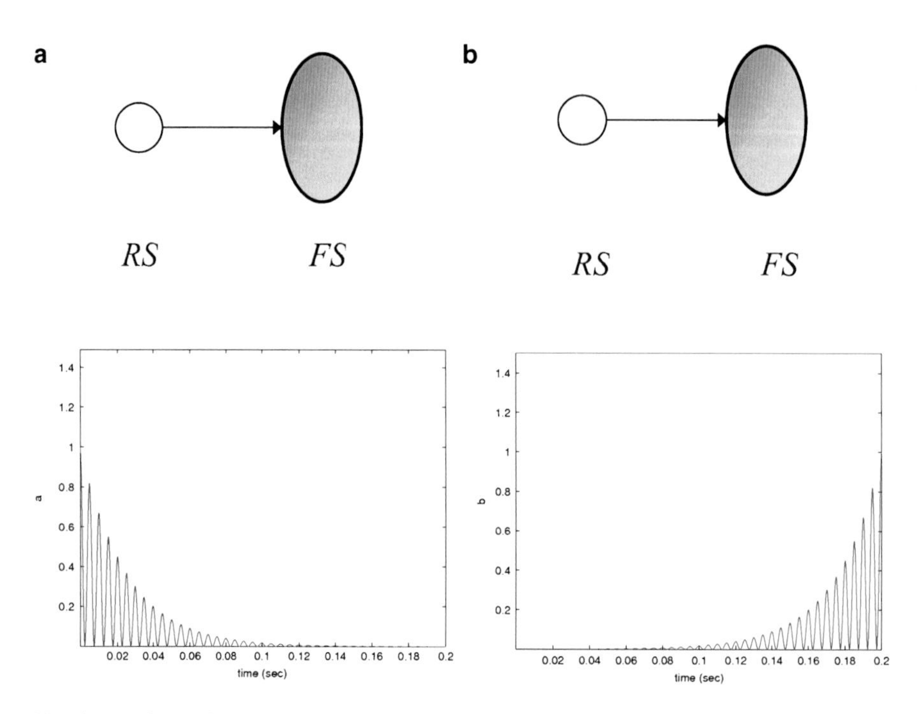

Fig. 4.5 (**a**) Synaptic depression: decrease in the amplitude of the post-synaptic current (**b**) synaptic facilitation: increase in the amplitude of the post-synaptic current

4.6 Short- and Long-Term Plasticity

Short-term plasticity is distinguished from long-term plasticity and denotes a change in the amplitude of the current appearing at the post-synaptic neuron, as a consequence of spikes generated at the pre-synaptic neuron. An example is given in the diagram of Fig. 4.5 and shows connection between cortical excitatory cells (RS) and cortical fast synaptic units (FS) [16, 65].

Next, it is explained how a pre-synaptic current affects the amplitude of the post-synaptic current. One defines

$$M(t) = q(t)f(t) \tag{4.20}$$

where $q(t) \in (0, 1)$ is the depression factor having as resting value q_0 and $f(t) \in (0, 1)$ is the facilitating factor having as resting value f_0. The variation of f and q is given by

$$\tau_f \tfrac{df}{dt} = f_0 - f, \qquad \tau_d \tfrac{dq}{dt} = d_0 - q \tag{4.21}$$

Each time a spike occurs $f(t)$ increases by $(1-f)$ and $q(t)$ decreases by $a_d d$. There are also more detailed models about the change of f, q, given by

$$\begin{aligned}\frac{df}{dt} &= \frac{f_0-f}{\tau_f} + \sum_j \delta(t-\tau_j)a_f(1-f)\\ \frac{dq}{dt} &= \frac{d_0-q}{\tau_d} - \sum_j \delta(t-\tau_j)a_d q\end{aligned} \tag{4.22}$$

Moreover, there are different models to describe the variation of f and q, which do not include the spiking times. For example,

$$\begin{aligned}\frac{dq}{dt} &= \frac{d_0-q}{\tau_d} - a_d(V)q\\ ad(V) &= \frac{a}{1+e^{-k(V-V_{\mathrm{thr}})}}\end{aligned} \tag{4.23}$$

Neurons plasticity affects memory. A basic assumption in neuroscience is that memories are stored in the strength of the synaptic connections (Hebbian Learning). Hebb rules enforce or weaken a synapse, depending on whether presynaptic or post-synaptic neurons are active. If both neurons are active, then one has enforcement of the neural synapse, otherwise the synapse is weakened (for both neurons to be active it is necessary the potential of each neuron's membrane to have a high value. The potential can be computed by the solution of the differential equations describing the propagation of the membrane's potential).

4.7 Synchronization of Coupled FitzHugh–Nagumo Neurons Using Differential Flatness Theory

4.7.1 The Problem of Synchronization of Coupled Neural Oscillators

The voltage of the neuron's membrane exhibits oscillatory variations after receiving suitable external excitation either when the neuron is independent from neighboring neural cells or when the neuron is coupled to neighboring neural cells through synapses or gap junctions [16, 98, 182]. In the latter case it is significant to analyze conditions under which synchronization between coupled neural oscillators takes place, which means that the neurons generate the same voltage variation pattern subject to a phase difference. The loss of synchronism between neurons can cause several neuro-degenerative disorders. Moreover, it can affect several basic functions of the body such as gait, respiration, and heart's rhythm. For this reason synchronization of coupled neural oscillators has become a subject of significant research during the last years. The associated results have been also used in several engineering applications, such as biomedical engineering and robotics. For example, synchronization between neural cells can result in a rhythm generator that controls joints motion in quadruped, multi-legged, and biped robots.

Dynamical models of coupled neural oscillators can serve as Central Pattern Generators (CPGs) [70, 212]. This means that they stand for higher level control elements in a multi-layered control scheme which provide the activation frequency and rhythm for controllers operating at the lower level, e.g. controllers that provide motion to robot's joints. CPG methods have been used to control various kinds of robots, such as crawling robots and legged robots and various modes of locomotion such as basic gait control, gait transitions control, dynamic adaptive locomotion control, etc. [85]. CPG models have been used with hexapod and octopod robots inspired by insect locomotion [11]. CPGs have been also used for controlling swimming robots, such as lamprey robots [40]. Quadruped walking robots controlled with the use of CPGs have been studied in [62]. Models of CPGs are also increasingly used for the control of biped locomotion in humanoid robots [79].

The problem of synchronization of coupled neural oscillators becomes more difficult when there is uncertainty about the parameters of the dynamical model of the neurons [8]. Thus, it is rather unlikely that coupled neurons will have identical dynamical models and that these models will be free of parametric variations and external disturbances. To synchronize neurons subject to model uncertainties and external disturbances several approaches have been proposed. In [130] by using the Lyapunov function method and calculating the largest Lyapunov exponent, respectively, a sufficient condition and a necessary condition of the coupling coefficient for achieving self-synchronization between two coupled FitzHugh–Nagumo neurons are established. In [5] matrix inequalities on the basis of Lyapunov stability theory are used to design a robust synchronizing controller for the model of the coupled FitzHugh–Nagumo neurons. In [217] robust control system combining backstepping and sliding mode control techniques is used to realize the synchronization of two gap junction coupled chaotic FitzHugh–Nagumo neurons under external electrical stimulation. In [207] a Lyapunov function-based control law is introduced, which transforms the FitzHugh–Nagumo neurons into an equivalent passive system. It is proved that the equivalent system can be asymptotically stabilized at any desired fixed state, namely, synchronization can be succeeded. In [144] synchronizing control for coupled FitzHugh–Nagumo neurons is succeeded using a Lyapunov function approach. The control signal is based on feedback of the synchronization error between the master and the slave neurons. Finally, the use of differential geometric methods in the modelling and control of FitzHugh–Nagumo neuron dynamics has been studied in [47, 220].

In this chapter, differential flatness theory has been proposed for the synchronization of coupled nonlinear oscillators of the FitzHugh–Nagumo type. Differential flatness theory is currently a main direction in nonlinear dynamical systems and enables linearization for a wider class of systems than the one succeeded with Lie-algebra methods [157, 173, 177]. To find out if a dynamical system is differentially flat, the following should be examined: (1) the existence of the so-called flat output, i.e. a new variable which is expressed as a function of the system's state variables. The flat output and its derivatives should not be coupled in the form of an ordinary differential equation (ODE), (2) the components of the system (i.e., state variables and control input) should be expressed as functions of the flat output and its derivatives [58, 101, 101, 109, 125, 172]. In certain cases the differential flatness theory enables transformation to a linearized form (canonical Brunovsky form) for which the design of the controller becomes easier. In other cases by showing that a system is differentially flat one can easily design a reference trajectory as a function of the so-called flat output and can find a control law that assures tracking of this desirable trajectory [58, 200].

This chapter is concerned with proving differential flatness of the model of the coupled FitzHugh–Nagumo neural oscillators and its resulting description in the Brunovksy (canonical) form [125]. By defining specific state variables as flat outputs an equivalent description of the coupled FitzHugh–Nagumo neurons in the Brunovksy (linear canonical) form is obtained. It is shown that for the linearized model of the coupled neural oscillators it is possible to design a feedback controller that succeeds their synchronization. At a second stage, a novel Kalman Filtering method, the Derivative-free nonlinear Kalman Filter, is proposed for estimating the non-directly measurable elements of the state vector of the linearized system. With the redesign of the proposed Kalman Filter as a disturbance observer, it becomes possible to estimate disturbance terms affecting the model of the coupled FitzHugh–Nagumo neurons and to use these terms in the feedback controller. By avoiding linearization approximations, the proposed filtering method improves the accuracy of estimation and results in smooth control signal variations and in minimization of the synchronization error [158, 159, 170].

4.7.2 Coupled Neural Oscillators as Coordinators of Motion

Coupled neural oscillators can provide as output signals that maintain a specific phase difference and which can be used for the coordination of robot's motion. These coupled neural oscillators are also known as CPGs and can synchronize the motion performed by the legs of quadruped and biped robots. The concept of CPGs comes from biological neurons, located at the spinal level, and which are able of generating rhythmic commands for the muscles. CPGs receive commands from higher levels of the central nervous system and also from peripheral afferents. Thus their functioning is the result of the interaction between central commands and local reflexes.

Neuronal mechanisms in the brain select which CPGs will be activated at every time instant. The basal ganglia play an important role in this. Under resting conditions the output layer of the basal ganglia (the pallidum) maintains different CPG neurons under inhibition. To achieve CPG activation, striatial neurons, the input layer of the basal ganglia, inhibit cells in the pallidum which keep under inhibition the CPG neurons. The striatial neurons can, in turn, be activated from either neocortex or directly from the thalamus. The responsiveness of striatial neurons to activation can be facilitated by dopaminergic inputs (Fig. 4.6). On

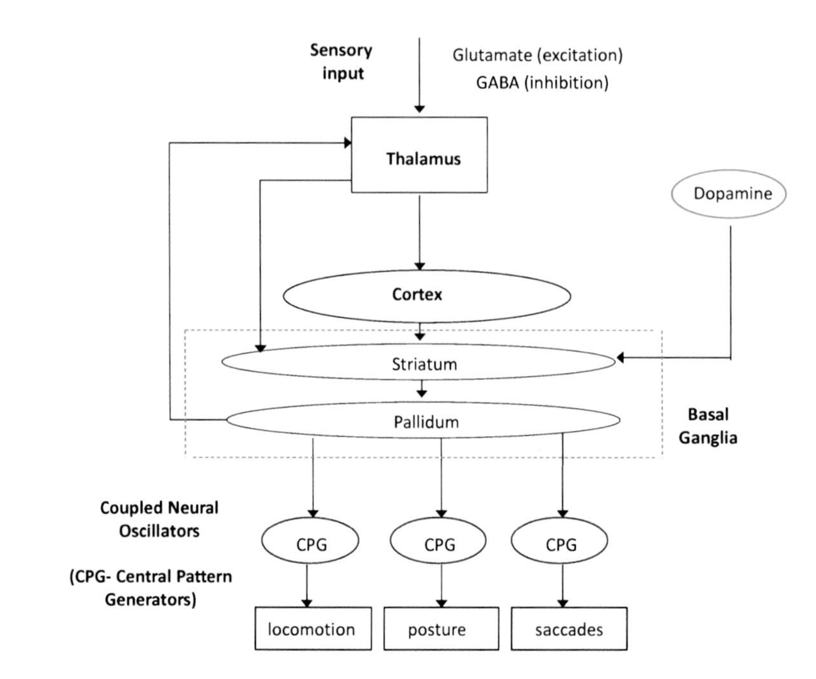

Fig. 4.6 Biological mechanism of activation of coupled neural oscillators (Central Patterns Generators)

the other hand, deficiencies or enhanced levels of dopaminergic inputs result in movements distortion and malfunction of the locomotion system [70, 212].

Coupled neuron models which have been used as CPGs are Hodkin–Huxley neurons and FitzHugh–Nagumo neurons (Fig. 4.7). CPG methods have been used to control various kinds of robots, such as crawling robots and legged robots and various modes of locomotion such as basic gait control, gait transitions control, dynamic adaptive locomotion control, etc. [85].

4.8 Differential Flatness Theory

4.8.1 Definition of Differentially Flat Systems

The main principles of differential flatness theory are as follows [58, 177, 200]: A finite dimensional system is considered. This can be written in the general form of an ODE, i.e. $S_i(w, \dot{w}, \ddot{w}, \cdots, w^{(i)}),\ \ i = 1, 2, \cdots, q$. The term w denotes the system variables (these variables are, for instance, the elements of the system's state vector and the control input) while $w^{(i)}$, $i = 1, 2, \cdots, q$ are the associated derivatives. Such a system is said to be differentially flat if there is a collection of m functions $y = (y_1, \cdots, y_m)$ of the system variables and of their time-

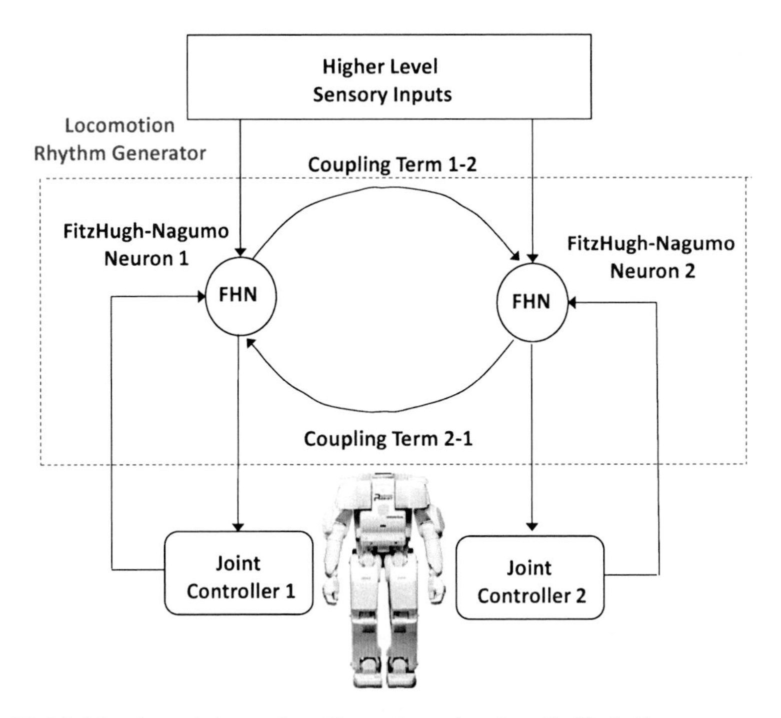

Fig. 4.7 Models of coupled neural oscillators in engineering: FitzHugh–Nagumo neurons for controlling robot's motion

derivatives, i.e. $y_i = \phi(w, \dot{w}, \ddot{w}, \cdots, w^{(\alpha_i)}),\ i = 1, \cdots, m$ satisfying the following two conditions [58, 157, 200]: (1) There does not exist any differential relation of the form $R(y, \dot{y}, \cdots, y^{(\beta)}) = 0$ which implies that the derivatives of the flat output are not coupled in the sense of an ODE, or equivalently it can be said that the flat output is differentially independent, (2) All system variables (i.e., the elements of the system's state vector w and the control input) can be expressed using only the flat output y and its time derivatives $w_i = \psi_i(y, \dot{y}, \cdots, y^{(\gamma_i)}),\ i = 1, \cdots, s$.

4.8.2 Conditions for Applying Differential Flatness Theory

The following generic class of nonlinear systems is considered

$$\dot{x} = f(x, u) \tag{4.24}$$

Such a system can be transformed to the form of an affine in the input system by adding an integrator to each input [25]

$$\dot{x} = f(x) + \sum_{i=1}^{m} g_i(x) u_i \tag{4.25}$$

If the system of Eq. (4.25) can be linearized by a diffeomorphism $z = \phi(x)$ and a static state feedback $u = \alpha(x) + \beta(x)v$ into the following form

$$\begin{aligned} \dot{z}_{i,j} &= z_{i+1,j} \text{ for } 1 \leq j \leq m \text{ and } 1 \leq i \leq v_j - 1 \\ \dot{z}_{v_{i,j}} &= v_j \end{aligned} \tag{4.26}$$

with $\sum_{j=1}^{m} v_j = n$, then $y_j = z_{1,j}$ for $1 \leq j \leq m$ are the 0-flat outputs which can be written as functions of only the elements of the state vector x (when the flat output of a system is only function of its states x, then this is called 0-flat). To define conditions for transforming the system of Eq. (4.25) into the canonical form described in Eq. (4.26) the following theorem holds [25].

Theorem. *For the nonlinear systems described by Eq. (4.25) the following variables are defined: (i)* $G_0 = span[g_1, \cdots, g_m]$*, (ii)* $G_1 = span[g_1, \cdots, g_m, ad_f g_1, \cdots, ad_f g_m]$*,* $\cdots$ *(k)* $G_k = span\{ad_f^j g_i \text{ for } 0 \leq j \leq k, 1 \leq i \leq m\}$ *(where* $ad_f^i g$ *stands for a Lie Bracket). Then, the linearization problem for the system of Eq. (4.25) can be solved if and only if: (1). The dimension of* G_i*,* $i = 1, \cdots, k$ *is constant for* $x \in X \subseteq R^n$ *and for* $1 \leq i \leq n-1$*, (2). The dimension of* G_{n-1} *is of order* n*, (3). The distribution* G_k *is involutive for each* $1 \leq k \leq n-2$.

4.8.3 Transformation of the Neurons' Model into the Linear Canonical Form

It is assumed now that after defining the flat outputs of the initial multi-input multi-output (MIMO) nonlinear system (this approach will be also shown to hold for the neuron model), and after expressing the system state variables and control inputs as functions of the flat output and of the associated derivatives, the system can be transformed in the Brunovsky canonical form:

$$\begin{aligned} \dot{x}_1 &= x_2 \\ &\cdots \\ \dot{x}_{r_1-1} &= x_{r_1} \\ \dot{x}_{r_1} &= f_1(x) + \sum_{j=1}^{p} g_{1_j}(x) u_j + d_1 \\ \\ \dot{x}_{r_1+1} &= x_{r_1+2} \\ &\cdots \\ \dot{x}_{p-1} &= x_p \\ \dot{x}_p &= f_p(x) + \sum_{j=1}^{p} g_{p_j}(x) u_j + d_p \\ \\ y_1 &= x_1 \\ &\cdots \\ y_p &= x_{n-r_p+1} \end{aligned} \tag{4.27}$$

where $x = [x_1, \cdots, x_n]^T$ is the state vector of the transformed system (according to the differential flatness formulation), $u = [u_1, \cdots, u_p]^T$ is the set of control inputs, $y = [y_1, \cdots, y_p]^T$ is the output vector, f_i are the drift functions, and $g_{i,j}$, $i, j = 1, 2, \cdots, p$ are smooth functions corresponding to the control input gains, while d_j is a variable associated with external disturbances. It holds that $r_1+r_2+\cdots+r_p = n$. Having written the initial nonlinear system into the canonical (Brunovsky) form it holds

$$y_i^{(r_i)} = f_i(x) + \sum_{j=1}^{p} g_{ij}(x)u_j + d_j \tag{4.28}$$

Next the following vectors and matrices can be defined $f(x)=[f_1(x), \cdots, f_n(x)]^T$, $g(x) = [g_1(x), \cdots, g_n(x)]^T$ with $g_i(x) = [g_{1i}(x), \cdots, g_{pi}(x)]^T$, $A = diag[A_1, \cdots, A_p]$, $B = diag[B_1, \cdots, B_p]$, $C^T = diag[C_1, \cdots, C_p]$, $d = [d_1, \cdots, d_p]^T$, where matrix A has the MIMO canonical form, i.e. with block-diagonal elements

$$A_i = \begin{pmatrix} 0 & 1 & \cdots & 0 \\ 0 & 0 & \cdots & 0 \\ \vdots & \vdots & \cdots & \vdots \\ 0 & 0 & \cdots & 1 \\ 0 & 0 & \cdots & 0 \end{pmatrix}_{r_i \times r_i}$$
$$B_i^T = \begin{pmatrix} 0 & 0 & \cdots & 0 & 1 \end{pmatrix}_{1 \times r_i}$$
$$C_i^T = \begin{pmatrix} 1 & 0 & \cdots & 0 & 0 \end{pmatrix}_{1 \times r_i} \tag{4.29}$$

Thus, Eq. (4.28) can be written in state-space form

$$\begin{aligned} \dot{x} &= Ax + Bv + B\tilde{d} \\ y &= Cx \end{aligned} \tag{4.30}$$

where the control input is written as $v = f(x) + g(x)u$.

4.9 Linearization of the FitzHugh–Nagumo Neuron

4.9.1 *Linearization of the FitzHugh–Nagumo Model Using a Differential Geometric Approach*

The model of the dynamics of the FitzhHugh Nagumo neuron is considered

$$\begin{aligned} \frac{dV}{dt} &= V(V-a)(1-V) - w + I \\ \frac{dw}{dt} &= \epsilon(V - \gamma w) \end{aligned} \tag{4.31}$$

By defining the state variables $x_1 = w$, $x_2 = V$ and setting current I as the external control input one has

$$\begin{array}{c}\frac{dx_1}{dt} = -\epsilon\gamma x_1 + \epsilon x_2 \\ \frac{dx_2}{dt} = x_2(x_2 - a)(1 - x_2) - x_1 + u \\ \\ y = h(x) = x_1\end{array} \quad (4.32)$$

Equivalently one has

$$\begin{pmatrix}\dot{x}_1 \\ \dot{x}_2\end{pmatrix} = \begin{pmatrix}-\epsilon\gamma x_1 + \epsilon x_2 \\ x_2(x_2 - a)(1 - x_2) - x_1\end{pmatrix} + \begin{pmatrix}0 \\ 1\end{pmatrix} u \quad (4.33)$$
$$y = h(x) = x_1$$

The state variable $z_1 = h_1(x) = x_1$ is defined. It holds that

$$z_2 = L_f h_1(x) = \frac{\partial h_1}{\partial x_1} f_1 + \frac{\partial h_2}{\partial x_2} f_2 = 1 \cdot f_1 + 0 \cdot f_2 = f_1 \quad (4.34)$$

$$\begin{array}{l}L_f^2 h_1(x) = \frac{\partial z_2}{\partial x_1} f_1 + \frac{\partial z_2}{\partial x_2} f_2 = \frac{\partial}{\partial x_1}(-\epsilon\gamma x_1 + \epsilon x_2) f_1 + \frac{\partial}{\partial x_2}(-\epsilon\gamma x_1 + \epsilon x_2) f_2 \Rightarrow \\ L_f^2 h_1(x) = (-\epsilon\gamma) f_1 + \epsilon f_2 \Rightarrow \\ L_f^2 h_1(x) = -\epsilon\gamma(-\epsilon\gamma x_1 + \epsilon x_2) + \epsilon[x_2(x_2 - a)(1 - x_2) - x_1]\end{array} \quad (4.35)$$

Moreover, it holds

$$L_g h_1(x) = \frac{\partial h_1}{\partial x_1} g_1 + \frac{\partial h_2}{\partial x_2} g_2 = 0 \cdot g_1 + 1 \cdot g_2 = g_2 = 1 \quad (4.36)$$

$$\begin{array}{l}L_g L_f h_1(x) = \frac{\partial z_2}{\partial x_1} g_1 + \frac{\partial z_2}{\partial x_2} g_2 = \frac{\partial}{\partial x_1}(-\epsilon\gamma x_1 + \epsilon x_2) g_1 + \frac{\partial}{\partial x_2}(-\epsilon\gamma x_1 + \epsilon x_2) g_2 \Rightarrow \\ L_g L_f h_1(x) = -\epsilon\gamma g_1 + \epsilon g_2 \Rightarrow L_g L_f h_1(x) = (-\epsilon\gamma)0 + \epsilon 1 \Rightarrow L_g L_f h_1(x) = \epsilon\end{array} \quad (4.37)$$

It holds that

$$\begin{array}{c}\dot{z}_1 = z_2 \\ \dot{z}_2 = L_f^2 h_1(x) + L_g L_f h_1(x) u\end{array} \quad (4.38)$$

or equivalently

$$\begin{array}{c}\dot{z}_1 = z_2 \\ \dot{z}_2 = \{-\epsilon\gamma x_1 + \epsilon x_2 + \epsilon[x_2(x_2 - a)(1 - x_2) - x_1]\} + \epsilon u\end{array} \quad (4.39)$$

which can be also written in the matrix canonical form

$$\begin{pmatrix} \dot{z}_1 \\ \dot{z}_2 \end{pmatrix} = \begin{pmatrix} 0 & 1 \\ 0 & 0 \end{pmatrix} \begin{pmatrix} z_1 \\ z_2 \end{pmatrix} + \begin{pmatrix} 0 \\ 1 \end{pmatrix} v \tag{4.40}$$

Next, the relative degree of the system is computed. It holds that $L_g h_1(x) = 0$, whereas

$$\begin{aligned} L_g L_f h_1(x) &= \tfrac{\partial f_1}{\partial x_1} g_1 + \tfrac{\partial f_1}{\partial x_2} g_2 \Rightarrow \\ L_g L_f h_1(x) &= -\epsilon\gamma 0 + \epsilon 1 = \epsilon \neq 0 \end{aligned} \tag{4.41}$$

and $L_g L_f^{n-1} h_1(x) \neq 0$ for $n = 2$. Therefore, the relative degree of the system is $n = 2$.

4.9.2 *Linearization of the FitzHugh–Nagumo Model Using Differential Flatness Theory*

Next, the FitzHugh–Nagumo model is linearized with the use of the differential flatness theory. The flat output of the system is taken to be $y = h(x) = x_1$. It holds that $\dot{y} = \dot{x}_1$. From the first row of the state space equation of the neuron given in Eq. (4.33) one gets

$$x_2 = \tfrac{\dot{x}_1 + \epsilon\gamma x_1}{\epsilon} \Rightarrow x_2 = \tfrac{\dot{y} + \epsilon\gamma y}{\epsilon} \tag{4.42}$$

Moreover, from the second row of Eq. (4.33) one gets

$$u = \dot{x}_2 - x_2(x_2 - \alpha)(1 - x_2) + x_1 \tag{4.43}$$

and since x_1, x_2 are functions of the flat output and its derivatives one has that the control input u is also a function of the flat output and its derivatives. Therefore, the considered neuron model stands for a differentially flat dynamical system.

From the computation of the derivatives of the flat output one obtains

$$\dot{y}_1 = \dot{x}_1 = f_1 \tag{4.44}$$

$$\begin{aligned} \ddot{y}_1 = \dot{f}_1 &= \tfrac{\partial}{\partial x_1}(-\epsilon\gamma x_1 + \epsilon x_2)\dot{x}_1 + \tfrac{\partial}{\partial x_1}(-\epsilon\gamma x_1 + \epsilon x_2)\dot{x}_2 \Rightarrow \\ \ddot{y}_1 = -\epsilon\gamma\dot{x}_1 + \epsilon\dot{x}_2 \Rightarrow \ddot{y}_1 &= -\epsilon\gamma(-\epsilon\gamma x_1 + \epsilon x_2) + \epsilon[x_2(x_2 - a)(1 - x_2) - x_1] + u. \end{aligned} \tag{4.45}$$

By denoting as $v = -\epsilon\gamma\dot{x}_1 + \epsilon\dot{x}_2 \Rightarrow \ddot{y}_1 = -\epsilon\gamma(-\epsilon\gamma x_1 + \epsilon x_2) + \epsilon[x_2(x_2 - a)(1 - x_2) - x_1] + u$, one arrives again at a state space description of the system in the linear canonical form. Thus setting $z_1 = y_1$ and $z_2 = \dot{y}_1 = y_2$ one gets Eq. (4.40)

$$\begin{pmatrix} \dot{z}_1 \\ \dot{z}_2 \end{pmatrix} = \begin{pmatrix} 0 & 1 \\ 0 & 0 \end{pmatrix} \begin{pmatrix} z_1 \\ z_2 \end{pmatrix} + \begin{pmatrix} 0 \\ 1 \end{pmatrix} u \tag{4.46}$$

For the linearized neuron model one can define a feedback control signal using that $\ddot{y} = v$.

$$v = \ddot{y}^d - k_d(\dot{y} - \dot{y}_d) - k_p(y - y_d) \Rightarrow \tag{4.47}$$

and using that $L_f^2 h(x) + L_g L_f h(x) u = v$ one has

$$u = (v - L_f^2 h(x))/L_g L_f h(x) \tag{4.48}$$

4.10 Linearization of Coupled FitzHugh–Nagumo Neurons Using Differential Geometry

The following system of coupled neurons is considered

$$\begin{aligned} \frac{dV_1}{dt} &= V_1(V_1 - \alpha)(1 - V_1) - w_1 + g(V_1 - V_2) + I_1 \\ \frac{dw_1}{dt} &= \epsilon(V_1 - \gamma w_1) \\ \frac{dV_2}{dt} &= V_2(V_2 - \alpha)(1 - V_2) - w_2 + g(V_2 - V_1) + I_2 \\ \frac{dw_2}{dt} &= \epsilon(V_2 - \gamma w_2) \end{aligned} \tag{4.49}$$

The following state variables are defined $x_1 = w_1$, $x_2 = V$, $x_3 = w_2$, $x_4 = V_2$, $u_1 = I_1$ and $u_2 = I_2$ (Fig. 4.8). Thus, one obtains the following state-space description

$$\begin{aligned} \frac{dx_1}{dt} &= -\epsilon\gamma x_1 + \epsilon x_2 \\ \frac{dx_2}{dt} &= x_2(x_2 - a)(1 - x_2) - x_1 + g(x_2 - x_1) + u_1 \\ \frac{dx_3}{dt} &= -\epsilon\gamma x_3 + \epsilon x_4 \\ \frac{dx_4}{dt} &= x_4(x_4 - a)(1 - x_4) - x_3 + g(x_4 - x_2) + u_2 \end{aligned} \tag{4.50}$$

while the measured output is taken to be

$$\begin{aligned} y_1 &= h_1(x) = x_1 \\ y_2 &= h_2(x) = x_3 \end{aligned} \tag{4.51}$$

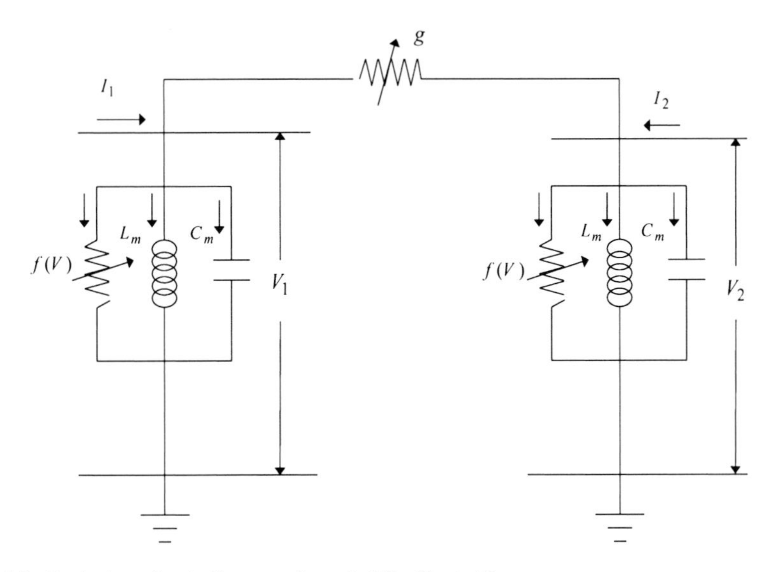

Fig. 4.8 Equivalent circuit diagram of coupled FitzHugh–Nagumo neurons

Equivalently, the state-space description of the system is given by

$$\begin{pmatrix} \dot{x}_1 \\ \dot{x}_2 \\ \dot{x}_3 \\ \dot{x}_4 \end{pmatrix} = \begin{pmatrix} -\epsilon\gamma x_1 + \epsilon x_2 \\ x_2(x_2 - a)(1 - x_2) - x_1 + g(x_2 - x_4) \\ -\epsilon\gamma x_3 + \epsilon x_4 \\ x_4(x_4 - \alpha)(1 - x_4) - x_3 + g(x_4 - x_2) \end{pmatrix} + \begin{pmatrix} 0 & 0 \\ 1 & 0 \\ 0 & 0 \\ 0 & 1 \end{pmatrix} \begin{pmatrix} u_1 \\ u_2 \end{pmatrix} \tag{4.52}$$

$$\begin{pmatrix} y_1 \\ y_2 \end{pmatrix} = \begin{pmatrix} h_1(x) \\ h_2(x) \end{pmatrix} = \begin{pmatrix} x_1 \\ x_3 \end{pmatrix} \tag{4.53}$$

It holds that

$$\begin{gathered} z_1^1 = h_1(x) = x_1 \\ z_2^1 = L_f h_1(x) = \frac{\partial h_1}{\partial x_1} f_1 + \frac{\partial h_1}{\partial x_2} f_2 + \frac{\partial h_1}{\partial x_3} f_3 + \frac{\partial h_1}{\partial x_4} f_4 \Rightarrow \\ z_2^1 = f_1 \Rightarrow z_2^1 = -\epsilon\gamma x_1 + \epsilon x_2 \end{gathered} \tag{4.54}$$

Moreover, it holds that

$$\begin{gathered} \dot{z}_2^1 = L_f^2 h_1(x) = \frac{\partial z_2^1}{\partial x_1} f_1 + \frac{\partial z_2^1}{\partial x_2} f_2 + \frac{\partial z_2^1}{\partial x_3} f_3 + \frac{\partial z_2^1}{\partial x_4} f_4 \Rightarrow \\ \dot{z}_2^1 = -\epsilon\gamma f_1 + \epsilon f_2 \Rightarrow \dot{z}_2^1 = \\ = -(\epsilon\gamma)(-\epsilon\gamma x_1 + \epsilon x_2) + \epsilon[x_2(x_2 - \alpha)(1 - x_2) - x_1 + g(x_2 - x_4)] \end{gathered} \tag{4.55}$$

In a similar manner one computes

$$\begin{array}{c} L_{g_1}h_1(x) = \frac{\partial h_1}{\partial x_1}g_{11} + \frac{\partial h_1}{\partial x_2}g_{12} + \frac{\partial h_1}{\partial x_3}g_{13} + \frac{\partial h_1}{\partial x_4}g_{14} \Rightarrow \\ L_{g_1}h_1(x) = 1g_{11} + 0g_{12} + 0g_{13} + 0g_{14} \Rightarrow L_{g_1}h_1(x) = 0 \end{array} \tag{4.56}$$

and

$$\begin{array}{c} L_{g_1}h_1(x) = \frac{\partial h_2}{\partial x_1}g_{21} + \frac{\partial h_2}{\partial x_2}g_{22} + \frac{\partial h_2}{\partial x_3}g_{23} + \frac{\partial h_2}{\partial x_4}g_{24} \Rightarrow \\ L_{g_1}h_1(x) = 1g_{21} + 0g22 + 0g_{23} + 0g_{24} \Rightarrow L_{g_1}h_1(x) = 0 \end{array} \tag{4.57}$$

Similarly one gets

$$\begin{array}{c} L_{g_1}L_fh_1(x) = \frac{\partial z_2^1}{\partial x_1}g_{11} + \frac{\partial z_2^1}{\partial x_2}g_{12} + \frac{\partial z_2^1}{\partial x_3}g_{13} + \frac{\partial z_2^1}{\partial x_4}g_{14} \Rightarrow \\ L_{g_1}L_fh_1(x) = (-\epsilon\gamma)g_{11} + \epsilon g12 + 0g_{13} + 0g_{14} \Rightarrow \\ L_{g_1}L_fh_1(x) = (-\epsilon\gamma)0 + \epsilon 1 + 0{\cdot}0 + 0{\cdot}0 \Rightarrow L_{g_1}L_fh_1(x) = \epsilon \end{array} \tag{4.58}$$

and

$$\begin{array}{c} L_{g_2}L_fh_1(x) = \frac{\partial z_2^1}{\partial x_1}g_{21} + \frac{\partial z_2^1}{\partial x_2}g_{22} + \frac{\partial z_2^1}{\partial x_3}g_{23} + \frac{\partial z_2^1}{\partial x_4}g_{24} \Rightarrow \\ L_{g_2}L_fh_1(x) = (-\epsilon\gamma)g_{21} + \epsilon g22 + 0g_{23} + 0g_{24} \Rightarrow \\ L_{g_2}L_fh_1(x) = (-\epsilon\gamma)0 + \epsilon 0 + 0{\cdot}0 + 0{\cdot}0 \Rightarrow L_{g_1}L_fh_1(x) = 0 \end{array} \tag{4.59}$$

Moreover, it holds

$$z_1^2 = h_2(x) = x_3 \tag{4.60}$$

and

$$\begin{array}{c} z_2^2 = L_fh_2(x) = \frac{\partial h_2}{\partial x_1}f_1 + \frac{\partial h_2}{\partial x_2}f_2 + \frac{\partial h_2}{\partial x_3}f_3 + \frac{\partial h_2}{\partial x_4}f_4 \Rightarrow \\ z_2^2 = f_3 \Rightarrow z_2^2 = -\epsilon\gamma x_3 + \epsilon x_4 \end{array} \tag{4.61}$$

Additionally, one has

$$\begin{array}{c} \dot{z}_2^2 = L_f^2h_2(x) = \frac{\partial z_2^1}{\partial x_1}f_1 + \frac{\partial z_2^1}{\partial x_2}f_2 + \frac{\partial z_2^1}{\partial x_3}f_3 + \frac{\partial z_2^1}{\partial x_4}f_4 \Rightarrow \\ \dot{z}_2^2 = 0f_1 + 0f_2 - \epsilon\gamma f_3 + \epsilon f_4 \Rightarrow \\ \dot{z}_2^2 = -\epsilon\gamma(-\epsilon\gamma x_3 + \epsilon x_4) + \epsilon[x_4(x_4 - a)(1 - x_4) - x_3 + g(x_4 - x_2)] \end{array} \tag{4.62}$$

Moreover one computes

$$\begin{array}{c} L_{g_1}h_2(x) = \frac{\partial h_2}{\partial x_1}g_{11} + \frac{\partial h_2}{\partial x_2}g_{12} + \frac{\partial h_2}{\partial x_3}g_{13} + \frac{\partial h_2}{\partial x_4}g_{14} \Rightarrow \\ L_{g_1}h_2(x) = 0g_{11} + 0g_{12} + 1g_{13} + 0g_{14} \Rightarrow L_{g_1}h_2(x) = g_{13} \Rightarrow L_{g_1}h_2(x) = 0 \end{array} \tag{4.63}$$

and

$$\begin{array}{c} L_{g_2}h_2(x) = \frac{\partial h_2}{\partial x_1}g_{21} + \frac{\partial h_2}{\partial x_2}g_{22} + \frac{\partial h_2}{\partial x_3}g_{23} + \frac{\partial h_2}{\partial x_4}g_{24} \Rightarrow \\ L_{g_2}h_2(x) = 1g_{21} + 0g_{22} + 0g_{23} + 0g_{24} \Rightarrow L_{g_2}h_2(x) = 0 \end{array} \tag{4.64}$$

Equivalently

$$\begin{array}{c} L_{g_1}L_fh_2(x) = \frac{\partial z_2^2}{\partial x_1}g_{11} + \frac{\partial z_2^2}{\partial x_2}g_{12} + \frac{\partial z_2^2}{\partial x_3}g_{13} + \frac{\partial z_2^2}{\partial x_4}g_{14} \Rightarrow \\ L_{g_1}L_fh_2(x) = 0g_{11} + 0g_{12} - \epsilon\gamma g_{13} + \epsilon g_{14} \Rightarrow \\ L_{g_1}L_fh_2(x) = 0{\cdot}0 + 0{\cdot}1 - \epsilon\gamma 0 + \epsilon 0 \Rightarrow L_{g_1}L_fh_2(x) = 0 \end{array} \tag{4.65}$$

and

$$\begin{array}{c} L_{g_2}L_fh_2(x) = \frac{\partial z_2^2}{\partial x_1}g_{21} + \frac{\partial z_2^2}{\partial x_2}g_{22} + \frac{\partial z_2^2}{\partial x_3}g_{23} + \frac{\partial z_2^2}{\partial x_4}g_{24} \Rightarrow \\ L_{g_2}L_fh_2(x) = 0g_{11} + 0g_{12} - \epsilon\gamma g_{23} + \epsilon g_{24} \\ L_{g_2}L_fh_2(x) = 0{\cdot}0 + 0{\cdot}0 - \epsilon\gamma 0 + \epsilon 1 \Rightarrow L_{g_2}L_fh_2(x) = \epsilon \end{array} \tag{4.66}$$

Therefore, it holds

$$\begin{array}{c} \dot{z}_1^1 = z_2^1 \\ \dot{z}_2^1 = L_f^2h_1(x) + L_{g_1}L_fh_1(x)u_1 + L_{g_2}L_fh_1(x)u_2 \\ \dot{z}_1^2 = z_2^2 \\ \dot{z}_2^2 = L_f^2h_2(x) + L_{g_1}L_fh_2(x)u_1 + L_{g_2}L_fh_2(x)u_2 \end{array} \tag{4.67}$$

or equivalently,

$$\begin{array}{c} \ddot{z}_2^1 = L_f^2h_1(x) + L_{g_1}L_fh_1(x)u_1 + L_{g_2}L_fh_1(x)u_2 \\ \ddot{z}_2^2 = L_f^2h_2(x) + L_{g_1}L_fh_2(x)u_1 + L_{g_2}L_fh_2(x)u_2 \end{array} \tag{4.68}$$

which can be also written in matrix form

$$\begin{pmatrix} \ddot{z}_2^1 \\ \ddot{z}_2^2 \end{pmatrix} = \begin{pmatrix} L_f^2h_1(x) \\ L_f^2h_2(x) \end{pmatrix} + \begin{pmatrix} L_{g_1}L_fh_1(x) & L_{g_2}L_fh_1(x) \\ L_{g_1}L_fh_2(x) & L_{g_2}L_fh_2(x) \end{pmatrix} \begin{pmatrix} u_1 \\ u_2 \end{pmatrix} \tag{4.69}$$

or

$$\ddot{z} = D_m + G_m u \tag{4.70}$$

Denoting

$$\begin{array}{c} v_1 = L_f^2h_1(x) + L_{g_1}L_fh_1(x)u_1 + L_{g_2}L_fh_1(x)u_2 \\ v_2 = L_f^2h_2(x) + L_{g_1}L_fh_2(x)u_1 + L_{g_2}L_fh_2(x)u_2 \end{array}$$

one also has

$$\begin{pmatrix} \ddot{z}_1^1 \\ \ddot{z}_1^2 \end{pmatrix} = \begin{pmatrix} v_1 \\ v_2 \end{pmatrix} \tag{4.71}$$

where the control inputs v_1 and v_2 are chosen as $v_1 = \ddot{z}_{1d}^1 - k_d^1(\dot{z}_{1d}^1 - \dot{z}_1^1) - k_p^1(z_{1d}^1 - z_1^1)$ and $v_2 = \ddot{z}_{1d}^2 - k_d^2(\dot{z}_{1d}^2 - \dot{z}_1^2) - k_p^1(z_{1d}^2 - z_1^2)$.

Therefore, the associated control signal is $u = G_m^{-1}(v - D_m(x))$.

4.11 Linearization of Coupled FitzHugh–Nagumo Neurons Using Differential Flatness Theory

4.11.1 Differential Flatness of the Model of the Coupled Neurons

It can be proven that the model of the coupled FitzHugh–Nagumo neurons described in Eq. (4.50) is a differentially flat one. The following flat outputs are defined

$$\begin{aligned} y_1 &= h_1(x) = x_1 \\ y_2 &= h_2(x) = x_3 \end{aligned} \tag{4.72}$$

From the first row of the dynamical model of the coupled neurons one has

$$\dot{y}_1 = -\epsilon\gamma y_1 + \epsilon x_2 \Rightarrow x_2 = \frac{\dot{y}_1 + \epsilon\gamma y_1}{\epsilon} \tag{4.73}$$

Similarly, from the third row of the dynamical model of the coupled neurons one has

$$\dot{y}_2 = -\epsilon\gamma y_2 + \epsilon x_4 \Rightarrow x_4 = \frac{\dot{y}_2 + \epsilon\gamma y_2}{\epsilon} \tag{4.74}$$

Thus all state variables x_i, $i = 1, \cdots, 4$ can be written as functions of the flat outputs and their derivatives.

From the second row of the dynamical model of the coupled neurons one has

$$u_1 = \dot{x}_2 - x_2(x_2 - a)(1 - x_2) + x_1 - g(x_2 - x_1) \Rightarrow u_1 = f_1(y_1, \dot{y}_1.y_2, \dot{y}_2) \tag{4.75}$$

Similarly, from the second row of the dynamical model of the coupled neurons one has

$$u_2 = \dot{x}_4 - x_4(x_4 - a)(1 - x_4) + x_3 - g(x_4 - x_2) \Rightarrow u_2 = f_2(y_1, \dot{y}_1.y_2, \dot{y}_2) \tag{4.76}$$

Consequently, all state variables x_i, $i = 1, \cdots, 4$ and all control inputs u_i $i = 1, 2$ of the model of the coupled FitzHugh–Nagumo neurons can be written as functions of the flat outputs y_i $i = 1, 2$ and their derivatives. Therefore, the model of the coupled neurons is a differentially flat one.

4.11.2 *Linearization of the Coupled Neurons Using Differential Flatness Theory*

The problem of linearization and decoupling of the interconnected FitzHugh–Nagumo neural oscillators can be efficiently solved using differential flatness theory. For the model of the coupled FitzHugh–Nagumo neurons described in Eq. (4.50) the flat outputs $y_1 = h_1(x) = x_1$ and $y_2 = h_2(x) = x_2$ have been defined. It holds that

$$\dot{y}_1 = \dot{x}_1 \Rightarrow \dot{y}_1 = f_1 \Rightarrow \dot{y}_1 = -\epsilon\gamma x_1 + \epsilon x_2 \tag{4.77}$$

and

$$\begin{gathered}\ddot{y}_1 = \tfrac{\partial f_1}{\partial x_1}\dot{x}_1 + \tfrac{\partial f_1}{\partial x_2}\dot{x}_2 + \tfrac{\partial f_1}{\partial x_3}\dot{x}_3 + \tfrac{\partial f_1}{\partial x_4}\dot{x}_4 \Rightarrow \\ \ddot{y}_1 = -(\epsilon\gamma)\dot{x}_1 + \epsilon\dot{x}_2 + 0\dot{x}_3 + 0\dot{x}_4 \Rightarrow \\ \ddot{y}_1 = -(\epsilon\gamma)f_1 + \epsilon(f_2 + u_1) \Rightarrow \\ \ddot{y}_1 = (-\epsilon\gamma)(-\epsilon\gamma x_1 + \epsilon x_2) + \epsilon[x_2(x_2 - a)(1 - x_2) - x_1 + g(x_2 - x_4)] + \epsilon u_1 \Rightarrow\end{gathered} \tag{4.78}$$

or equivalently

$$\ddot{y}_1 = L_f^2 h_1(x) + L_{g_1} L_f h_1(x) u_1 + L_{g_2} L_f h_1(x) u_2 \tag{4.79}$$

where

$$\begin{gathered}L_f^2 h_1(x) = (-\epsilon\gamma)(-\epsilon\gamma x_1 + \epsilon x_2) + \epsilon[x_2(x_2 - a)(1 - x_2) - x_1 + g(x_2 - x_4)] \\ L_{g_1} L_f h_1(x) = \epsilon \quad L_{g_2} L_f h_1(x) = 0\end{gathered} \tag{4.80}$$

Equivalently, one computes

$$\dot{y}_2 = \dot{x}_3 \Rightarrow \dot{y}_2 = f_3 \Rightarrow \dot{y}_2 = -\epsilon\gamma x_3 + \epsilon x_4 \tag{4.81}$$

and

$$\begin{gathered}\ddot{y}_2 = \tfrac{\partial f_3}{\partial x_1}\dot{x}_1 + \tfrac{\partial f_3}{\partial x_2}\dot{x}_2 + \tfrac{\partial f_3}{\partial x_3}\dot{x}_3 + \tfrac{\partial f_3}{\partial x_4}\dot{x}_4 \Rightarrow \\ \ddot{y}_2 = 0\dot{x}_1 + 0\dot{x}_2 + (-\epsilon\gamma)\dot{x}_3 + \epsilon\dot{x}_4 \Rightarrow \\ \ddot{y}_2 = (-\epsilon\gamma)f_3 + \epsilon(f_4 + u_2) \Rightarrow \\ \ddot{y}_2 = (-\epsilon\gamma)(-\epsilon\gamma x_3 + \epsilon x_4) + \epsilon[x_4(x_4 - a)(1 - x_4) - x_3 + g(x_4 - x_2)] + \epsilon u_2 \Rightarrow \\ \ddot{y}_2 = L_f^2 h_2(x) + L_{g_1} L_f h_2(x) u_1 + L_{g_2} L_f h_2(x) u_2\end{gathered} \tag{4.82}$$

where

$$L_f^2 h_2(x) = (-\epsilon\gamma)(-\epsilon\gamma x_3 + \epsilon x_4) + \epsilon[x_4(x_4 - a)(1 - x_4) - x_3 + g(x_4 - x_2)]$$
$$L_{g_1} L_f h_2(x) = 0 \quad L_{g_2} L_f h_2(x) = \epsilon \tag{4.83}$$

By defining the state variables $z_1^1 = y_1$, $\dot{z}_1^1 = \dot{y}_1$, $z_1^2 = y_2$ and $\dot{z}_1^2 = \dot{y}_2$ one obtains a description of the coupled neurons in the state-space form

$$\begin{pmatrix} \ddot{z}_2^1 \\ \ddot{z}_2^2 \end{pmatrix} = \begin{pmatrix} L_f^2 h_1(x) \\ L_f^2 h_2(x) \end{pmatrix} + \begin{pmatrix} L_{g_1} L_f h_1(x) & L_{g_2} L_f h_1(x) \\ L_{g_1} L_f h_2(x) & L_{g_2} L_f h_2(x) \end{pmatrix} \begin{pmatrix} u_1 \\ u_2 \end{pmatrix} \tag{4.84}$$

$$\ddot{z} = D_m + G_m u \tag{4.85}$$

Denoting

$$v_1 = L_f^2 h_1(x) + L_{g_1} L_f h_1(x) u_1 + L_{g_2} L_f h_1(x) u_2$$
$$v_2 = L_f^2 h_2(x) + L_{g_1} L_f h_2(x) u_1 + L_{g_2} L_f h_2(x) u_2$$

one also has

$$\begin{pmatrix} \ddot{z}_1^1 \\ \ddot{z}_1^2 \end{pmatrix} = \begin{pmatrix} v_1 \\ v_2 \end{pmatrix} \tag{4.86}$$

where the control inputs v_1 and v_2 are chosen as $v_1 = \ddot{z}_{1d}^1 - k_d^1(\dot{z}_{1d}^1 - \dot{z}_1^1) - k_p^1(z_{1d}^1 - z_1^1)$ and $v_2 = \ddot{z}_{1d}^2 - k_d^2(\dot{z}_{1d}^2 - \dot{z}_1^2) - k_p^1(z_{1d}^2 - z_1^2)$.

Therefore, the associated control signal is $u = G_m^{-1}(v - D_m(x))$.

4.12 State and Disturbances Estimation with the Derivative-Free Nonlinear Kalman Filter

4.12.1 *Kalman and Extended Kalman Filtering*

It will be shown that a new nonlinear filtering method, the so-called Derivative-free nonlinear Kalman Filter, can be used for estimating wave-type dynamics in the neuron's membrane. This can be done through the processing of noisy measurements and without knowledge of boundary conditions. Other results on the application of Kalman Filtering in the estimation of neuronal dynamics can be found in [77, 99, 134, 161]

An overview of Kalman and Extended Kalman Filtering is given first. In the discrete-time case a dynamical system is assumed to be expressed in the form of a discrete-time state model [166]:

$$\begin{aligned} x(k+1) &= A(k)x(k) + B(k)u(k) + w(k) \\ z(k) &= Cx(k) + v(k) \end{aligned} \tag{4.87}$$

where the state $x(k)$ is a m-vector, $w(k)$ is an m-element process noise vector, and A is an $m \times m$ real matrix. Moreover the output measurement $z(k)$ is a p-vector, C is a $p \times m$-matrix of real numbers, and $v(k)$ is the measurement noise. It is assumed that the process noise $w(k)$ and the measurement noise $v(k)$ are uncorrelated. The process and measurement noise covariance matrices are denoted as $Q(k)$ and $R(k)$, respectively. Now the problem is to estimate the state $x(k)$ based on the measurements $z(1), z(2), \cdots, z(k)$. This can be done with the use of Kalman Filtering. The discrete-time Kalman filter can be decomposed into two parts: (1) time update (prediction stage), and (2) measurement update (correction stage).
Measurement update:

$$\begin{aligned} K(k) &= P^{-}(k)C^{T}[C \cdot P^{-}(k)C^{T} + R]^{-1} \\ \hat{x}(k) &= \hat{x}^{-}(k) + K(k)[z(k) - C\hat{x}^{-}(k)] \\ P(k) &= P^{-}(k) - K(k)CP^{-}(k) \end{aligned} \tag{4.88}$$

Time update:

$$\begin{aligned} P^{-}(k+1) &= A(k)P(k)A^{T}(k) + Q(k) \\ \hat{x}^{-}(k+1) &= A(k)\hat{x}(k) + B(k)u(k) \end{aligned} \tag{4.89}$$

Next, the following nonlinear state-space model is considered:

$$\begin{aligned} x(k+1) &= \phi(x(k)) + L(k)u(k) + w(k) \\ z(k) &= \gamma(x(k)) + v(k) \end{aligned} \tag{4.90}$$

The operators $\phi(x)$ and $\gamma(x)$ are $\phi(x) = [\phi_1(x), \phi_2(x), \cdots, \phi_m(x)]^T$, and $\gamma(x) = [\gamma_1(x), \gamma_2(x), \cdots, \gamma_p(x)]^T$, respectively. It is assumed that ϕ and γ are sufficiently smooth in x so that each one has a valid series Taylor expansion. Following a linearization procedure, about the current state vector estimate $\hat{x}(k)$ the linearized version of the system is obtained:

$$\begin{aligned} x(k+1) &= \phi(\hat{x}(k)) + J_{\phi}(\hat{x}(k))[x(k) - \hat{x}(k)] + w(k) \\ z(k) &= \gamma(\hat{x}^{-}(k)) + J_{\gamma}(\hat{x}^{-}(k))[x(k) - \hat{x}^{-}(k)] + v(k) \end{aligned} \tag{4.91}$$

where $J_{\phi}(\hat{x}(k))$ and $J_{\gamma}(\hat{x}(k))$ are the associated Jacobian matrices of ϕ and γ, respectively. Now, the EKF recursion is as follows [157].
Measurement update. Acquire $z(k)$ and compute:

$$\begin{aligned} K(k) &= P^{-}(k)J_{\gamma}^{T}(\hat{x}^{-}(k)) \cdot [J_{\gamma}(\hat{x}^{-}(k))P^{-}(k)J_{\gamma}^{T}(\hat{x}^{-}(k)) + R(k)]^{-1} \\ \hat{x}(k) &= \hat{x}^{-}(k) + K(k)[z(k) - \gamma(\hat{x}^{-}(k))] \\ P(k) &= P^{-}(k) - K(k)J_{\gamma}(\hat{x}^{-}(k))P^{-}(k) \end{aligned} \tag{4.92}$$

Time update. Compute:

$$\begin{aligned} P^{-}(k+1) &= J_{\phi}(\hat{x}(k))P(k)J_{\phi}^{T}(\hat{x}(k)) + Q(k) \\ \hat{x}^{-}(k+1) &= \phi(\hat{x}(k)) + L(k)u(k) \end{aligned} \tag{4.93}$$

4.12.2 Design of a Disturbance Observer for the Model of the Coupled Neurons

Using Eq. (4.69), the nonlinear model of the coupled neurons can be written in the MIMO canonical form that was described in Sect. 4.8, that is

$$\begin{pmatrix} \dot{z}_1 \\ \dot{z}_2 \\ \dot{z}_3 \\ \dot{z}_4 \end{pmatrix} = \begin{pmatrix} 0 & 1 & 0 & 0 \\ 0 & 0 & 0 & 0 \\ 0 & 0 & 1 & 0 \\ 0 & 0 & 0 & 0 \end{pmatrix} \begin{pmatrix} z_1 \\ z_2 \\ z_3 \\ z_4 \end{pmatrix} + \begin{pmatrix} 0 & 0 \\ 1 & 0 \\ 0 & 0 \\ 0 & 1 \end{pmatrix} \begin{pmatrix} v_1 \\ v_2 \end{pmatrix} \tag{4.94}$$

where $z_1 = y_1$, $z_2 = \dot{y}_1$, $z_3 = y_2$ and $z_4 = \dot{y}_2$, while $v_1 = L_f^2 h_1(x) + L_{g_1} L_f h_1(x) u_1 + L_{g_2} L_f h_1(x) u_2$ and $v_2 = L_f^2 h_2(x) + L_{g_1} L_f h_2(x) u_1 + L_{g_2} L_f h_2(x) u_2$. Thus one has a MIMO linear model of the form

$$\begin{aligned} \dot{z} &= Az + Bv \\ z_m &= Cz \end{aligned} \tag{4.95}$$

where $z = [z_1, z_2, z_3, z_4]^T$ and matrices A, B, C are in the MIMO canonical form

$$A = \begin{pmatrix} 0 & 1 & 0 & 0 \\ 0 & 0 & 0 & 0 \\ 0 & 0 & 1 & 0 \\ 0 & 0 & 0 & 0 \end{pmatrix} \quad B = \begin{pmatrix} 0 & 0 \\ 1 & 0 \\ 0 & 0 \\ 0 & 1 \end{pmatrix} \quad C^T = \begin{pmatrix} 1 & 0 \\ 0 & 0 \\ 0 & 1 \\ 0 & 0 \end{pmatrix} \tag{4.96}$$

Assuming now the existence of additive input disturbances in the linearized model of the coupled neurons described in Eq. (4.69) one gets

$$\begin{aligned} \begin{pmatrix} \ddot{z}_2^1 \\ \ddot{z}_2^2 \end{pmatrix} &= \begin{pmatrix} L_f^2 h_1(x) \\ L_f^2 h_2(x) \end{pmatrix} \\ &+ \begin{pmatrix} L_{g_1} L_f h_1(x) & L_{g_2} L_f h_1(x) \\ L_{g_1} L_f h_2(x) & L_{g_2} L_f h_2(x) \end{pmatrix} \begin{pmatrix} u_1 + d_1 \\ u_2 + d_2 \end{pmatrix} \end{aligned} \tag{4.97}$$

These disturbances can be due to external perturbations affecting the coupled neurons and can also represent parametric uncertainty for the neurons' model. It can be assumed that the additive disturbances d_i, $i = 1, 2$ are described by the

n-th order derivatives of d_i and the associated initial conditions. Without loss of generality it will be considered that $n = 2$. This means

$$\begin{aligned} \ddot{d}_1 &= f_a(y_1, \dot{y}_1, y_2, \dot{y}_2) \\ \ddot{d}_2 &= f_b(y_1, \dot{y}_1, y_2, \dot{y}_2) \end{aligned} \tag{4.98}$$

By defining the additional state variables $z_5 = d_1$, $z_6 = \dot{d}_1$, $z_7 = d_2$, $z_8 = \dot{d}_2$ one has $\dot{z}_1 = z_2$, $\dot{z}_2 = v_1 + d_1$, $\dot{z}_3 = z_4$, $\dot{z}_4 = v_2 + d_2$, $\dot{z}_5 = z_6$, $\dot{z}_6 = f_a$, $\dot{z}_7 = z_8$ and $\dot{z}_8 = f_b$. Thus, in the case of additive input disturbances the state-space equations of the model of the coupled neurons can be written in the following matrix form

$$\begin{aligned} \dot{z} &= Az + Bv \\ z_m &= Cz \end{aligned} \tag{4.99}$$

where the state vector is $z = [z_1, z_2, z_3, z_4, z_5, z_6, z_7, z_8]^T$, the control input vector is $u = [v_1, v_2, f_a, f_b]^T$, and the measured output is $z_m = [z_1, z_3]^T$, while matrices A, B, and C are defined as

$$A = \begin{pmatrix} 0&1&0&0&0&0&0&0 \\ 0&0&0&0&1&0&0&0 \\ 0&0&0&1&0&0&0&0 \\ 0&0&0&0&0&0&1&0 \\ 0&0&0&0&0&1&0&0 \\ 0&0&0&0&0&0&0&0 \\ 0&0&0&0&0&0&0&1 \\ 0&0&0&0&0&0&0&0 \end{pmatrix} \quad B = \begin{pmatrix} 0&0&0&0 \\ 1&0&0&0 \\ 0&0&0&0 \\ 0&1&0&0 \\ 0&0&0&0 \\ 0&0&1&0 \\ 0&0&0&0 \\ 0&0&0&1 \end{pmatrix} \tag{4.100}$$

$$C = \begin{pmatrix} 1&0&0&0&0&0&0&0 \\ 0&0&1&0&0&0&0&0 \end{pmatrix} \tag{4.101}$$

The associated disturbance observer is

$$\begin{aligned} \dot{\hat{z}} &= A_o\hat{z} + B_o u + K(z_m - \hat{z}_m) \\ \hat{z}_m &= C_o\hat{z} \end{aligned} \tag{4.102}$$

where $A_o = A$, $C_o = C$ and

$$B_o^T = \begin{pmatrix} 0&1&0&0&0&0&0&0 \\ 0&0&0&1&0&0&0&0 \end{pmatrix} \tag{4.103}$$

where now $u = [u_1, u_2]^T$.

4.12.3 Disturbances Compensation for the Model of the Coupled Neurons

For the aforementioned model, and after carrying out discretization of matrices A_o, B_o, and C_o with common discretization methods one can perform Kalman filtering [157–159, 166, 169, 170]. This is *Derivative-free nonlinear Kalman filtering* which, unlike EKF, is performed without the need to compute Jacobian matrices and does not introduce numerical errors due to approximative linearization with Taylor series expansion.

In the design of the associated disturbances' estimator one has the dynamics defined in Eq. (5.85), where $K \in R^{8\times 2}$ is the state estimator's gain and matrices A_o, B_o, and C_o have been defined in Eqs. (4.100) and (4.101). The discrete-time equivalents of matrices A_o, B_o, and C_o are denoted as $\tilde{A}_d$, $\tilde{B}_d$, and $\tilde{C}_d$, respectively, and are computed with the use of common discretization methods. Next, a Derivative-free nonlinear Kalman Filter can be designed for the aforementioned representation of the system dynamics [157, 158]. The associated Kalman Filter-based disturbance estimator is given by the recursion [159, 166, 169].
Measurement update:

$$
\begin{aligned}
K(k) &= P^{-}(k)\tilde{C}_d^T[\tilde{C}_d \cdot P^{-}(k)\tilde{C}_d^T + R]^{-1} \\
\hat{z}(k) &= \hat{z}^{-}(k) + K(k)[\tilde{C}_d z(k) - \tilde{C}_d \hat{z}^{-}(k)] \\
P(k) &= P^{-}(k) - K(k)\tilde{C}_d P^{-}(k)
\end{aligned} \tag{4.104}
$$

Time update:

$$
\begin{aligned}
P^{-}(k+1) &= \tilde{A}_d(k)P(k)\tilde{A}_d^T(k) + Q(k) \\
\hat{z}^{-}(k+1) &= \tilde{A}_d(k)\hat{z}(k) + \tilde{B}_d(k)\tilde{v}(k)
\end{aligned} \tag{4.105}
$$

To compensate for the effects of the disturbance inputs it suffices to use in the control loop the estimates of the disturbance terms that is $\hat{z}_5 = d_1$ and $\hat{z}_7 = d_2$ and the modified control input vector

$$
v = \begin{pmatrix} v_1 - \hat{f}_a \\ v_2 - \hat{f}_b \end{pmatrix} \text{ or } v = \begin{pmatrix} v_1 - \hat{z}_5 \\ v_2 - \hat{z}_7 \end{pmatrix} \tag{4.106}
$$

4.13 Simulation Tests

Results on the synchronization between the master and the slave FitzHugh–Nagumo neuron are provided for three different cases, each one associated with difference disturbance and uncertainty terms affecting the neuron model as well as with different setpoints for the state vector elements of the coupled neurons' model.

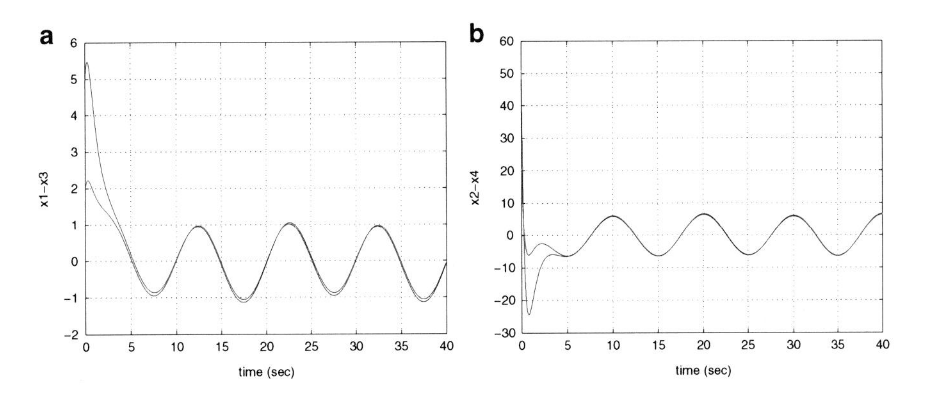

Fig. 4.9 Synchronization in the model of the two coupled FitzHugh–Nagumo neurons (**a**) between state variables x_1 (master neuron) and x_3 (slave neuron), (**b**) between state variables x_2 (master neuron) and x_4 (slave neuron)

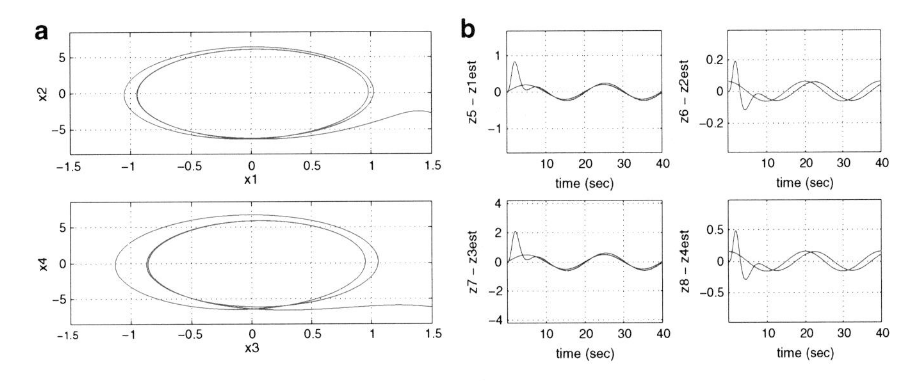

Fig. 4.10 Phase diagrams of the synchronized FitzHugh–Nagumo neurons (**a**) between state variables x_1–x_2 (master neuron) and between x_3–x_4 (slave neuron), (**b**) estimation of disturbances affecting the master neuron (state variables z_5 and z_6) and the slave neuron (state variables z_7 and z_8)

The results obtained in the first test case are shown in Figs. 4.9 and 4.10. The results obtained in the second case are shown in Figs. 4.11 and 4.12. Finally, the results obtained in the third case are shown in Figs. 4.13 and 4.14.

In Figs. 4.9a, 4.11a, and 4.13a it is shown how the proposed Kalman Filter-based feedback control scheme succeeded the convergence of state variable x_1 of the master neuron to state variable x_3 of the second neuron. Similarly, in Figs. 4.9b, 4.11b, and 4.13b it is shown how the proposed Kalman Filter-based feedback control scheme succeeded the convergence of state variable x_2 of the master neuron to state variable x_3 of the second neuron. In Figs. 4.10a, 4.12a, and 4.14a the phase diagrams between the state variables x_1, x_2 of the master neuron and the phase diagrams between the state variables x_3, x_4 of the slave neuron are presented. The phase diagrams for the master and slave neuron become identical and this is an indication

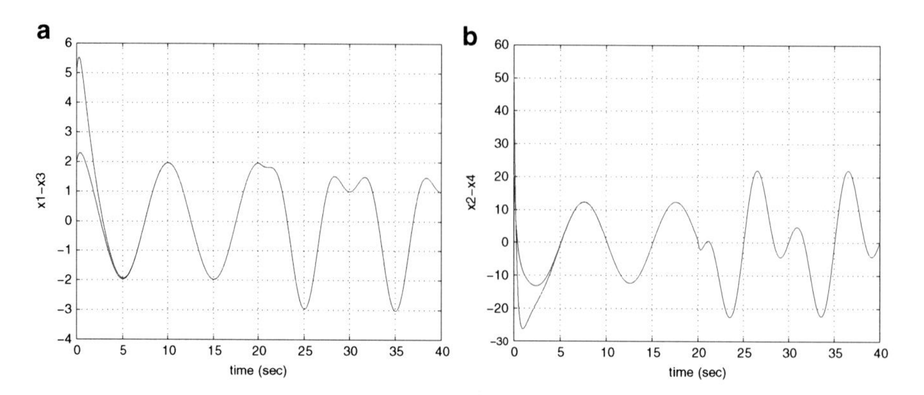

Fig. 4.11 Synchronization in the model of the two coupled FitzHugh–Nagumo neurons (**a**) between state variables x_1 (master neuron) and x_3 (slave neuron), (**b**) between state variables x_2 (master neuron) and x_4 (slave neuron)

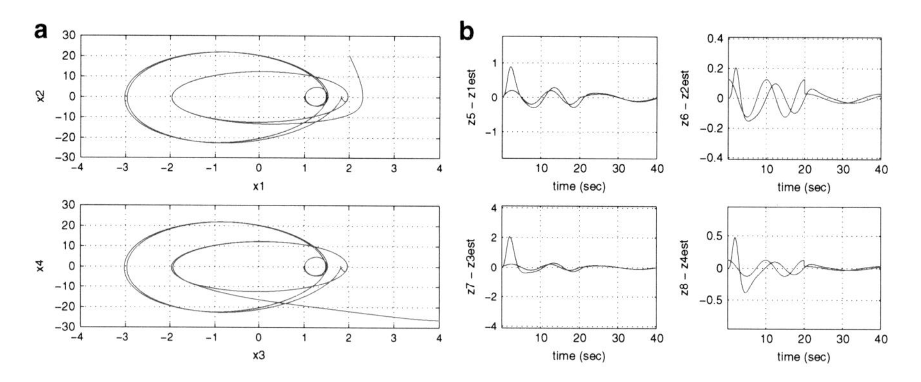

Fig. 4.12 Phase diagrams of the synchronized FitzHugh–Nagumo neurons (**a**) between state variables x_1–x_2 (master neuron) and between x_3–x_4 (slave neuron), (**b**) estimation of disturbances affecting the master neuron (state variables z_5 and z_6) and the slave neuron (state variables z_7 and z_8)

about the succeeded synchronization. Finally, in Figs. 4.10b, 4.12b, and 4.14b the estimates of the disturbances terms affecting the neurons' model are provided. It can be noticed that the Kalman Filter-based observer provides accurate estimates about the non-measurable disturbances and this information enables the efficient compensation of the perturbations' effects.

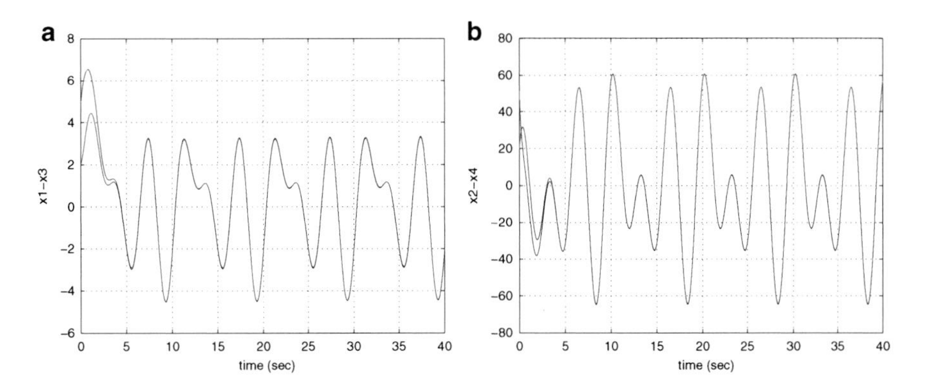

Fig. 4.13 Synchronization in the model of the two coupled FitzHugh–Nagumo neurons (**a**) between state variables x_1 (master neuron) and x_3 (slave neuron), (**b**) between state variables x_2 (master neuron) and x_4 (slave neuron)

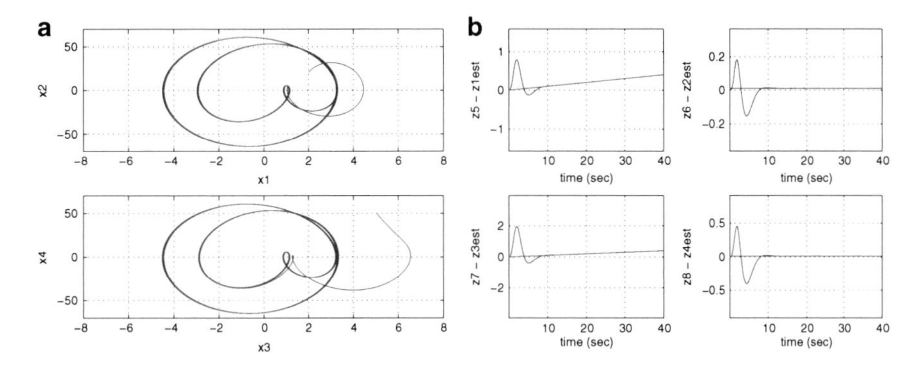

Fig. 4.14 Phase diagrams of the synchronized FitzHugh–Nagumo neurons (**a**) between state variables x_1–x_2 (master neuron) and between x_3–x_4 (slave neuron), (**b**) estimation of disturbances affecting the master neuron (state variables z_5 and z_6) and the slave neuron (state variables z_7 and z_8)

4.14 Conclusions

A new method for robust synchronization of coupled neural oscillators has been developed. The model of the FitzHugh–Nagumo neural oscillators has been considered which represents efficiently voltage variations on the neuron's membrane, due to ionic currents and external activation currents. It was pointed out that synchronism between the coupled neurons is responsible for rhythm generation and control of several functions in living species such as gait, breathing, heart's pulsing, etc. Moreover, models of synchronized neurons can be used in several robotic applications to control locomotion of multilegged, quadruped, and biped robots.

With the application of differential geometric methods and differential flatness theory it was shown that the nonlinear model of the FitzHugh–Nagumo neuron can be written in the linear canonical form. It was also shown that by applying differential geometric methods and by computing Lie derivatives exact linearization of the model of the coupled FitzHugh–Nagumo neurons (connected through a gap junction) can be succeeded. Moreover, it was proven that the model of the coupled FitzHugh–Nagumo neuron is a differentially flat one and by defining appropriate flat outputs it can be transformed to the MIMO linear canonical form. For the linearized representation of the coupled neuron's model the design of a feedback control is possible and synchronization between the two neurons can be attained.

Next, the problem of synchronization of the coupled neurons under external perturbations and parametric uncertainties was examined. To obtain simultaneous state and disturbances estimation a disturbance observer based on the Derivative-free nonlinear Kalman Filter has been used. The Derivative-free nonlinear Kalman Filter consists of the standard Kalman Filter recursion on the linearized equivalent model of the coupled neurons and on computation of state and disturbance estimates using the diffeomorphism (relations about state variables transformation) provided by differential flatness theory. After estimating the disturbance terms in the neurons' model their compensation has become possible.

The performance of the synchronizing control loop has been tested through simulation experiments. It was shown that the proposed method assures that the neurons will remain synchronized, despite parametric variations and uncertainties in the associated dynamical model and despite the existence of external perturbations. The method can be extended to multiple interconnected neurons, thus enabling to succeed more complicated patterns in robot's motion.

Chapter 5
Synchronization of Circadian Neurons and Protein Synthesis Control

Abstract The chapter proposes a new method for synchronization of coupled circadian cells and for nonlinear control of the associated protein synthesis process using differential flatness theory and the derivative-free nonlinear Kalman Filter. By proving that the dynamic model of the FRQ protein synthesis is a differentially flat one its transformation to the linear canonical (Brunovsky) form becomes possible. For the transformed model one can find a state feedback control input that makes the oscillatory characteristics in the concentration of the FRQ protein vary according to desirable setpoints. To estimate nonmeasurable elements of the state vector the Derivative-free nonlinear Kalman Filter is used. The Derivative-free nonlinear Kalman Filter consists of the standard Kalman Filter recursion on the linearized equivalent model of the coupled circadian cells and on computation of state and disturbance estimates using the diffeomorphism (relations about state variables transformation) provided by differential flatness theory. Moreover, to cope with parametric uncertainties in the model of the FRQ protein synthesis and with stochastic disturbances in measurements, the Derivative-free nonlinear Kalman Filter is redesigned in the form of a disturbance observer. The efficiency of the proposed Kalman Filter-based control scheme is tested through simulation experiments.

5.1 Overview

The chapter studies synchronization of networks of circadian oscillators and control of the protein synthesis procedure performed by them. Circadian oscillators are cells performing protein synthesis within a feedback loop. The concentration of the produced proteins exhibits periodic variations in time. In humans, circadian oscillators are neural cells which are found in the suprachiasmatic nucleus of the hypothalamus and which are coupled through the exchange of neurotransmitters. There are also peripheral circadian oscillators which are responsible for rhythmicity in the functioning of various organs in the human body. The peripheral

G.G. Rigatos, *Advanced Models of Neural Networks*,
DOI 10.1007/978-3-662-43764-3_5, © Springer-Verlag Berlin Heidelberg 2015

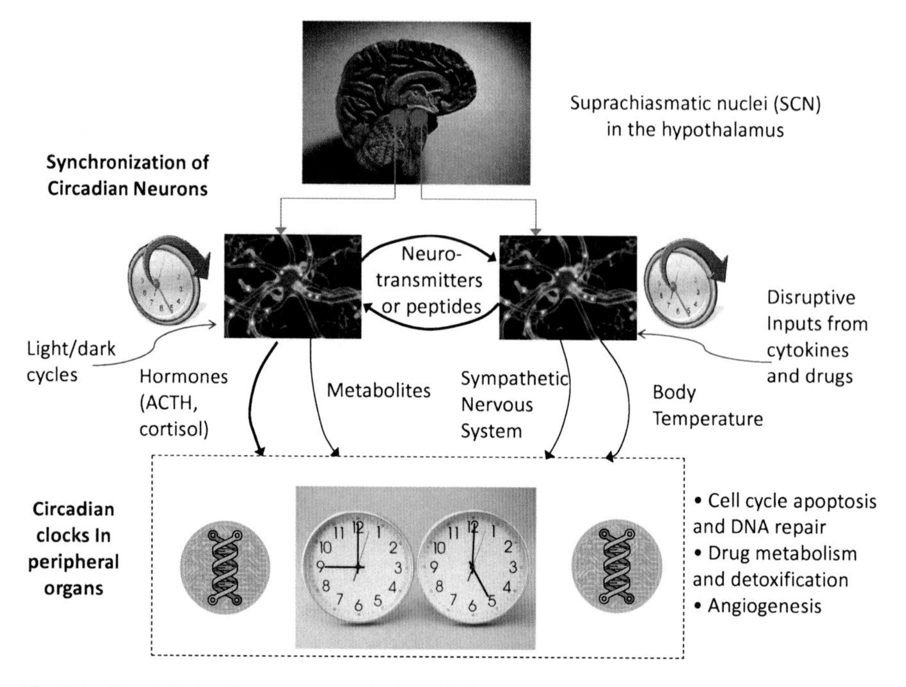

Fig. 5.1 Coupled circadian neurons at the hypothalamus and synchronization of circadian oscillators in peripheral organs

circadian oscillators are controlled by the circadian neurons through messages sent by the latter, in the form of hormones (ACTH, cortisol, etc.), metabolites, inputs from the sympathetic neural system, or variations in body temperature [61, 67, 68, 104] (see Fig. 5.1). Circadian rhythms affect cell cycle apoptosis and DNA repair, drug metabolism, and detoxification, as well as angiogenesis. The disruption of periodicity and synchronism in circadian oscillators is known to be responsible for the appearance of malignancies [32,33,105,106]. On the other hand, pharmacological treatment of tumors when taking into account the variations of proteins concentration controlled by the circadian oscillators can be more efficient. Mathematical models of circadian oscillators consist of sets of ordinary differential equations describing the feedback loop of the generation of proteins in the circadian cells. The main stages of these feedback loops are transcription of genes into mRNA, translation of mRNA into proteins, and inhibition of the transcription stage by the generated protein. The number of ordinary differential equations in such models is moderate in simple organisms (e.g., neurospora and drosophila) whereas it becomes elevated in the case of mammals. In this chapter the model of the neurospora circadian oscillators will be considered for testing the method of feedback control and synchronization of the circadian cells. However, the method is generic and can be applied to more elaborated and increased order models of circadian oscillators.

Previous results on the control of circadian oscillators have been presented in [14, 53, 102, 142, 182, 188–190, 219]. It appears that this control problem is a nontrivial one when taking into account uncertainty and stochasticity of the oscillators model. Up to now the problem of robust synchronization of coupled circadian oscillators has not been solved efficiently. The approach followed in this chapter for succeeding control of the protein synthesis procedure in the circadian cells as well as for synchronizing coupled circadian cells when performing such processes is based on transformation of the dynamical model of the circadian oscillators into the linear canonical (Brunovsky) form. This transformation makes use of differential flatness theory [58, 101, 109, 172].

The problem of synchronization of coupled circadian oscillators becomes more difficult when there is uncertainty about the parameters of the dynamical model of the circadian cells and when the parameters of the dynamical models of these cells are uneven. This chapter is concerned with proving differential flatness of the model of the coupled circadian oscillators and its resulting description in the Brunovksy (canonical) form [125]. By defining specific state variables as flat outputs an equivalent description of the coupled circadian oscillators in the Brunovksy (linear canonical) form is obtained. It is shown that for the linearized model of the coupled oscillators it is possible to design a feedback controller that succeeds their synchronization. At a second stage, the novel Kalman Filtering method proposed in Chap. 4, that is the Derivative-free nonlinear Kalman Filter, is used as a disturbance observer, thus making possible to estimate disturbance terms affecting the model of the coupled oscillators and to use these terms in the feedback controller. By avoiding linearization approximations, the proposed filtering method improves the accuracy of estimation and results in smooth control signal variations and in minimization of the synchronization error [158, 159, 170].

5.2 Modelling of Circadian Oscillators Dynamics

5.2.1 The Functioning of the Circadian Oscillators

Circadian neurons are a small cluster of 2×10^4 neurons in the ventral hypothalamus and provide the body with key time-keeping signals, known as circadian rhythms. Circadian neurons mainly receive input from the retina, can be coupled through neurotransmitters, and can stimulate other cells in the body also exhibiting an oscillatory behavior that provides time keeping for other functions of the body.

Circadian oscillators affect cell division. A model that describes the cell cycle is given in [32]. The model comprises several stages of the cell division cycle each one denoted by a different index i $(1\leq i\leq I)$, whereas the primary variable is denoted as $\eta_i = \eta_i(t,a)$ and represents the density of cells of age a in phase i at time instant t. During each phase, the density of cells varies either due to spontaneous death rate $d_i = d_i(t,a)$ or due to a transition rate $K_{i+1,i}(t,a)$ from phase i to phase $i+1$.

Actually, the transition rate $K_{i+1,i}$ stands for a control input to the cells proliferation model and is dependent on excitation received from circadian oscillators.

The PDE model of the cell cycle is given by

$$\begin{array}{l} \frac{\partial}{\partial t}\eta_i + \frac{\partial}{\partial a}[v_i(a)\eta_i] + [d_i + K_{i,i+1}]\eta_i = 0 \\ v_i(0)\eta_i(t,a=0) = \int_{a\geq 0} K_{i-1,i}(t,a)\eta_{i-1}(t,a)da \ \ 2\leq i\leq I \\ \eta_1(t,a=0)2\int_{a\geq 0} K_{I,1}(t,a)\eta_I(t,a)da \end{array} \tag{5.1}$$

where $\sum_{i=1}^{I}\int_{a\geq 0}\eta_i(t,a)da = 1$, $v_i(a)$ denotes a speed function of age a in phase i with respect to time t and $K_{i,i+1}(t,a) = \psi_i(t)\mathbf{1}_{\{a\geq a_i\}}(a)$. The transition rates $\psi_i(t)\mathbf{1}_{\{a\geq a_i\}}(a)$ have the following significance: a minimum age a_i must be spent in phase i, and a dynamic control $t\rightarrow\psi_i(t)$ is exerted on the transition from phase i to phase $i+1$. Functionals ψ_i are dependent on physiological control, that is hormonal or circadian, as well as on pharmacological factors. From the above, it is concluded that circadian oscillators indeed affect the cells' cycle. They may also have an effect on cell's cycle through the death rate d_i.

It has been verified that the cells's growth is controlled by proteins known as cyclines and cycline-dependent kinases (CDK) and in turn the concentration of cyclines is controlled by circadian oscillators. Cyclines and CDKs, apart from circadian control can be pharmacologically controlled. Anticancer drugs, such as alcylating agents act by damaging DNA and thus triggering p53 protein which provokes cell cycle arrest. It has been shown that p53 protein, on the one hand, and many cellular drug detoxification mechanisms (such as reduced glutathione), on the other hand, are dependent on the cell circadian clock, showing 24 h periodicity. Consequently, the circadian neurons affect cell cycle's transitions in both physiological and pathological conditions. By disrupting the circadian clock cells' proliferation and tumor development can be triggered. Moreover, by infusing anticancer drugs in time periods which are in-phase with high levels of proteins which arrest the cell cycle and which are controlled by the circadian mechanism one can anticipate a more efficient anticancer treatment. Finally, altering the period of the oscillations of the circadian cells with disruptive inputs from cytokines and drugs is an approach to improving the efficacy of chemotherapy [106].

It is also known that light through the retino-hypothalamic tract modulates equally for all neurons. A higher risk of cancer is related with circadian disruption induced by dark/light rhythm perturbations. Disruptive inputs can be due to circulating molecules such as cytokines (e.g., interferon and interleukin) either secreted by the immune system in the presence of cancer cells or delivered by therapy. These inputs are known to have a disruptive effect on circadian neurons mostly at the central pacemaking revel. It is also known that elevated levels of cytokines are associated with fatigue.

Finally a note is given about transmission pathways from central to peripheral circadian cells. These pathways maybe provided by the autonomic nervous system, neurohormonal ways using the hypothalamocorticosurrenal axis. In a simple way communication between circadian neurons may be based on intercentral, hormonal, such as adrenocorticotropic hormone (ACTH) and peripheral tissue terminal inputs (such as cortisol).

5.2.2 Mathematical Model of the Circadian Oscillator

The considered model of the circadian oscillator comes from Neurospora Crassa (fungus), which is a benchmark biological system. This is a third order oscillator model. Oscillator models of higher order describe the functioning of human circadian neurons.

The considered state variables are: x_1: describing the concentration of mRNA, x_2: describing the concentration of the FRQ (frequency) protein outside the nucleus, and x_3: describing the concentration of the FRQ protein inside the nucleus. The Neurospora oscillator model is described by a set of three differential equations:

$$\dot{x}_1 = v_s \frac{K_i^n}{K_i^n + x_3^n} - v_m \frac{x_1}{K_M + x_1} \tag{5.2}$$

$$\dot{x}_2 = K_s x_1 - v_d \frac{x_2}{K_d + x_2} - K_1 x_2 + K_2 x_3 \tag{5.3}$$

$$\dot{x}_3 = K_1 x_2 - K_2 x_3 \tag{5.4}$$

This model describes the molecular mechanism of circadian rhythms in Neurospora. The molecular mechanism is based on a negative feedback loop that regulates the synthesis of FRQ protein which is extracted from the frq gene. Transcription of the *frq* gene produces messenger RNA (mRNA). Transcription occurs in the nucleus. At a next step, the translation of the mRNA produces FRQ protein in the cytoplasm. Once translation process takes place in the cytoplasm, the generated FRQ protein is transferred in the nucleus and this has as a result to inhibit further transcription of the *frq* gene into mRNA. This completes the feedback loop.

The rest of the model's parameters are defined as follows: v_s denotes the rate of transcription of the *frq* gene into FRQ protein inside the nucleus. K_i represents the threshold constant beyond which nuclear FRQ protein inhibits transcription of the *frq* gene (Fig. 5.2). n is the Hill coefficient showing the degree of cooperativity of the inhibition process, v_m is the maximum rate of *frq* mRNA degradation, K_M is the Michaelis constant related to the transcription process. Parameter K_s defines the rate of FRQ synthesis and stands for the control input to the model. Parameter K_1 denotes the rate of transfer of FRQ protein from the cytoplasm to the nucleus and K_2 denotes the rate of transfer of the FRQ protein from the nucleus into the cytoplasm. Parameter v_d denotes the maximum rate of degradation of the FRQ protein and K_d is the Michaelis constant associated with this process.

Typical values for the aforementioned parameters are: $v_s = 1.6nMh^{-1}$, $K_i = 1nM$, $n = 4$, $v_m = 0.7nMh^{-1}$, $K_M = 0.4nM$, $k_s = 1h^{-1}$ (when no control is exerted to the protein synthesis), $v_d = 4nMh^{-1}$, $K_d = 1.4nM$, $k_1 = 0.3h^{-1}$, and $k_2 = 0.15h^{-1}$.

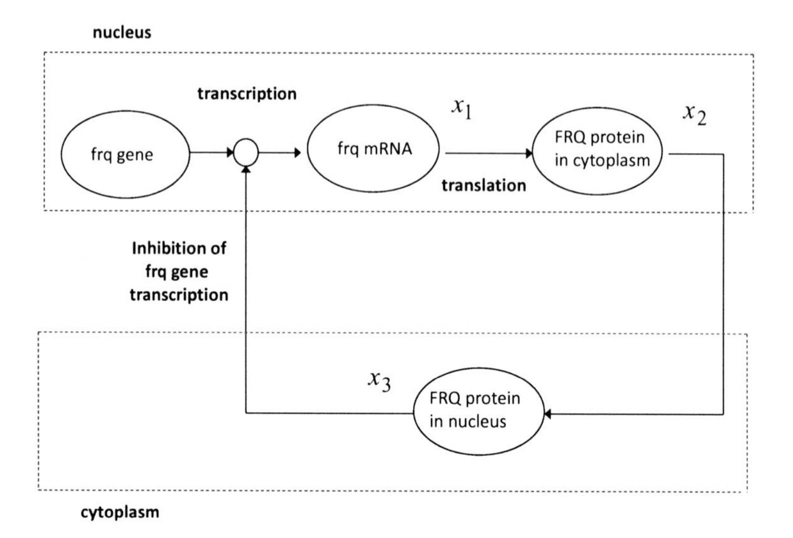

Fig. 5.2 Feedback loop of FRQ protein synthesis in circadian oscillators

5.3 Protein Synthesis Control Using Differential Geometry Methods

It will be shown that using a differential geometric approach and the computation of Lie derivatives one can arrive into the linear canonical form for the dynamical model of the circadian oscillator.

First, the nonlinear model of the circadian oscillator is written in the following state-space form:

$$\begin{pmatrix} \dot{x}_1 \\ \dot{x}_2 \\ \dot{x}_3 \end{pmatrix} = \begin{pmatrix} v_s \frac{K_i^n}{K_i^n + x_3^n} - v_m \frac{x_1}{K_m + x_1} \\ -v_d \frac{x_2}{K_d + x_2} - K_1 x_2 + K_2 x_3 \\ K_1 x_2 - K_2 x_3 \end{pmatrix} + \begin{pmatrix} 0 \\ x_1 \\ 0 \end{pmatrix} K_s \tag{5.5}$$

Next, the state variable $z_1 = h_1(x) = x_1$ is defined. Next, the variable $z_2 = L_f h_1(x)$ is computed. It holds that

$$\begin{aligned} z_2 &= L_f h_1(x) = \tfrac{\partial h_1}{\partial x_1} f_1 + \tfrac{\partial h_1}{\partial x_2} f_2 + \tfrac{\partial h_1}{\partial x_3} f_3 \Rightarrow \\ z_2 &= L_f h_1(x) = f_1 \Rightarrow L_f h_1(x) = v_s \frac{K_i^n}{K_i^n + x_3^n} - v_m \frac{x_1}{K_m + x_1} \end{aligned} \tag{5.6}$$

Equivalently

$$\begin{aligned} z_3 &= L_f^2 h_1(x) = L_f z_2 \Rightarrow z_3 = \tfrac{\partial z_2}{\partial x_1} f_1 + \tfrac{\partial z_2}{\partial x_2} f_2 + \tfrac{\partial z_2}{\partial x_3} f_3 \Rightarrow \\ z_3 &= -v_m \frac{K_m}{(K_m + x_1)^2} \left[v_s \frac{K_i^n}{K_i^n + x_3^n} - v_m \frac{x_1}{K_m + x_1} \right] - v_s \frac{K_i^n n x_3^{n-1}}{(K_i^n + x_3^n)^2} (K_1 x_2 - K_2 x_3) \end{aligned} \tag{5.7}$$

For the derivative of z_3 with respect to time it holds

$$\dot{z}_3 = L_f^3 h_1(x) + L_g L_f^2 h_1(x) u \tag{5.8}$$

For the computation of $L_f^3 h_1(x)$ one has

$$L_f^3 h_1(x) = \frac{\partial z_3}{\partial x_1} f_1 + \frac{\partial z_3}{\partial x_2} f_2 + \frac{\partial z_3}{\partial x_3} f_3 \tag{5.9}$$

which gives

$$\begin{aligned} &L_f^3 h_1(x) \\ &= \left\{ v_m \frac{K_m 2(K_m + x_1)}{(K_m + x_1)^4} [v_s \frac{K_i^n}{K_i^n + x_3^n} - v_m \frac{x_1}{K_m + x_1}] + v_m \frac{K_m}{(K_m + x_1)^2} v_m \frac{K_m}{(K_m + x_1)^2} \right\} \left[v_s \frac{K_i^n}{K_i^n + x_3^n} - v_m \frac{x_1}{K_m + x_1} \right] \\ &- \left\{ v_s \frac{K_i^n n x_3^{n-1}}{(K_i^n + x_3^n)^2} K_1 \right\} \left[-v_d \frac{x_2}{K_d + x_2} - K_1 x_2 + K_2 x_3 \right] + \left\{ v_m \frac{K_m}{(K_m + x_1)^2} v_s \frac{-K_i^n n x_3^{n-1}}{(K_i^n + x_3^n)^2} \right. \\ &\left. - v_s \frac{K_i^n n(n-1) x_3^{n-2} (K_i^n + x_3^n) - K_i^n n x_3^{n-1} 2(K_i^n + x_3^n)}{(K_i^n + x_3^n)^4} (K_1 x_2 - K_2 x_3) + v_s \frac{K_i^n n x_3^{n-1}}{(K_i^n + x_3^n)^2} K_2 \right\} [K_1 x_2 - K_2 x_3] \end{aligned} \tag{5.10}$$

For the computation of $L_g L_f^2 h_1(x)$ one has

$$L_g L_f^2 h_1(x) = \frac{\partial z_3}{\partial x_1} g_1 + \frac{\partial z_3}{\partial x_2} g_2 + \frac{\partial z_3}{\partial x_3} g_3 \tag{5.11}$$

which gives

$$L_g L_f^2 h_1(x) = \left\{ v_s \frac{K_i^n n x_3^{n-1}}{(K_i^n + x_3^n)^2} K_1 \right\} x_1 \tag{5.12}$$

It can be noticed that $L_f^3 h_1(x)$ computed in Eq. (5.37) is the same as function $f(y, \dot{y})$ computed using Eq. (5.24), while $L_g L_f^2 h_1(x)$ computed in Eq. (5.12) is the same as function $g(y, \dot{y})$ that was computed using Eq. (5.25).

Finally, the relative degree of the system is computed. It holds that $L_g h_1(x) = 0$ and $L_g L_f h_1(x) = 0$, while as shown above in Eq. (5.12) it holds that $L_g L_f^2 h_1(x) \neq 0$. Therefore, the relative degree of the system is $n = 3$. This confirms the appearance of the control input in the equation obtained after computing the third order derivative of x_1 with respect to time.

Next, by defining the new control variables $z_1 = h_1(x)$, $z_2 = L_f h_1(x)$, and $z_3 = L_f^2 h_1(x)$ and using the relation $\dot{z}_3 = L_f^3 h_1(x) + L_g L_f^2 h_1(x) u = v$ one arrives at a description of the circadian oscillator dynamics in the linear canonical form, that is

$$\begin{pmatrix} \dot{z}_1 \\ \dot{z}_2 \\ \dot{z}_3 \end{pmatrix} = \begin{pmatrix} 0 & 1 & 0 \\ 0 & 0 & 1 \\ 0 & 0 & 0 \end{pmatrix} \begin{pmatrix} z_1 \\ z_2 \\ z_3 \end{pmatrix} + \begin{pmatrix} 0 \\ 0 \\ 1 \end{pmatrix} v \tag{5.13}$$

and the associated measurement equation is

$$z_m = \begin{pmatrix} 1 & 0 & 0 \end{pmatrix} \begin{pmatrix} z_1 \\ z_2 \\ z_3 \end{pmatrix} \quad (5.14)$$

For the linearized model of the circadian oscillator dynamics the control law is computed using Eqs. (5.31) and (5.32), as in the case of differential flatness theory.

5.4 Protein Synthesis Control Using Differential Flatness Theory

5.4.1 *Differential Flatness of the Circadian Oscillator*

Considering the model of the circadian oscillator described in Eqs. (5.2)–(5.4) the flat output is taken to be $y = x_1$.

From Eq. (5.2) it holds

$$\begin{aligned} \dot{y} &= v_s \frac{K_i^n}{K_i^n + x_3^n} - v_m \frac{y}{K_M + y} \Rightarrow \\ {x_3}^n &= \left\{ \left(\frac{(K_m + y)K_i^n}{(K_m + y)\dot{y} + v_m y} \right) \left[v_s - v_m \frac{y}{K_m + y} - \dot{y} \right] \right\} \Rightarrow \\ x_3 &= \left\{ \left(\frac{(K_m + y)K_i^n}{(K_m + y)\dot{y} + v_m y} \right) \left[v_s - v_m \frac{y}{K_m + y} - \dot{y} \right] \right\}^{1/n} \end{aligned} \quad (5.15)$$

Therefore, state variable x_3 is a function of the flat output and its derivatives that is $x_3 = f_1(y, \dot{y})$.

From Eq. (5.4) one obtains

$$\begin{aligned} \dot{x}_3 &= k_1 x_2 - k_2 x_3 \Rightarrow \\ x_2 &= \frac{1}{K_1}[\dot{x}_3 + k_2 x_3] \end{aligned} \quad (5.16)$$

Therefore, state variable x_2 is a function of the flat output and its derivatives that is $x_2 = f_3(y, \dot{y})$.

From Eq. (5.3) it holds

$$\begin{aligned} \dot{x}_2 &= K_s x_1 - v_d \frac{x_2}{K_d + x_2} - K_1 x_2 + K_2 x_3 \Rightarrow \\ K_s &= \frac{1}{x_1}[\dot{x}_2] - v_d \frac{x_2}{K_d + x_2} - K_1 x_2 - K_2 x_3 \end{aligned} \quad (5.17)$$

Taking into account that $x_1 = y$, $x_3 = f_1(y, \dot{y})$, and $x_2 = f_2(y, \dot{y})$ one obtains that the control input K_s is also a function of the flat output y and of the associated derivatives, that is $K_s = f_3(y, \dot{y})$.

Consequently, all state variables and the control input of the system are written as functions of the flat output and its derivatives and this proves that the model of the circadian oscillator is a differentially flat one.

5.4.2 Transformation into a Canonical Form

From Eq. (5.2) it holds that

$$\dot{x}_1 = v_s \frac{K_i^n}{K_i^n + x_3^n} - v_m \frac{x_1}{K_M + x_1} \tag{5.18}$$

Deriving the aforementioned relation with respect to time one has

$$\ddot{x}_1 = v_s \frac{-K_i^n n x_3^{n-1}}{(K_i^n + x_3^n)^2} \dot{x}_3 - v_m \frac{K_M}{(K_M + x_1)^2} \dot{x}_1 \tag{5.19}$$

Substituting $\dot{x}_2$ from Eq. (5.3) and $\dot{x}_3$ from Eqs. (5.4) into (5.19) one has

$$\ddot{x}_1 = v_s \frac{-K_i^n n x_3^{n-1}}{(K_i^n + x_3^n)^2} (K_1 K_2 - k_2 x_3) - v_m \frac{K_M}{(K_M + x_1)^2} \left(v_s \frac{K_i^n}{K_i^n + x_3^n} - v_m \frac{x_1}{K_m + x_1} \right) \tag{5.20}$$

which can be also written as

$$\begin{aligned} \ddot{x}_1 = {} & v_s \frac{-K_i^n n x_3^{n-1}}{(K_i^n + x_3^n)^2} K_1 x_2 - v_s \frac{-K_i^n n x_3^{n-1}}{(K_i^n + x_3^n)^2} K_2 x_3 \\ & - v_m \frac{K_M}{(K_M + x_1)^2} v_s \frac{K_i^n}{K_i^n + x_3^n} + v_m \frac{K_M}{(K_M + x_1)^2} v_m \frac{x_1}{K_m + x_1} \end{aligned} \tag{5.21}$$

By deriving once more with respect to time one obtains

$$\begin{aligned} x_1^{(3)} = {} & v_s \frac{[-K_i^n n(n-1) x_3^{n-2} (K_i^n + x_3^n)^2 + K_i^n n x_3^{n-1} 2 (K_i^n + x_3^n)]}{(K_i^n + x_3^n)^4} \dot{x}_3 K_1 x_2 \\ & + v_s \frac{-K_i^n n x_3^{n-1}}{(K_i^n + x_3^n)^2} K_1 \dot{x}_2 \\ & - v_s \frac{[-K_i^n n(n-1) x_3^{n-2} (K_i^n + x_3^n)^2 + K_i^n n x_3^{n-1} 2 (K_i^n + x_3^n)]}{(K_i^n + x_3^n)^4} \dot{x}_3 K_2 x_3 \\ & - v_s \frac{-K_i^n n x_3^{n-1}}{(K_i^n + x_3^n)^2} K_2 \dot{x}_3 \\ & - v_m \frac{[-K_m 2 (K_m + x_1)]}{(K_m + x_1)^4} \dot{x}_1 v_s \frac{K_i^n}{K_i^n + x_3^n} \end{aligned}$$

$$-v_m\frac{-K_m}{(K_m+x_1)^2}v_s\frac{-K_i^n n x_3^{n-1}}{(K_i^n+x_3^n)^2}\dot{x}_3 - v_m\frac{[-K_m 2(K_m+x_1)]}{(K_m+x_1)^4}\dot{x}_1 v_m\frac{x_1}{K_m+x_1}$$
$$+v_m\frac{-K_m}{(K_m+x_1)^2}v_m\frac{K_m}{(K_m+x_1)^2}\dot{x}_1 \quad (5.22)$$

By substituting the relations about the derivatives $\dot{x}_1$, $\dot{x}_2$, and $\dot{x}_3$ and after performing intermediate operations one finally obtains

$$\begin{aligned} x_1^{(3)} = & \left\{v_m\frac{K_m 2(K_m+x_1)}{(K_m+x_1)^4}\left[v_s\frac{K_i^n}{K_i^n+x_3^n}-v_m\frac{x_1}{K_m+x_1}\right]\right. \\ & \left.+v_m\frac{K_m}{(K_m+x_1)^2}v_m\frac{K_m}{(K_m+x_1)^2}\right\}\times\left[v_s\frac{K_i^n}{K_i^n+x_3^n}-\right] \\ & -v_m\frac{x_1}{K_m+x_1}-\left\{v_s\frac{K_i^n n x_3^{n-1}}{(K_i^n+x_3^n)^2}K_1\right\}\left[-v_d\frac{x_2}{K_d+x_2}-K_1x_2+K_2x_3\right] \\ & +\left\{v_m\frac{K_m}{(K_m+x_1)^2}v_s\frac{-K_i^n n x_3^{n-1}}{(K_i^n+x_3^n)^2}\right. \\ & -v_s\frac{K_i^n n(n-1)x_3^{n-2}(K_i^n+x_3^n)-K_i^n n x_3^{n-1}2(K_i^n+x_3^n)}{(K_i^n+x_3^n)^4}(K_1x_2-K_2x_3) \\ & \left.+v_s\frac{K_i^n n x_3^{n-1}}{(K_i^n+x_3^n)^2}K_2\right\}[K_1x_2-K_2x_3] \\ & -\left\{v_s\frac{K_i^n n x_3^{n-1}}{(K_i^n+x_3^n)^2}K_1\right\}x_1K_s \end{aligned} \quad (5.23)$$

By defining the control input $u = K_s$ and the following two functions

$$\begin{aligned} f(x) = & \left\{v_m\frac{K_m 2(K_m+x_1)}{(K_m+x_1)^4}\left[v_s\frac{K_i^n}{K_i^n+x_3^n}-v_m\frac{x_1}{K_m+x_1}\right]+v_m\frac{K_m}{(K_m+x_1)^2}v_m\frac{K_m}{(K_m+x_1)^2}\right\}\left[v_s\frac{K_i^n}{K_i^n+x_3^n}-\right] \\ & -v_m\frac{x_1}{K_m+x_1}-\left\{v_s\frac{K_i^n n x_3^{n-1}}{(K_i^n+x_3^n)^2}K_1\right\}\left[-v_d\frac{x_2}{K_d+x_2}-K_1x_2+K_2x_3\right] \\ & +\left\{v_m\frac{K_m}{(K_m+x_1)^2}v_s\frac{-K_i^n n x_3^{n-1}}{(K_i^n+x_3^n)^2}\right. \\ & -v_s\frac{K_i^n n(n-1)x_3^{n-2}(K_i^n+x_3^n)-K_i^n n x_3^{n-1}2(K_i^n+x_3^n)}{(K_i^n+x_3^n)^4}(K_1x_2-K_2x_3) \\ & \left.+v_s\frac{K_i^n n x_3^{n-1}}{(K_i^n+x_3^n)^2}K_2\right\}[K_1x_2-K_2x_3] \end{aligned} \quad (5.24)$$

$$g(x) = \left\{v_s\frac{K_i^n n x_3^{n-1}}{(K_i^n+x_3^n)^2}K_1\right\}x_1 \quad (5.25)$$

one has the dynamics of the circadian oscillator in the form

$$x^{(3)} = f(x) + g(x)u \tag{5.26}$$

or equivalently

$$y^{(3)} = f(y, \dot{y}) + g(y, \dot{y})u \tag{5.27}$$

where function $f(y, \dot{y})$ comprises all terms in the right part of Eq. (5.23) that do not multiply the control input K_s, whereas function $g(y, \dot{y})$ comprises those terms in the right part of Eq. (5.23) that multiply the control input K_s.

After transforming the circadian oscillator in the form $y^{(3)} = f(y, \dot{y}) + g(y, \dot{y})u$ and by defining the new control input $v = f(y, \dot{y}) + g(y, \dot{y})u$ one has the dynamics

$$y^{(3)} = v \tag{5.28}$$

Next, by defining the new state variables $z_1 = y$, $z_2 = \dot{y}$, and $z_3 = \ddot{y}$ one obtains the following state-space description for the circadian oscillator

$$\begin{pmatrix} \dot{z}_1 \\ \dot{z}_2 \\ \dot{z}_3 \end{pmatrix} = \begin{pmatrix} 0 & 1 & 0 \\ 0 & 0 & 1 \\ 0 & 0 & 0 \end{pmatrix} \begin{pmatrix} z_1 \\ z_2 \\ z_3 \end{pmatrix} + \begin{pmatrix} 0 \\ 0 \\ 1 \end{pmatrix} v \tag{5.29}$$

and the associated measurement equation is

$$z_m = \begin{pmatrix} 1 & 0 & 0 \end{pmatrix} \begin{pmatrix} z_1 \\ z_2 \\ z_3 \end{pmatrix} \tag{5.30}$$

Knowing that $\dot{z}_3 = y^{(3)} = v$ the control law for the synthesis of the FRQ protein and consequently the control law of the circadian rhythms is

$$y^{(3)} = v = y_d^{(3)} - K_1(\ddot{y} - \ddot{y}_d) - K_2(\dot{y} - \dot{y}_d) - K_3(y - y_d) \tag{5.31}$$

Since $v = f(y, \dot{y}) + g(y, \dot{y})u$, the control input that is finally applied to the model of the circadian oscillator is

$$u = g^{-1}(y, \dot{y})[v - f(y, \dot{y})] \tag{5.32}$$

The above control law assures that $\lim_{t\to\infty} y(t) = y_d(t)$ which implies $\lim_{t\to\infty} x_1(t) = x_{1,d}(t)$ (concentration of frq mRNA), $\lim_{t\to\infty} x_2(t) = x_{2,d}(t)$ (concentration of FRQ protein in cytoplasm), and $\lim_{t\to\infty} x_3(t) = x_{3,d}(t)$ (concentration of FRQ protein in nucleus).

5.5 Robust Synchronization of Coupled Circadian Oscillators Using Differential Geometry Methods

The nonlinear state-space model of the coupled circadian oscillators is given by

$$\begin{pmatrix} \dot{x}_1 \\ \dot{x}_2 \\ \dot{x}_3 \\ \dot{x}_4 \\ \dot{x}_5 \\ \dot{x}_6 \end{pmatrix} = \begin{pmatrix} v_s \frac{K_i^n}{K_i^n + x_3^n} - v_m \frac{x_1}{K_M + x_1} \\ -v_d \frac{x_2}{K_d + x_2} - K_1 x_2 + K_2 x_3 + K_c(x_2 - x_5) \\ K_1 x_2 - K_2 x_3 \\ v_s \frac{K_i^n}{K_i^n + x_6^n} - v_m \frac{x_4}{K_M + x_4} \\ -v_d \frac{x_5}{K_d + x_5} - K_1 x_5 + K_2 x_6 + K_c(x_5 - x_2) \\ K_1 x_5 - K_2 x_6 \end{pmatrix} + \begin{pmatrix} 0 & 0 \\ x_1 & 0 \\ 0 & 0 \\ 0 & 0 \\ 0 & x_4 \\ 0 & 0 \end{pmatrix} \begin{pmatrix} u_1 \\ u_2 \end{pmatrix} \tag{5.33}$$

The following two functions are defined: $z_1 = h_1(x) = x_1$ and $z_4 = h_2(x) = x_4$. It holds that

$$z_2 = L_f h_1(x) = v_3 \frac{K_i^n}{K_i^n + x_3^n} - v_m \frac{x_1}{K_M + x_1} \tag{5.34}$$

$$\begin{aligned} z_3 = L_f^2 h_1(x) &= v_3 \frac{-K_i^n n x_3^{n-1}}{(K_i^n + x_3^n)^2}(K_1 x_2 - K_2 x_3) \\ &\quad - v_m \frac{K_m}{(K_M + x_1)^2}\left(v_s \frac{K_i^n}{K_i^n + x_3^n} - v_m \frac{x_1}{K_m + x_1}\right) \end{aligned} \tag{5.35}$$

Moreover, it holds that

$$\dot{z}_3 = L_f^3 h_1(x) + L_{g_1} L_f^2 h_1(x) u_1 + L_{g_2} L_f^2 h_1(x) u_2 \tag{5.36}$$

where

$$\begin{aligned} L_f^3 h_1(x) = &\left\{ v_m \frac{K_m 2(K_m + x_1)}{(K_m + x_1)^4} \left[v_s \frac{K_i^n}{K_i^n + x_3^n} - v_m \frac{x_1}{K_m + x_1} \right] \right. \\ &\left. + v_m \frac{K_m}{(K_m + x_1)^2} v_m \frac{K_m}{(K_m + x_1)^2} \right\} \left[v_s \frac{K_i^n}{K_i^n + x_3^n} - v_m \frac{x_1}{K_m + x_1} \right] \\ &- \left\{ v_s \frac{K_i^n n x_3^{n-1}}{(K_i^n + x_3^n)^2} K_1 \right\} \left[-v_d \frac{x_2}{K_d + x_2} - K_1 x_2 + K_2 x_3 + K_c(x_2 - x_5) \right] \\ &+ \left\{ v_m \frac{K_m}{(K_m + x_1)^2} v_s \frac{-K_i^n n x_3^{n-1}}{(K_i^n + x_3^n)^2} \right. \\ &- v_s \frac{K_i^n n(n-1) x_3^{n-2}(K_i^n + x_3^n) - K_i^n n x_3^{n-1} 2(K_i^n + x_3^n)}{(K_i^n + x_3^n)^4}(K_1 x_2 - K_2 x_3) \\ &\left. + v_s \frac{K_i^n n x_3^{n-1}}{(K_i^n + x_3^n)^2} K_2 \right\} [K_1 x_2 - K_2 x_3] \end{aligned} \tag{5.37}$$

$$L_{g_1} L_f^2 h_1(x) = v_s \frac{-K_i^n n x_3^{(n-1)}}{(K_i^n + x_3^n)^2} K_1 x_1 \tag{5.38}$$

$$L_{g_2} L_f^2 h_2(x) = 0 \tag{5.39}$$

Equivalently, it holds that

$$\dot{z}_6 = L_f^3 h_2(x) + L_{g_1} L_f^2 h_2(x) u_1 + L_{g_2} L_f^2 h_2(x) u_2 \tag{5.40}$$

where

$$\begin{aligned} L_f^3 h_2(x) = & \left\{ v_m \frac{K_m 2(K_m+x_1)}{(K_m+x_4)^4} \left[v_s \frac{K_i^n}{K_i^n+x_6^n} - v_m \frac{x_4}{K_m+x_4} \right] + v_m \frac{K_m}{(K_m+x_4)^2} v_m \frac{K_m}{(K_m+x_4)^2} \right\} \\ & \left[v_s \frac{K_i^n}{K_i^n+x_6^n} - \right. \\ & \left. -v_m \frac{x_4}{K_m+x_4} \right] - \left\{ v_s \frac{K_i^n n x_3^{n-1}}{(K_i^n+x_6^n)^2} K_1 \right\} \left[-v_d \frac{x_5}{K_d+x_5} - K_1 x_5 + K_2 x_6 + K_c(x_5 - x_2) \right] \\ & + \left\{ v_m \frac{K_m}{(K_m+x_4)^2} v_s \frac{-K_i^n n x_6^{n-1}}{(K_i^n+x_6^n)^2} \right. \\ & -v_s \frac{K_i^n n(n-1) x_6^{n-2} (K_i^n+x_6^n) - K_i^n n x_6^{n-1} 2(K_i^n+x_6^n)}{(K_i^n+x_6^n)^4} (K_1 x_5 - K_2 x_6) \\ & \left. + v_s \frac{K_i^n n x_6^{n-1}}{(K_i^n+x_6^n)^2} K_2 \right\} [K_1 x_2 - K_2 x_6] \end{aligned} \tag{5.41}$$

$$L_{g_1} L_f^2 h_2(x) = 0 \tag{5.42}$$

$$L_{g_2} L_f^2 h_2(x) = v_s \frac{-K_i^n n x_6^{(n-1)}}{(K_i^n + x_6^n)^2} K_1 x_4 \tag{5.43}$$

Therefore, for the system of the coupled circadian neurons one obtains an input output linearized form given by

$$\begin{pmatrix} z_1^{(3)} \\ z_4^{(3)} \end{pmatrix} = \begin{pmatrix} L_f^3 h_1(x) \\ L_f^3 h_2(x) \end{pmatrix} + \begin{pmatrix} L_{g_1} L_f^2 h_1(x) & L_{g_2} L_f^2 h_1(x) \\ L_{g_1} L_f^2 h_2(x) & L_{g_2} L_f^2 h_2(x) \end{pmatrix} \begin{pmatrix} u_1 \\ u_2 \end{pmatrix} \tag{5.44}$$

The new control inputs are defined as

$$\begin{aligned} v_1 &= L_f^3 h_1(x) + L_{g_1} L_f^2 h_1(x) u_1 + L_{g_2} L_f^2 h_1(x) u_2 \\ v_2 &= L_f^3 h_2(x) + L_{g_1} L_f^2 h_2(x) u_1 + L_{g_2} L_f^2 h_2(x) u_2 \end{aligned} \tag{5.45}$$

Therefore, the system's dynamics becomes

$$\begin{aligned} x_1^{(3)} &= v_1 \\ x_4^{(3)} &= v_2 \end{aligned} \tag{5.46}$$

By defining the new state variables $z_1 = x_1, z_2 = x_2, z_3 = x_3, z_4 = x_4, z_5 = x_5$, and $z_6 = x_6$ one obtains a description for the system of the coupled circadian oscillators in the linear canonical (Brunovsky) form

$$\begin{pmatrix} \dot{z}_1 \\ \dot{z}_2 \\ \dot{z}_3 \\ \dot{z}_4 \\ \dot{z}_5 \\ \dot{z}_6 \end{pmatrix} = \begin{pmatrix} 0 & 1 & 0 & 0 & 0 & 0 \\ 0 & 0 & 1 & 0 & 0 & 0 \\ 0 & 0 & 0 & 0 & 0 & 0 \\ 0 & 0 & 0 & 0 & 1 & 0 \\ 0 & 0 & 0 & 0 & 0 & 1 \\ 0 & 0 & 0 & 0 & 0 & 0 \end{pmatrix} \begin{pmatrix} z_1 \\ z_2 \\ z_3 \\ z_4 \\ z_5 \\ z_6 \end{pmatrix} + \begin{pmatrix} 0 & 0 \\ 0 & 0 \\ 1 & 0 \\ 0 & 0 \\ 0 & 0 \\ 0 & 1 \end{pmatrix} \begin{pmatrix} v_1 \\ v_2 \end{pmatrix} \tag{5.47}$$

$$\begin{pmatrix} z_{m_1} \\ z_{m_2} \end{pmatrix} = \begin{pmatrix} 1 & 0 & 0 & 0 & 0 & 0 \\ 0 & 0 & 0 & 1 & 1 & 0 \end{pmatrix} \begin{pmatrix} z_1 \\ z_2 \\ z_3 \\ z_4 \\ z_5 \\ z_6 \end{pmatrix} \tag{5.48}$$

The control law that allows FRQ protein concentration converge to desirable levels is given by

$$\begin{aligned} v_1 &= x_{1,d}^{(3)} - K_1^1(\ddot{x}_1 - \ddot{x}_1^d) - K_2^1(\dot{x}_1 - \dot{x}_1^d) - K_3^1(x_1 - x_1^d) \\ v_2 &= x_{1,d}^{(3)} - K_1^2(\ddot{x}_4 - \ddot{x}_4^d) - K_2^2(\dot{x}_4 - \dot{x}_4^d) - K_3^2(x_4 - x_4^d) \end{aligned} \tag{5.49}$$

By knowing the control inputs $(v_1, v_2)^T$ one can compute the control inputs $(u_1, u_2)^T$ that are finally applied to the system of the coupled circadian neurons. By setting

$$\tilde{f} = \begin{pmatrix} L_f^3 h_1(x) \\ L_f^3 h_2(x) \end{pmatrix} \; \tilde{G} = \begin{pmatrix} L_{g_1} L_f^2 h_1(x) & L_{g_2} L_f^2 h_1(x) \\ L_{g_1} L_f^2 h_2(x) & L_{g_2} L_f^2 h_2(x) \end{pmatrix} \tag{5.50}$$

5.6 Robust Synchronization of Coupled Circadian Oscillators Using Differential Flatness Theory

In case that there is coupling between circadian oscillators (neurons) it is also possible to find a linearizing and decoupling control law based on differential flatness theory. Without loss of generality the model of two coupled circadian neurons is considered. The state vector of this model now comprises the following state variables: x_1: mRNA concentration for the first neuron, x_2: concentration of the FRQ protein in the cytoplasm of the first neuron, x_3: concentration of FRQ protein in the nucleus of the first neuron, x_4: concentration of mRNA of the second neuron, x_5: concentration of the FRQ protein in the cytoplasm of the second neuron, x_6: concentration of FRQ protein in the nucleus of the second neuron.

$$\dot{x}_1 = v_s \frac{K_i^n}{K_i^n + x_3^n} - v_m \frac{x_1}{K_M + x_1} \tag{5.51}$$

$$\dot{x}_2 = K_{s_1} x_1 - v_d \frac{x_2}{K_d + x_2} - K_1 x_2 + K_2 x_3 + K_c (x_2 - x_5) \tag{5.52}$$

$$\dot{x}_3 = K_1 x_2 - K_2 x_3 \tag{5.53}$$

$$\dot{x}_4 = v_s \frac{K_i^n}{K_i^n + x_6^n} - v_m \frac{x_4}{K_M + x_4} \tag{5.54}$$

$$\dot{x}_5 = K_{s_2} x_4 - v_d \frac{x_5}{K_d + x_5} - K_1 x_5 + K_2 x_6 + K_c (x_5 - x_2) \tag{5.55}$$

$$\dot{x}_6 = K_1 x_5 - K_2 x_6 \tag{5.56}$$

It can be proven that the model of the coupled circadian neurons is differentially flat. The flat output is defined as

$$y = [y_1, y_2] = [x_1, x_4] \tag{5.57}$$

As it has been shown in Eq. (5.51) in the case of the independent circadian oscillator it holds

$$x_3 = \left\{ \left(\frac{(K_m + y_1) K_i^n}{(K_m + y_1)\dot{y} + v_m y_1} \right) \left[v_s - v_m \frac{y_1}{K_m + y_1} - \dot{y}_1 \right] \right\}^{1/n} \tag{5.58}$$

therefore it holds $x_3 = f_1(y_1, \dot{y}_1, y_2, \dot{y}_2)$. Moreover from Eq. (5.53) it holds

$$\dot{x}_3 = K_1 x_2 - K_2 x_3 \Rightarrow x_2 = \frac{1}{K_1} [\dot{x}_3 + K_2 x_3] \tag{5.59}$$

therefore it holds $x_2 = f_2(y_1, \dot{y}_1, y_2, \dot{y}_2)$. Additionally, from Eq. (5.54) it holds

$$x_6 = \left\{ \left(\frac{(K_m + y_2) K_i^n}{(K_m + y_2)\dot{y}_2 + v_m y_2} \right) \left[v_s - v_m \frac{y_2}{K_m + y_2} - \dot{y}_2 \right] \right\}^{1/n} \tag{5.60}$$

therefore it holds $x_6 = f_3(y_1, \dot{y}_1, y_2, \dot{y}_2)$. Furthermore, from Eq. (5.56) one obtains

$$\dot{x}_5 = K_1 x_5 - K_2 x_6 \Rightarrow x_5 = \frac{1}{K_1} [\dot{x}_6 + K_2 x_6] \tag{5.61}$$

therefore it holds $x_5 = f_4(y_1, \dot{y}_1, y_2, \dot{y}_2)$. Next, using Eq. (5.52) one finds

$$K_{s_1} = \frac{1}{x_1} \left[\dot{x}_2 - v_d \frac{x_2}{K_d + x_2} + K_1 x_2 - K_2 x_3 - K_c (x_2 - x_5) \right] \tag{5.62}$$

Since, x_1, x_2, x_3, and x_5 are functions of the flat output and of its derivatives it holds that $K_{s_1} = f_5(y_1, \dot{y}_1, y_2, \dot{y}_2)$. Similarly, from Eq. (5.55) one obtains for the second control input

$$K_{s_2} = \frac{1}{x_4} \left[\dot{x}_5 - v_d \frac{x_5}{K_d + x_5} + K_1 x_5 - K_2 x_6 - K_c (x_5 - x_2) \right] \tag{5.63}$$

Since x_2, x_4, x_5, and x_6 are functions of the flat output and of its derivatives it holds that $K_{s_2} = f_6(y_1, \dot{y}_1, y_2, \dot{y}_2)$.

From the above analysis, for the model of the coupled circadian oscillators it holds that all state variables and the control inputs are functions of the flat output and of its derivatives. Therefore, this system is a differentially flat one.

As in the case of the isolated circadian neuron here it holds

$$\dot{x}_1 = v_3 \frac{K_i^n}{K_i^n + x_3^n} - v_m \frac{x_1}{K_M + x_1} \tag{5.64}$$

and deriving with respect to time gives

$$\begin{aligned}
\ddot{x}_1 = {} & v_3 \frac{-K_i^n n x_3^{n-1}}{(K_i^n + x_3^n)^2}(K_1 x_2 - K_2 x_3) \\
& - v_m \frac{K_m}{(K_M + x_1)^2} \left(v_s \frac{K_i^n}{K_i^n + x_3^n} - v_m \frac{x_1}{K_m + x_1} \right)
\end{aligned} \tag{5.65}$$

while the relation for $x_1^{(3)}$ is modified by including in it the coupling term $K_c(x_2 - x_5)$ which appears now in $\dot{x}_2$ as described in Eq. (5.52). This results into

$$\begin{aligned}
x_1^{(3)} = {} & f_A(y_1, \dot{y}_1, y_2, \dot{y}_2) \\
& + g_{A_1}(y_1, \dot{y}_1, y_2, \dot{y}_2) u_1 + g_{A_2}(y_1, \dot{y}_1, y_2, \dot{y}_2) u_2
\end{aligned} \tag{5.66}$$

$u_1 = K_{s_1}$ and $u_2 = K_{s_2}$. For $f_A(y_1, \dot{y}_1, y_2, \dot{y}_2)$, $g_{A_1}(y_1, \dot{y}_1, y_2, \dot{y}_2)$, and $g_{A_2}(y_1, \dot{y}_1, y_2, \dot{y}_2)$ one has

$$\begin{aligned}
& f_A(y_1, \dot{y}_1, y_2, \dot{y}_2) \\
& = \left\{ v_m \frac{K_m 2(K_m + x_1)}{(K_m + x_1)^4} \left[v_s \frac{K_i^n}{K_i^n + x_3^n} - v_m \frac{x_1}{K_m + x_1} \right] + v_m \frac{K_m}{(K_m + x_1)^2} v_m \frac{K_m}{(K_m + x_1)^2} \right\} \\
& \left[v_s \frac{K_i^n}{K_i^n + x_3^n} - v_m \frac{x_1}{K_m + x_1} \right] \\
& - \left\{ v_s \frac{K_i^n n x_3^{n-1}}{(K_i^n + x_3^n)^2} K_1 \right\} \left[-v_d \frac{x_2}{K_d + x_2} - K_1 x_2 + K_2 x_3 + K_c(x_2 - x_5) \right] \\
& + \left\{ v_m \frac{K_m}{(K_m + x_1)^2} v_s \frac{-K_i^n n x_3^{n-1}}{(K_i^n + x_3^n)^2} \right. \\
& - v_s \frac{K_i^n n(n-1) x_3^{n-2} (K_i^n + x_3^n) - K_i^n n x_3^{n-1} 2(K_i^n + x_3^n)}{(K_i^n + x_3^n)^4} (K_1 x_2 - K_2 x_3) \\
& \left. + v_s \frac{K_i^n n x_3^{n-1}}{(K_i^n + x_3^n)^2} K_2 \right\} [K_1 x_2 - K_2 x_3]
\end{aligned} \tag{5.67}$$

$$g_{A_1}(y_1, \dot{y}_1, y_2, \dot{y}_2) = v_s \frac{-K_i^n n x_3^{(n-1)}}{(K_i^n + x_3^n)^2} K_1 x_1 \tag{5.68}$$

$$g_{A_2}(y_1, \dot{y}_1, y_2, \dot{y}_2) = 0 \tag{5.69}$$

Similarly, for the second neuron one has

$$x_4^{(3)} = f_B(y_1, \dot{y}_1, y_2, \dot{y}_2) + g_{B_1}(y_1, \dot{y}_1, y_2, \dot{y}_2)u_1 + g_{B_2}(y_1, \dot{y}_1, y_2, \dot{y}_2)u_2 \quad (5.70)$$

$u_1 = K_{s_1}$ and $u_2 = K_{s_2}$. For $f_B(y_1, \dot{y}_1, y_2, \dot{y}_2)$, $g_{B_1}(y_1, \dot{y}_1, y_2, \dot{y}_2)$, and $g_{B_2}(y_1, \dot{y}_1, y_2, \dot{y}_2)$ one has

$$\begin{aligned}
&f_B(y_1, \dot{y}_1, y_2, \dot{y}_2) \\
&= \left\{ v_m \frac{K_m 2(K_m+x_1)}{(K_m+x_4)^4} \left[v_s \frac{K_i^n}{K_i^n+x_6^n} - v_m \frac{x_4}{K_m+x_4} \right] + v_m \frac{K_m}{(K_m+x_4)^2} v_m \frac{K_m}{(K_m+x_4)^2} \right\} \\
&\left[v_s \frac{K_i^n}{K_i^n+x_6^n} - v_m \frac{x_4}{K_m+x_4} \right] \\
&- \left\{ v_s \frac{K_i^n n x_3^{n-1}}{(K_i^n+x_6^n)^2} K_1 \right\} \left[-v_d \frac{x_5}{K_d+x_5} - K_1 x_5 + K_2 x_6 + K_c(x_5 - x_2) \right] \\
&+ \left\{ v_m \frac{K_m}{(K_m+x_4)^2} v_s \frac{-K_i^n n x_6^{n-1}}{(K_i^n+x_6^n)^2} \right. \\
&-v_s \frac{K_i^n n(n-1) x_6^{n-2}(K_i^n+x_6^n) - K_i^n n x_6^{n-1} 2(K_i^n+x_6^n)}{(K_i^n+x_6^n)^4} (K_1 x_5 - K_2 x_6) \\
&\left. +v_s \frac{K_i^n n x_6^{n-1}}{(K_i^n+x_6^n)^2} K_2 \right\} [K_1 x_2 - K_2 x_6]
\end{aligned} \quad (5.71)$$

$$g_{B_1}(y_1, \dot{y}_1, y_2, \dot{y}_2) = 0 \quad (5.72)$$

$$g_{B_2}(y_1, \dot{y}_1, y_2, \dot{y}_2) = v_s \frac{-K_i^n n x_6^{(n-1)}}{(K_i^n + x_6^n)^2} K_1 x_4 \quad (5.73)$$

Therefore, the system of the coupled circadian neurons can be written in the input–output linearized form

$$\begin{pmatrix} x_1^{(3)} \\ x_4^{(3)} \end{pmatrix} = \begin{pmatrix} f_A(y_1, \dot{y}_1, y_2, \dot{y}_2) \\ f_B(y_1, \dot{y}_1, y_2, \dot{y}_2) \end{pmatrix} + \begin{pmatrix} g_{A_1}(y_1, \dot{y}_1, y_2, \dot{y}_2) & g_{A_2}(y_1, \dot{y}_1, y_2, \dot{y}_2) \\ g_{B_1}(y_1, \dot{y}_1, y_2, \dot{y}_2) & g_{B_2}(y_1, \dot{y}_1, y_2, \dot{y}_2) \end{pmatrix} \begin{pmatrix} u_1 \\ u_2 \end{pmatrix} \quad (5.74)$$

The new control inputs are defined as

$$\begin{aligned}
v_1 &= f_A(y_1, \dot{y}_1, y_2, \dot{y}_2) + g_{A_1}(y_1, \dot{y}_1, y_2, \dot{y}_2)u_1 + g_{A_2}(y_1, \dot{y}_1, y_2, \dot{y}_2)u_2 \\
v_2 &= f_B(y_1, \dot{y}_1, y_2, \dot{y}_2) + g_{B_1}(y_1, \dot{y}_1, y_2, \dot{y}_2)u_1 + g_{B_2}(y_1, \dot{y}_1, y_2, \dot{y}_2)u_2
\end{aligned} \quad (5.75)$$

Therefore, the system's dynamics becomes

$$\begin{aligned}
x_1^{(3)} &= v_1 \\
x_4^{(3)} &= v_2
\end{aligned} \quad (5.76)$$

By defining the new state variables $z_1 = x_1, z_2 = x_2, z_3 = x_3, z_4 = x_4, z_5 = x_5$, and $z_6 = x_6$ one obtains a description for the system of the coupled circadian oscillators in the linear canonical (Brunovsky) form

$$\begin{pmatrix} \dot{z}_1 \\ \dot{z}_2 \\ \dot{z}_3 \\ \dot{z}_4 \\ \dot{z}_5 \\ \dot{z}_6 \end{pmatrix} = \begin{pmatrix} 0 & 1 & 0 & 0 & 0 & 0 \\ 0 & 0 & 1 & 0 & 0 & 0 \\ 0 & 0 & 0 & 0 & 0 & 0 \\ 0 & 0 & 0 & 0 & 1 & 0 \\ 0 & 0 & 0 & 0 & 0 & 1 \\ 0 & 0 & 0 & 0 & 0 & 0 \end{pmatrix} \begin{pmatrix} z_1 \\ z_2 \\ z_3 \\ z_4 \\ z_5 \\ z_6 \end{pmatrix} + \begin{pmatrix} 0 & 0 \\ 0 & 0 \\ 1 & 0 \\ 0 & 0 \\ 0 & 0 \\ 0 & 1 \end{pmatrix} \begin{pmatrix} v_1 \\ v_2 \end{pmatrix} \quad (5.77)$$

$$\begin{pmatrix} z_{m_1} \\ z_{m_2} \end{pmatrix} = \begin{pmatrix} 1 & 0 & 0 & 0 & 0 & 0 \\ 0 & 0 & 0 & 1 & 1 & 0 \end{pmatrix} \begin{pmatrix} z_1 \\ z_2 \\ z_3 \\ z_4 \\ z_5 \\ z_6 \end{pmatrix} \quad (5.78)$$

The control law that allows FRQ protein concentration converge to desirable levels is given by

$$\begin{aligned} v_1 &= x_{1,d}^{(3)} - K_1^1(\ddot{x}_1 - \ddot{x}_1^d) - K_2^1(\dot{x}_1 - \dot{x}_1^d) - K_3^1(x_1 - x_1^d) \\ v_2 &= x_{1,d}^{(3)} - K_1^2(\ddot{x}_4 - \ddot{x}_4^d) - K_2^2(\dot{x}_4 - \dot{x}_4^d) - K_3^2(x_4 - x_4^d) \end{aligned} \quad (5.79)$$

By knowing the control inputs $(v_1, v_2)^T$ one can compute the control inputs $(u_1, u_2)^T$ that are finally applied to the system of the coupled circadian neurons. By setting

$$\tilde{f} = \begin{pmatrix} f_A(y_1, \dot{y}_1, y_2, \dot{y}_2) \\ f_B(y_1, \dot{y}_1, y_2, \dot{y}_2) \end{pmatrix} \quad \tilde{G} = \begin{pmatrix} g_{A_1}(y_1, \dot{y}_1, y_2, \dot{y}_2) & g_{A_2}(y_1, \dot{y}_1, y_2, \dot{y}_2) \\ g_{B_1}(y_1, \dot{y}_1, y_2, \dot{y}_2) & g_{B_2}(y_1, \dot{y}_1, y_2, \dot{y}_2) \end{pmatrix} \quad (5.80)$$

It holds $v = \tilde{f} + \tilde{G}u \Rightarrow u = \tilde{G}^{-1}(v - \tilde{f})$.

5.7 State Estimation and Disturbances Compensation with the Derivative-Free Nonlinear Kalman Filter

The parametric uncertainty terms and the external disturbance terms affecting $x^{(3)} = v$ or $y^{(3)} = v$ are denoted with variable $\tilde{d}$. In that case the dynamics of the coupled circadian oscillators takes the form

$$y_1^{(3)} = v_1 + \tilde{d}_1 \qquad y_2^{(3)} = v_2 + \tilde{d}_2 \quad (5.81)$$

It is also assumed that the dynamics of the disturbance and model uncertainty terms is described by the associated n-th order derivative, where without loss of generality here it is assumed that $n = 3$

$$\tilde{d}_1^{(3)} = f_{d_1}(y, \dot{y})\tilde{d}_2^{(3)} = f_{d_2}(y, \dot{y}) \tag{5.82}$$

The state vector of the circadian oscillator is extended by including in it state variables that describe the disturbance's variations. Thus one has $z_1 = y_1$, $z_2 = \dot{y}_1$, $z_3 = \ddot{y}_1$, $z_4 = y_2$, $z_5 = \dot{y}_2$, $z_6 = \ddot{y}_2$, $z_7 = \tilde{d}_1$, $z_8 = \dot{\tilde{d}}_1$, and $z_9 = \ddot{\tilde{d}}_1$, $z_{10} = \tilde{d}_2$, $z_{11} = \dot{\tilde{d}}_2$, and $z_{12} = \ddot{\tilde{d}}_2$. The associated description of the system in the form of state-space equations becomes

$$\begin{aligned} \dot{z}_e &= A z_e + B u_e \\ z_m &= C_e z_e \end{aligned} \tag{5.83}$$

where z_e is the extended state vector $z_e = [z_1, \cdots, z_{12}]^T$, $v_e = [v_1, v_2, f_{d_1}, f_{d_2}]^T$, is the extended control input vector, and matrices A_e, B_e, and C_e are defined as

$$A_e = \begin{pmatrix} 0&1&0&0&0&0&0&0&0&0&0&0 \\ 0&0&1&0&0&0&0&0&0&0&0&0 \\ 0&0&0&0&0&0&1&0&0&0&0&0 \\ 0&0&0&0&1&0&0&0&0&0&0&0 \\ 0&0&0&0&0&1&0&0&0&0&0&0 \\ 0&0&0&0&0&0&0&0&0&1&0&0 \\ 0&0&0&0&0&0&0&1&0&0&0&0 \\ 0&0&0&0&0&0&0&0&1&0&0&0 \\ 0&0&0&0&0&0&0&0&0&0&0&0 \\ 0&0&0&0&0&0&0&0&0&0&1&0 \\ 0&0&0&0&0&0&0&0&0&0&0&1 \\ 0&0&0&0&0&0&0&0&0&0&0&0 \end{pmatrix} \quad B_e = \begin{pmatrix} 0&0&0&0 \\ 0&0&0&0 \\ 1&0&0&0 \\ 0&0&0&0 \\ 0&0&0&0 \\ 0&1&0&0 \\ 0&0&0&0 \\ 0&0&0&0 \\ 0&0&1&0 \\ 0&0&0&0 \\ 0&0&0&0 \\ 0&0&0&1 \end{pmatrix} \quad C_e^T = \begin{pmatrix} 1&0 \\ 0&0 \\ 0&0 \\ 0&1 \\ 0&0 \\ 0&0 \\ 0&0 \\ 0&0 \\ 0&0 \\ 0&0 \\ 0&0 \\ 0&0 \end{pmatrix} \tag{5.84}$$

The associated disturbance observer is

$$\begin{aligned} \dot{\hat{z}} &= A_o \hat{z} + B_o u + K(z_m - \hat{z}_m) \\ \hat{z}_m &= C_o \hat{z} \end{aligned} \tag{5.85}$$

where $A_o = A$, $C_o = C$, and

$$B_o^T = \begin{pmatrix} 0&0&1&0&0&0&0&0&0&0&0&0 \\ 0&0&0&0&0&1&0&0&0&0&0&0 \end{pmatrix} \tag{5.86}$$

where now $u = [u_1, u_2]^T$. For the aforementioned model, and after carrying out discretization of matrices A_o, B_o, and C_o with common discretization methods one can perform Kalman filtering [157, 169]. This is *Derivative-free nonlinear Kalman filtering*. As explained in Chap. 4, unlike EKF, the filtering method is performed without the need to compute Jacobian matrices and does not introduce numerical errors due to approximative linearization with Taylor series expansion.

The design of a disturbance observer based on the Derivative-free nonlinear Kalman Filter enables simultaneous estimation of the nonmeasurable elements of the state vector, that is y_2, y_3 and y_5, y_6 and estimation of the disturbance term $\tilde{d}_1 = z_7$ and $\tilde{d}_2 = z_{10}$. For the compensation of the disturbance term the control input that is applied to the circadian oscillator model is modified as follows:

$$\begin{array}{c} v_1(k) = v(k) - \hat{\tilde{d}}_1(k) \text{ or} \\ v_1(k) = v(k) - \hat{z}_7(k) \\ \\ v_2(k) = v(k) - \hat{\tilde{d}}_2(k) \text{ or} \\ v_2(k) = v(k) - \hat{z}_{10}(k) \end{array} \tag{5.87}$$

In the design of the associated disturbances' estimator one has the dynamics defined in Eq. (5.85), where $K \in R^{12\times 1}$ is the state estimator's gain and matrices A_o, B_o, and C_o have been given in Eqs. (5.84)–(5.86). The discrete-time equivalents of matrices A_o, B_o, and C_o are denoted as $\tilde{A}_d$, $\tilde{B}_d$, and $\tilde{C}_d$ respectively, and are computed with the use of common discretization methods. Next, a Derivative-free nonlinear Kalman Filter can be designed for the aforementioned representation of the system dynamics [157, 158]. The associated Kalman Filter-based disturbance estimator is given by the recursion [166, 169]

Measurement update:

$$\begin{array}{l} K(k) = P^-(k)\tilde{C}_d^T[\tilde{C}_d \cdot P^-(k)\tilde{C}_d^T + R]^{-1} \\ \hat{z}(k) = \hat{z}^-(k) + K(k)[\tilde{C}_d z(k) - \tilde{C}_d \hat{z}^-(k)] \\ P(k) = P^-(k) - K(k)\tilde{C}_d P^-(k) \end{array} \tag{5.88}$$

Time update:

$$\begin{array}{l} P^-(k+1) = \tilde{A}_d(k)P(k)\tilde{A}_d^T(k) + Q(k) \\ \hat{z}^-(k+1) = \tilde{A}_d(k)\hat{z}(k) + \tilde{B}_d(k)\tilde{v}(k) \end{array} \tag{5.89}$$

5.8 Simulation Tests

The nonlinear dynamical model of the coupled circadian oscillators has been described in Eqs. (5.51)–(5.56). Using the canonical form model control law for the FRQ protein synthesis was computed according to the stages described in Sect. 5.6. The Derivative-free nonlinear Kalman Filter used for model uncertainty, measurement noise, and external disturbances compensation was designed according to Sect. 5.7.

Results on the synchronization between a primary and a secondary circadian cell are provided for six different cases, each one associated with different disturbance

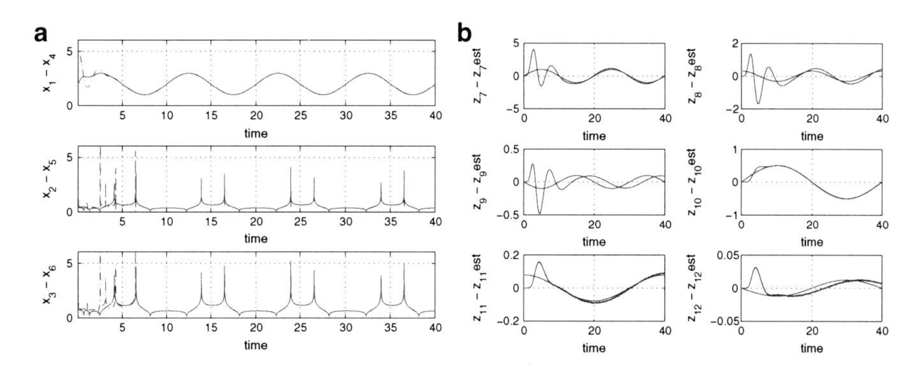

Fig. 5.3 Test 1: (**a**) Synchronization of state variables x_i ($i = 1, 4$: frq mRNA concentration, $i = 2, 5$: FRQ protein concentration in cytoplasm, $i = 3, 6$: FRQ protein concentration in nucleus) between the two circadian cells (*red continuous line* denotes concentration in cell 1 whereas the *dashed blue line* denotes concentration in cell 2) (**b**) Estimation of disturbance inputs and of their derivatives (*blue lines*) with the use of the Derivative-free nonlinear Kalman Filter

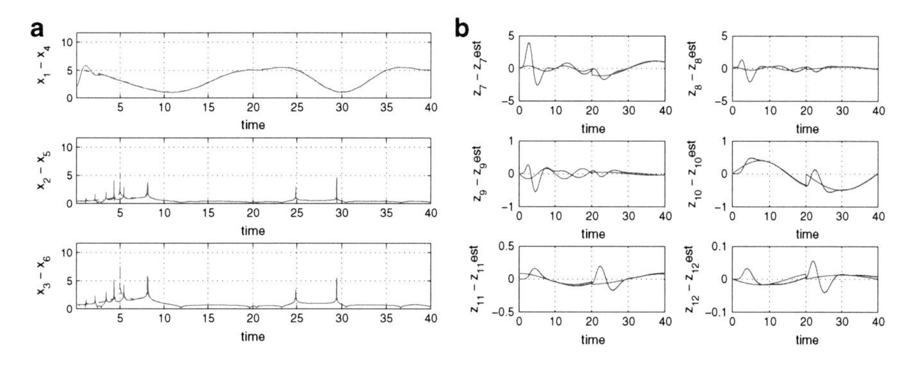

Fig. 5.4 Test 2: (**a**) Synchronization of state variables x_i ($i = 1, 4$: frq mRNA concentration, $i = 2, 5$: FRQ protein concentration in cytoplasm, $i = 3, 6$: FRQ protein concentration in nucleus) between the two circadian cells (*red continuous line* denotes concentration in cell 1 whereas the *dashed blue line* denotes concentration in cell 2) (**b**) Estimation of disturbance inputs and of their derivatives (*blue lines*) with the use of the Derivative-free nonlinear Kalman Filter

and uncertainty terms affecting the cell as well as with different setpoints for the state vector elements of the coupled circadian cell model. The results obtained are shown in Figs. 5.3a, 5.4a, 5.5a, 5.6a, 5.7a, and 5.8a. It can be observed that the proposed control scheme that was based on linearization and decoupling of the circadian's oscillator model with the use of differential flatness theory enabled complete synchronization between the state variables of the first cell (x_i, $i = 1, 2, 3$ denoted with the continuous red line) and the state variables of the second cell (x_i, $i = 4, 5, 6$ denoted with the dashed blue line).

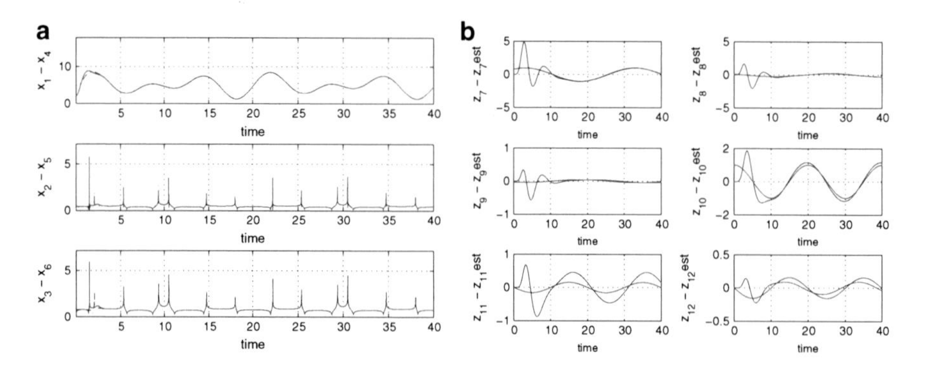

Fig. 5.5 Test 3: (**a**) Synchronization of state variables x_i ($i = 1, 4$: frq mRNA concentration, $i = 2, 5$: FRQ protein concentration in cytoplasm, $i = 3, 6$: FRQ protein concentration in nucleus) between the two circadian cells (*red continuous line* denotes concentration in cell 1 whereas the *dashed blue line* denotes concentration in cell 2) (**b**) Estimation of disturbance inputs and of their derivatives (*blue lines*) with the use of the Derivative-free nonlinear Kalman Filter

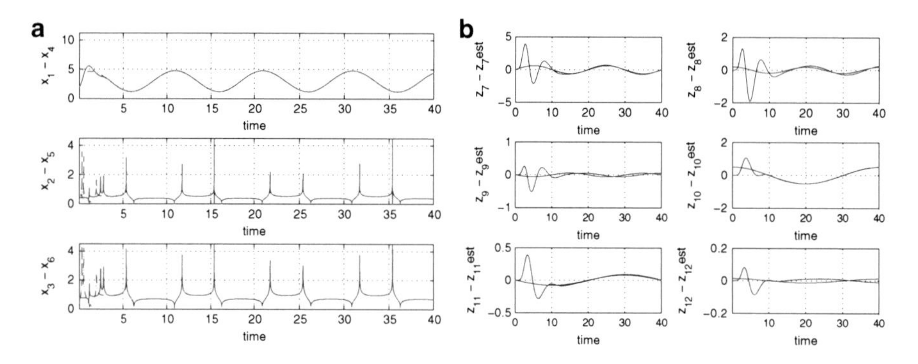

Fig. 5.6 Test 4: (**a**) Synchronization of state variables x_i ($i = 1, 4$: frq mRNA concentration, $i = 2, 5$: FRQ protein concentration in cytoplasm, $i = 3, 6$: FRQ protein concentration in nucleus) between the two circadian cells (*red continuous line* denotes concentration in cell 1 whereas the *dashed blue line* denotes concentration in cell 2) (**b**) Estimation of disturbance inputs and of their derivatives (*blue lines*) with the use of the Derivative-free nonlinear Kalman Filter

In Figs. 5.3b, 5.4b, 5.5b, 5.6b, 5.7b, and 5.8b the estimates of the disturbance inputs affecting the model of the coupled circadian oscillators are presented. The real value of the disturbance variable is denoted with the red line while the estimated value is denoted with the blue line. The disturbance terms affecting the inputs of the model are variables z_7 and z_{10}. Their first and second order derivatives are variables z_8, z_9 and z_{11}, z_{12}, respectively. It can be noticed that the Kalman Filter-based observer provides accurate estimates about the non-measurable disturbances and this information enables the efficient compensation of the perturbations' effects.

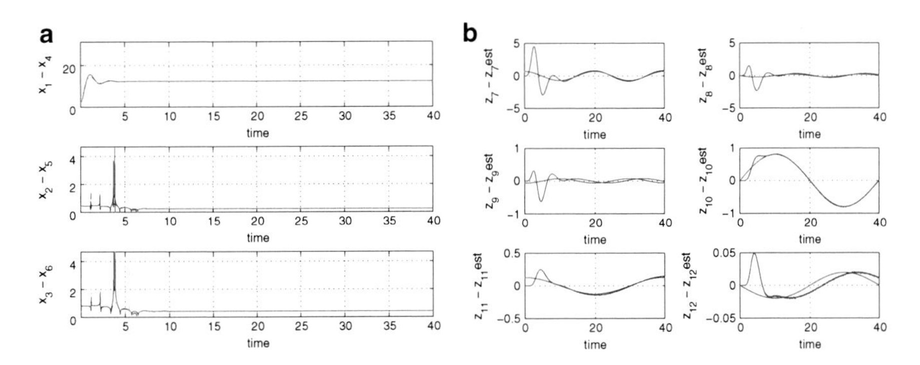

Fig. 5.7 Test 5: (**a**) Synchronization of state variables x_i ($i = 1, 4$: frq mRNA concentration, $i = 2, 5$: FRQ protein concentration in cytoplasm, $i = 3, 6$: FRQ protein concentration in nucleus) between the two circadian cells (*red continuous line* denotes concentration in cell 1 whereas the *dashed blue line* denotes concentration in cell 2) (**b**) Estimation of disturbance inputs and of their derivatives (*blue lines*) with the use of the Derivative-free nonlinear Kalman Filter

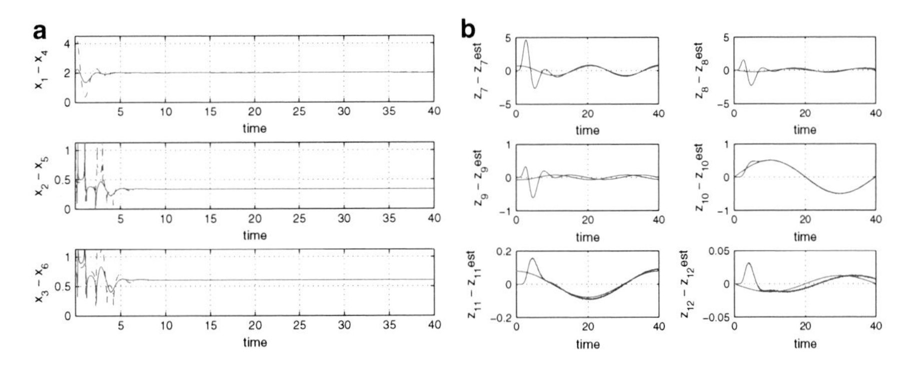

Fig. 5.8 Test 6: (**a**) Synchronization of state variables x_i ($i = 1, 4$: frq mRNA concentration, $i = 2, 5$: FRQ protein concentration in cytoplasm, $i = 3, 6$: FRQ protein concentration in nucleus) between the two circadian cells (*red continuous line* denotes concentration in cell 1 whereas the *dashed blue line* denotes concentration in cell 2) (**b**) Estimation of disturbance inputs and of their derivatives (*blue lines*) with the use of the Derivative-free nonlinear Kalman Filter

5.9 Conclusions

The problem of robust synchronization of coupled circadian oscillators and the problem of the nonlinear control of the associated protein synthesis has been studied. Control of the periodic variations of protein levels in circadian cells is important because it affects several functions performed by living organisms. When the periodicity of circadian cells is disrupted pathological situations such as tumor growth may appear. On the other hand, knowing the effect of the circadian cycle on the variation of proteins levels can help in administering more efficiently anticancer treatment. The chapter has proposed a synchronizing control method for coupled circadian cells using differential flatness theory and the Derivative-free nonlinear Kalman Filter.

With the application of differential geometric methods and differential flatness theory it was shown that the nonlinear model of the coupled circadian cells can be written in the linear canonical form. It was also shown that by applying differential geometric methods and by computing Lie derivatives exact linearization of the model of the coupled circadian cells can be succeeded. Moreover, it was proven that the model of the coupled circadian cells is a differentially flat one and by defining appropriate flat outputs it can be transformed to the MIMO (multi-input multi-output) linear canonical form. For the linearized representation of the coupled neuron's model the design of a feedback control is possible and synchronization between the two neurons can be attained.

Next, the problem of synchronization of the coupled circadian cells under external perturbations and parametric uncertainties was examined. As explained, synchronization of coupled circadian oscillators becomes more difficult when there is uncertainty about the parameters of the dynamical model of the circadian cells and when the parameters of the dynamical models of these cells are uneven. To obtain simultaneous state and disturbances estimation, a disturbance observer based on the Derivative-free nonlinear Kalman Filter has been used. The Derivative-free nonlinear Kalman Filter consists of the standard Kalman Filter recursion on the linearized equivalent model of the coupled cells and on computation of state and disturbance estimates using the diffeomorphism (relations about state variables transformation) provided by differential flatness theory. After estimating the disturbance terms in the model of the coupled circadian cells their compensation has become possible.

The performance of the synchronizing control loop has been tested through simulation experiments. It was shown that the proposed method assures that the circadian cells will remain synchronized, despite parametric variations and uncertainties in the associated dynamical model and despite the existence of external perturbations. Robust synchronizing control of biological oscillators is a developing research field while the application of elaborated control and estimation methods in biological models is anticipated to give nontrivial results for further advancements in the fields of biophysics and biomedical engineering.

Chapter 6
Wave Dynamics in the Transmission of Neural Signals

Abstract The chapter analyzes wave-type partial differential equations (PDEs) that describe the transmission of neural signals and proposes filtering for estimating the spatiotemporal variations of voltage in the neurons' membrane. It is shown that in specific neuron models the spatiotemporal variations of the membrane's voltage follow PDEs of the wave type while in other models such variations are associated with the propagation of solitary waves in the membrane. To compute the dynamics of the membrane PDE model without knowledge of initial conditions and through the processing of noisy measurements, a new filtering method, under the name Derivative-free nonlinear Kalman Filtering, is proposed. The PDE of the membrane is decomposed into a set of nonlinear ordinary differential equations with respect to time. Next, each one of the local models associated with the ordinary differential equations is transformed into a model of the linear canonical (Brunovsky) form through a change of coordinates (diffeomorphism) which is based on differential flatness theory. This transformation provides an extended model of the nonlinear dynamics of the membrane for which state estimation is possible by applying the standard Kalman Filter recursion. The proposed filtering method is tested through numerical simulation tests.

6.1 Outline

As explained in Chap. 1, spatiotemporal variations of voltage in neurons' membrane are usually described by partial differential equations (PDEs) of the wave or cable type [1, 38, 176, 185, 194]. Certain models of the voltage dynamics consider also the propagation of solitary waves through the membrane. Thus, voltage dynamics in the membrane can be viewed as a distributed parameters system [23, 211]. State estimation in distributed parameter systems and in infinite dimensional systems described by PDEs is a much more complicated problem than state estimation in lumped parameter systems [46, 72, 74, 174, 213, 218]. Previous results on state estimation of wave-type nonlinear dynamics can be found in [29, 81, 201, 208].

G.G. Rigatos, *Advanced Models of Neural Networks*,
DOI 10.1007/978-3-662-43764-3_6,

To compute the dynamics of the membrane's PDE model without knowledge of initial conditions and through the processing of noisy measurements, the following stages are followed: Using the method for numerical solution of the PDE through discretization the initial PDE is decomposed into a set of nonlinear ordinary differential equations with respect to time [139]. Next, each one of the local models associated with the ordinary differential equations is transformed into a model of the linear canonical (Brunovsky) form through a change of coordinates (diffeomorphism) which is based on differential flatness theory. This transformation provides an extended model of the nonlinear system for which state estimation is possible by application of the standard Kalman Filter recursion [122, 200]. Unlike other nonlinear estimation methods (e.g., Extended Kalman Filter) the application of the standard Kalman Filter recursion to the linearized equivalent of the nonlinear PDE does not need the computation of Jacobian matrices and partial derivatives and does not introduce cumulative numerical errors [157–160, 170].

6.2 Propagating Action Potentials

The Hodgkin–Huxley equation, in the cable PDE form, is re-examined. It holds that:

$$\begin{aligned} C_m \frac{\partial V}{\partial t} &= \frac{4d}{R_i}\frac{\partial^2 V}{\partial x^2} - I_{\text{ion}} + I \\ \frac{\partial \Gamma}{\partial t} &= \alpha_\Gamma(V)(1-\Gamma) - \beta_\Gamma(V)\Gamma \end{aligned} \tag{6.1}$$

where $\Gamma = m, h, \eta$, that is parameters that affect conductivity and consequently the input currents of the membrane. If c is the velocity of propagation of the wave solution, then one can use the notation

$$V(x, ct, t) = \hat{V}(x) \tag{6.2}$$

By performing a change of coordinates $\xi = x - ct$ one has

$$\begin{aligned} C_m \frac{\partial V}{\partial t} &= C_m c \frac{\partial V}{\partial \xi} + \frac{4d}{R_i}\frac{\partial^2 V}{\partial \xi^2} + I - I_{\text{ion}} \\ \frac{\partial \Gamma}{\partial t} &= c\frac{\partial \Gamma}{\partial \xi} + \alpha_\Gamma(V)(1-\Gamma) - \beta_\Gamma(V)\Gamma \end{aligned} \tag{6.3}$$

where again $\Gamma = m, h, \eta$. The following boundary condition holds $V(\xi = \pm\infty) = V_{\text{rest}}$. Equation (6.3) is decomposed into a set of equivalent ordinary differential equations

$$\begin{aligned} \frac{dV}{d\xi} &= U \\ \frac{dU}{d\xi} &= \frac{R_i}{4d}(I_{\text{ion}} - I - cC_m U) \\ \frac{d\Gamma}{d\xi} &= -(\alpha_{\text{Gamma}}(V)(1-\Gamma) - \beta_\Gamma(V)\Gamma)/c \end{aligned} \tag{6.4}$$

where $\Gamma = m, h, \eta$. The solution should satisfy the following boundary conditions

$$(V, U, m, h, n)(\pm\infty) = (V_{\text{rest}}, 0, m_{\infty}(V_{\text{rest}}), h_{\infty}(V_{\text{rest}}), n_{\infty}(V_{\text{rest}})) \qquad (6.5)$$

6.3 Dynamics of the Solution of the Wave PDE

Neurons communicate over long distances and this is succeeded by electric signals (action potentials) which are transmitted along the neuron's axis. Diffusion phenomena describing nonlinear flow of ions across the membrane generate changes in the potential of the membrane which are transmitted on its longitudinal axis. The transmitted potential takes the form of a traveling wave. In several cases the wave equation is computed from the solution of a reaction-diffusion PDE. The model of propagating action potential in the form of a wave equation holds both for the neuron's membrane and for dendrites.

For example, the neuron described by FitzHugh–Nagumo equation can be considered to have dynamics of the Shilnikov type that is

$$\begin{aligned} \frac{\partial V}{\partial t} &= \frac{\partial^2 V}{\partial x^2} + f(V) - w + I \\ \frac{\partial w}{\partial t} &= \epsilon(v - kw) \end{aligned} \qquad (6.6)$$

where $f(w) = v(1-v)(v-a)$ with $0 < a < 1$, $\epsilon > 0$ and $k \geq 0$. If $\epsilon > 0$ is sufficiently small, then the wave differential equation has a solution (homoclinic orbit). In particular, if $f(v, w) = I - v + H(v - a) - w$ where H is a Heaviside function, then the so-called McKean model is obtained.

The dynamics of the neuron described by Eq. (6.6) is called type-II neuron dynamics. There is also the type-I neuron dynamics described by the diffusion PDE

$$\frac{\partial V}{\partial t} = \frac{\partial^2 V}{\partial x^2} + f(V) \qquad (6.7)$$

where $f(v)$ is a periodic function of period 2π.

6.4 Myelinated Axons and Discrete Diffusion

It is assumed that the membrane's axis that is covered by myelin (a fatty substant which one the one side serves as insulator of the axis and on the other side reduces the capacitance of the membrane). It is also assumed that in regular intervals, known as Ranvier nodes, the axis contacts the extracellular mean, and additionally that there is high density of Na^+ channels (Fig. 6.1).

The diameter of the axis is denoted as a_1, while the diameter of the myelinated axis is denoted as a_2.

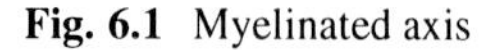
Fig. 6.1 Myelinated axis

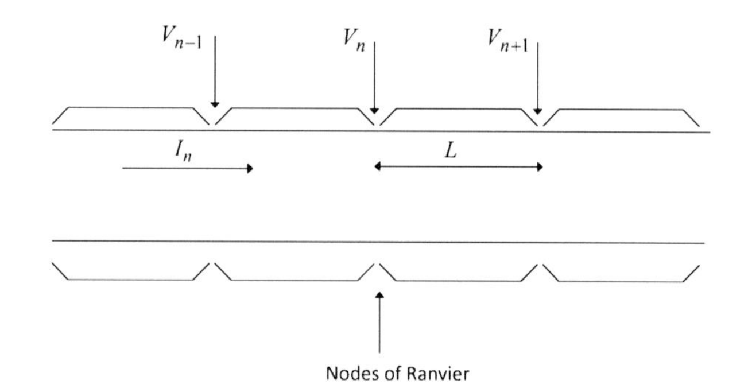

The capacitance due to myelination is defined from the following relation

$$\frac{1}{c_m} = \frac{\ln(a_2/a_1)}{2C_m \pi d_m L} \tag{6.8}$$

where L is the length of the myelinated area, C_m is the capacitance of the material, and d_m is the thickness of the cellular wall. The potential on the myelinated membrane satisfies a diffusion PDE of the form

$$\frac{c_m}{L}\frac{\partial V}{\partial t} = \frac{4\pi a_1^2}{R_L}\frac{\partial^2 V}{\partial t^2} \tag{6.9}$$

where R_L is the transmembrane resistivity. The above diffusion equation can be also written as

$$\frac{\partial V}{\partial t} = D\frac{\partial^2 V}{\partial t^2} \tag{6.10}$$

where $D = \frac{4\pi a_1^2 L}{c_m R_1}$. It holds that the transmembrane conductance and the capacitance of the myelinated areas is about 1,000 times smaller comparing to unmyelinated areas. The membrane's potential varies linearly between the nodes

$$V(x) = V_{n-1} + \frac{(V_n - V_{n-1})x}{L} \tag{6.11}$$

The current at the node is proportional to the gradient of the voltage at the myelinated segments. Thus, at node n voltage satisfies the relation

$$AC_m \frac{dV_n}{dt} = -AI_{\text{ionic}}(V_n \cdots) + I_n - I_{n-1} \tag{6.12}$$

where A is the area of the membrane which corresponds to the node and the longitudinal current is

$$I_n = -\frac{4a_1^2}{R_l}\frac{\partial V}{\partial x} = 4\pi a_1^2 \frac{(V_n - V_{n-1})}{R_1 L} \tag{6.13}$$

Area A is the surface of the membrane emerging at the node and μ is the length of the node. By dividing with surface A Eq. (6.12) becomes

$$C_m \frac{dV_n}{dt} = -I_{\text{ionic}}(V_n \cdots) + D(V_{n+1} - 2V_n + V_{n-1}) \tag{6.14}$$

where $D = \frac{4a_1}{R_l L\mu}$. Thus the PDE of the voltage along the myelinated axis is turned into a system of equivalent ordinary differential equations.

$$\begin{aligned} C_m \frac{dV}{dt} &= D[V(t+\tau) - 2V(t) + V(t-\tau)] - I_{\text{ionic}}(V, w, \cdots) \\ \frac{dw}{dt} &= g(V, w) \end{aligned} \tag{6.15}$$

where parameter w represents gating variables, calcium, etc.

It can be set $f(V) = -I_{\text{ionic}}(V)$ and it can be assumed that $f(V)$ has three roots, V_{rest}, V_{thr}, and V_{ex}, which mean resting state, threshold, and excited state, respectively. As boundary conditions, one can assume $V(-\infty) = V_{\text{ex}}$ and $V_{\infty} = V_{\text{ex}}$. Using the approximation about the second derivative with respect to time

$$\tau^2 V'' = V(t+\tau) - 2V(t) + V(t-\tau) \tag{6.16}$$

one arrives at the ordinary differential equation

$$C_m V' = f(V) + \frac{\tau^2}{L\mu} \frac{4a_1}{R_1} V'' \tag{6.17}$$

where τ is the discretization step. By performing the change of variables $\xi = \frac{\sqrt{L\mu}}{\tau}$ the wave equation is written as

$$C_m \frac{\sqrt{L\mu}}{\tau} V_{\xi} = f(V) + \frac{4a_1}{R_l} V_{\xi\xi} \tag{6.18}$$

Assume that c is the velocity of the wave traveling along the myelinated axis. It holds that $c = \frac{\sqrt{L\mu}}{\tau}$, and consequently for the myelinated part it holds

$$c_{\text{myelin}} = \frac{L}{\tau} \simeq \sqrt{\frac{L}{\mu}} c \tag{6.19}$$

For the values $\mu = 1\,\mu\text{m}$, $L = 100\,\mu\text{m}$ one obtains $c_{\text{myelin}} \simeq 10c$.

6.4.1 Solitons in Neuron's Dynamical Model

In the following approach of modelling for the neuron's membrane dynamics, the neuron's axon is considered to be a one-dimensional cylinder with lateral excitations moving along the coordinate x. The dynamics of this cylinder is described by a wave-type equation [10, 80, 86, 103]

$$\frac{\partial^2 \Delta\rho^A}{\partial t^2} = \frac{\partial}{\partial x} \left(c^2 \frac{\partial}{\partial x} \Delta\rho^A \right) \tag{6.20}$$

where $\Delta\rho^A = \rho^A - \rho_0^A$ is the charge in the area of the membrane as a function of x and t and $c = \sqrt{1/(\kappa_s^A \rho^A)}$ is the density-dependent velocity of the wave. Here, ρ_0^A is the density of the membrane at physiological conditions. To the extent that the compressibility is independent of the density and the amplitude of the propagating density wave $\Delta\rho^A << \rho_0^A$ the wave's velocity is approximately constant ($c = c_0$). Thus, the wave equation is simplified in the form

$$\frac{\partial^2}{\partial t^2}\Delta\rho^A = c_0^2 \frac{\partial^2}{\partial x^2}\Delta\rho^A \tag{6.21}$$

The compressibility κ_s^A depends on temperature and on density of the lipid membranes. The lipids of the membrane can change from the liquid to the gel state. At densities near the phase transition, where the two phases co-exist, a small increase in pressure can cause a significant increase in density by converting lipids from liquid to gel. Near this phase transition, the compression modulus becomes significantly smaller. In the latter case, the wave's velocity is approximated by

$$c^2 = \frac{1}{\rho^A \kappa_s^A} = c_0^2 + p\Delta\rho^A + q(\Delta\rho^A)^2 \tag{6.22}$$

with $p < 0$ and $q > 0$.

A more complicated form of the wave's propagation in the nerve is obtained by including the term $-h\partial^4\Delta\rho^A/\partial x^4$ with $h > 0$. This term shows that compressibility decreases at higher frequencies and leads to a propagation velocity which increases with increasing frequency. The term also shows a linear dependence on frequency of the pulse propagation velocity. Changes in coefficient h affect the spatial size of the solitary waves but not their functional form. Substituting Eqs. (6.22) in (6.21) and including also the abovementioned compressibility factor $-h\partial^4\Delta\rho^A/\partial x^4$, the equation describing the wave propagation becomes

$$\frac{\partial^2}{\partial t^2}\Delta\rho^A = \frac{\partial}{\partial x}\{[c_0^2 + p\Delta\rho^A + q(\Delta\rho^A)^2]\frac{\partial}{\partial x}\Delta\rho^A\} - h\frac{\partial^4}{\partial x^4}\Delta\rho^A \tag{6.23}$$

In the above equation $\Delta\rho^A$ is the change in lateral density of the membrane $\Delta\rho^A = \rho^A - \rho_0^A$, ρ^A is the lateral density of the membrane, ρ_0^A is the equilibrium lateral density of the membrane, c_0 is the velocity of small amplitude wave, p and q are parameters defining the dependence of the wave velocity on the membrane's density. Equation (6.23) is closely related to Boussinesq equation.

6.4.2 *Comparison Between the Hodgkin–Huxley and the Soliton Model*

In the following, some major features of the two neuron models (the Hodgkin–Huxley model and the soliton model) are summarized [10, 80].

The Hodgkin–Huxley model is exclusively of electric nature: (1) The action potential is based on the electric cable theory in which the pulse is a consequence of voltage and time-dependent changes of the conductance for sodium and potassium, (2) The nerve pulse consists of a voltage signal that changes both in space and time and this change is driven by the flow of ions, through channels formed by proteins, (3) The model is consistent with the channel-blocking effects of several poisons, such as tetrodotoxin, but does not provide an explication of anesthesia (4) Reversible changes in heat and mechanical changes are not explicitly addressed, and according to this model heat generation should be expected.

The soliton model is of both electric and mechanical nature: (1) The propagation of solitons is a consequence of the nonlinearity of the elastic constants close to the melting transition of the lipid's membrane, (2) The model does not consider an explicit role of protein and ion channels, (3) The propagating pulse is associated with changes in all variables of the membrane including temperature, lateral pressure, thickness, and length (4) In accordance with experimental findings, the propagating pulse in this model does not dissipate heat, (5) the model provides an explanation of the mechanism of anesthesia.

6.5 Estimation of Nonlinear Wave Dynamics

It will be shown that the new nonlinear filtering method, the so-called Derivative-free nonlinear Kalman Filter, can be used for estimating wave-type dynamics in the neuron's membrane. This can be done through the processing of noisy measurements and without knowledge of boundary conditions. Previous results on the application of Kalman Filtering in the estimation of neuronal dynamics can be found in [77, 134, 208]

The following nonlinear wave equation is considered

$$\frac{\partial^2\phi}{\partial t^2} = K\frac{\partial^2\phi}{\partial x^2} + f(\phi) \tag{6.24}$$

Using the approximation for the partial derivative

$$\frac{\partial^2\phi}{\partial x^2} \simeq = \frac{\phi_{i+1}-2\phi_i+\phi_{i-1}}{\Delta x^2} \tag{6.25}$$

and considering spatial measurements of variable ϕ along axis x at points $x_0 + i\Delta x,\ i = 1, 2, \cdots, N$ one has

$$\frac{\partial^2\phi_i}{\partial t^2} = \frac{K}{\Delta x^2}\phi_{i+1} - \frac{2K}{\Delta x^2}\phi_i + \frac{1}{\Delta x^2}\phi_{i-1} + f(\phi_i) \tag{6.26}$$

By considering the associated samples of ϕ given by $\phi_0, \phi_1, \cdots, \phi_N, \phi_{N+1}$ one has

$$\begin{aligned}
\frac{\partial^2 \phi_1}{\partial t^2} &= \frac{K}{\Delta x^2}\phi_2 - \frac{2K}{\Delta x^2}\phi_1 + \frac{K}{\Delta x^2}\phi_0 + f(\phi_1) \\
\frac{\partial^2 \phi_2}{\partial t^2} &= \frac{K}{\Delta x^2}\phi_3 - \frac{2K}{\Delta x^2}\phi_2 + \frac{K}{\Delta x^2}\phi_1 + f(\phi_2) \\
\frac{\partial^2 \phi_3}{\partial t^2} &= \frac{K}{\Delta x^2}\phi_4 - \frac{2K}{\Delta x^2}\phi_3 + \frac{K}{\Delta x^2}\phi_2 + f(\phi_3) \\
&\cdots \\
\frac{\partial^2 \phi_{N-1}}{\partial t^2} &= \frac{K}{\Delta x^2}\phi_N - \frac{2K}{\Delta x^2}\phi_{N-1} + \frac{K}{\Delta x^2}\phi_{N-2} + f(\phi_{N-1}) \\
\frac{\partial^2 \phi_N}{\partial t^2} &= \frac{K}{\Delta x^2}\phi_{N+1} - \frac{2K}{\Delta x^2}\phi_N + \frac{K}{\Delta x^2}\phi_{N-1} + f(\phi_N)
\end{aligned} \tag{6.27}$$

By defining the following state vector

$$x^T = (\phi_1, \phi_2, \cdots, \phi_N) \tag{6.28}$$

one obtains the following state-space description

$$\begin{aligned}
\ddot{x}_1 &= \frac{K}{\Delta x^2}x_2 - \frac{2K}{\Delta x^2}x_1 + \frac{K}{\Delta x^2}\phi_0 + f(x_1) \\
\ddot{x}_2 &= \frac{K}{\Delta x^2}x_3 - \frac{2K}{\Delta x^2}x_2 + \frac{K}{\Delta x^2}x_1 + f(x_2) \\
\ddot{x}_3 &= \frac{K}{\Delta x^2}x_4 - \frac{2K}{\Delta x^2}x_3 + \frac{K}{\Delta x^2}x_2 + f(x_3) \\
&\cdots \\
\ddot{x}_{N-1} &= \frac{K}{\Delta x^2}x_N - \frac{2K}{\Delta x^2}x_{N-1} + \frac{K}{\Delta x^2}x_{N-2} + f(x_{N-1}) \\
\ddot{x}_N &= \frac{K}{\Delta x^2}\phi_{N+1} - \frac{2K}{\Delta x^2}x_N + \frac{K}{\Delta x^2}x_{N-1} + f(x_N)
\end{aligned} \tag{6.29}$$

Next, the following state variables are defined

$$\begin{aligned}
y_{1,i} &= x_i \\
y_{2,i} &= \dot{x}_i
\end{aligned} \tag{6.30}$$

and the state-space description of the system becomes as follows

$$\begin{aligned}
\dot{y}_{1,1} &= y_{2,1} \\
\dot{y}_{2,1} &= \frac{K}{\Delta x^2}y_{1,2} - \frac{2K}{\Delta x^2}y_{1,1} + \frac{K}{\Delta x^2}\phi_0 + f(y_{1,1}) \\
\dot{y}_{1,2} &= y_{2,2} \\
\dot{y}_{2,2} &= \frac{K}{\Delta x^2}y_{1,3} - \frac{2K}{\Delta x^2}y_{1,2} + \frac{K}{\Delta x^2}y_{1,1} + f(y_{1,2}) \\
\dot{y}_{1,3} &= y_{2,3} \\
\dot{y}_{2,3} &= \frac{K}{\Delta x^2}y_{1,4} - \frac{2K}{\Delta x^2}y_{1,3} + \frac{K}{\Delta x^2}y_{1,2} + f(y_{1,3}) \\
&\cdots \\
&\cdots \\
\dot{y}_{1,N-1} &= y_{2,N-1} \\
\dot{y}_{2,N-1} &= \frac{K}{\Delta x^2}y_{1,N} - \frac{2K}{\Delta x^2}y_{1,N-1} + \frac{K}{\Delta x^2}y_{1,N-2} + f(y_{1,N-1}) \\
\dot{y}_{1,N} &= y_{2,N} \\
\dot{y}_{2,N} &= \frac{K}{\Delta x^2}\phi_{N+1} - \frac{2K}{\Delta x^2}y_{1,N} + \frac{K}{\Delta x^2}y_{1,N-1} + f(y_{1,N})
\end{aligned} \tag{6.31}$$

The dynamical system described in Eq. (6.31) is a differentially flat one with flat output defined as the vector $\tilde{y} = [y_{1,1}, y_{1,2}, \cdots, y_{1,N}]$. Indeed all state variables can be written as functions of the flat output and its derivatives.

Moreover, by defining the new control inputs

$$\begin{aligned}
v_1 &= \frac{K}{\Delta x^2} y_{1,2} - \frac{2K}{\Delta x^2} y_{1,1} + \frac{K}{\Delta x^2}\phi_0 + f(y_{1,1}) \\
v_2 &= \frac{K}{\Delta x^2} y_{1,3} - \frac{2K}{\Delta x^2} y_{1,2} + \frac{K}{\Delta x^2} y_{1,1} + f(y_{1,2}) \\
v_3 &= \frac{K}{\Delta x^2} y_{1,4} - \frac{2K}{\Delta x^2} y_{1,3} + \frac{K}{\Delta x^2} y_{1,2} + f(y_{1,3}) \\
&\cdots \\
v_{N-1} &= \frac{K}{\Delta x^2} y_{1,N} - \frac{2K}{\Delta x^2} y_{1,N-1} + \frac{K}{\Delta x^2} y_{1,N-2} + f(y_{1,N-1}) \\
v_N &= \frac{K}{\Delta x^2}\phi_{N+1} - \frac{2K}{\Delta x^2} y_{1,N} + \frac{K}{\Delta x^2} y_{1,N-1} + f(y_{1,N})
\end{aligned} \tag{6.32}$$

the following state-space description is obtained

$$\begin{pmatrix} \dot{y}_{1,1} \\ \dot{y}_{2,1} \\ \dot{y}_{1,2} \\ \dot{y}_{2,2} \\ \cdots \\ \dot{y}_{1,N-1} \\ \dot{y}_{2,N-1} \\ \dot{y}_{1,N} \\ \dot{y}_{2,N} \end{pmatrix} = \begin{pmatrix} 0\,1\,0\,0 \cdots 0\,0\,0\,0 \\ 0\,0\,0\,0 \cdots 0\,0\,0\,0 \\ 0\,0\,0\,1 \cdots 0\,0\,0\,0 \\ 0\,0\,0\,0 \cdots 0\,0\,0\,0 \\ 0\,0\,0\,0 \cdots 0\,0\,0\,0 \\ 0\,0\,0\,0 \cdots 0\,0\,0\,0 \\ \cdots\cdots\cdots\cdots\cdots\cdots \\ 0\,0\,0\,0 \cdots 0\,1\,0\,0 \\ 0\,0\,0\,0 \cdots 0\,0\,0\,0 \\ 0\,0\,0\,0 \cdots 0\,0\,0\,1 \\ 0\,0\,0\,0 \cdots 0\,0\,0\,0 \end{pmatrix} \begin{pmatrix} y_{1,1} \\ y_{2,1} \\ y_{1,2} \\ y_{2,2} \\ \cdots \\ y_{1,N-1} \\ y_{2,N-1} \\ y_{1,N} \\ y_{2,N} \end{pmatrix} + \begin{pmatrix} 0\,0\,0 \cdots 0\,0 \\ 1\,0\,0 \cdots 0\,0 \\ 0\,0\,0 \cdots 0\,0 \\ 0\,1\,0 \cdots 0\,0 \\ 0\,0\,0 \cdots 0\,0 \\ 0\,0\,1 \cdots 0\,0 \\ \cdots\cdots\cdots\cdots \\ 0\,0\,0 \cdots 0\,0 \\ 0\,0\,0 \cdots 1\,0 \\ 0\,0\,0 \cdots 0\,0 \\ 0\,0\,0 \cdots 0\,1 \end{pmatrix} \begin{pmatrix} v_1 \\ v_2 \\ v_3 \\ \cdots \\ v_{N-1} \\ v_N \end{pmatrix} \tag{6.33}$$

By selecting measurements from a subset of points x_j $j \in [1, 2, \cdots, m]$, the associated observation (measurement) equation becomes

$$\begin{pmatrix} z_1 \\ z_2 \\ \cdots \\ z_m \end{pmatrix} = \begin{pmatrix} 1\,0\,0 \cdots 0\,0 \\ 0\,0\,0 \cdots 0\,0 \\ \cdots\cdots\cdots\cdots \\ 0\,0\,0 \cdots 1\,0 \\ 0\,0\,0 \cdots 0\,0 \end{pmatrix} \begin{pmatrix} y_{1,1} \\ y_{2,1} \\ y_{1,2} \\ y_{2,2} \\ \cdots \\ y_{1,N} \\ y_{2,N} \end{pmatrix} \tag{6.34}$$

Thus, in matrix form one has the following state-space description of the system

$$\begin{aligned}
\dot{\tilde{y}} &= A\tilde{y} + Bv \\
\tilde{z} &= C\tilde{y}
\end{aligned} \tag{6.35}$$

Denoting $a = \frac{K}{Dx^2}$ and $b = -\frac{2K}{Dx^2}$, the initial description of the system given in Eq. (6.33) is rewritten as follows:

$$\begin{pmatrix} \dot{y}_{1,1} \\ \dot{y}_{2,1} \\ \dot{y}_{1,2} \\ \dot{y}_{2,2} \\ \cdots \\ \dot{y}_{1,N-1} \\ \dot{y}_{2,N-1} \\ \dot{y}_{1,N} \\ \dot{y}_{2,N} \end{pmatrix} = \begin{pmatrix} 0\,1\,0\,0\,0\,0\,0 \cdots 0\,0\,0\,0\,0\,0 \\ b\,0\,a\,0\,0\,0\,0 \cdots 0\,0\,0\,0\,0\,0 \\ 0\,0\,0\,1\,0\,0\,0 \cdots 0\,0\,0\,0\,0\,0 \\ a\,0\,b\,0\,a\,0\,0 \cdots 0\,0\,0\,0\,0\,0 \\ 0\,0\,0\,0\,0\,1\,0 \cdots 0\,0\,0\,0\,0\,0 \\ 0\,0\,a\,0\,b\,0\,a \cdots 0\,0\,0\,0\,0\,0 \\ \cdots\cdots\cdots\cdots\cdots\cdots\cdots \\ 0\,0\,0\,0\,0\,0\,0 \cdots 0\,0\,0\,1\,0\,0 \\ 0\,0\,0\,0\,0\,0\,0 \cdots a\,0\,b\,0\,a\,0 \\ 0\,0\,0\,0\,0\,0\,0 \cdots 0\,0\,0\,0\,0\,1 \\ 0\,0\,0\,0\,0\,0\,0 \cdots 0\,0\,a\,0\,b\,0 \end{pmatrix} \begin{pmatrix} y_{1,1} \\ y_{2,1} \\ y_{1,2} \\ y_{2,2} \\ \cdots \\ y_{1,N-1} \\ y_{2,N-1} \\ y_{1,N} \\ y_{2,N} \end{pmatrix} + \begin{pmatrix} 0\,0\,0 \cdots 0\,0 \\ 1\,0\,0 \cdots 0\,0 \\ 0\,0\,0 \cdots 0\,0 \\ 0\,1\,0 \cdots 0\,0 \\ 0\,0\,0 \cdots 0\,0 \\ 0\,0\,1 \cdots 0\,0 \\ \cdots\cdots\cdots\cdots \\ 0\,0\,0 \cdots 0\,0 \\ 0\,0\,0 \cdots 1\,0 \\ 0\,0\,0 \cdots 0\,0 \\ 0\,0\,0 \cdots 0\,1 \end{pmatrix} \begin{pmatrix} v_1 \\ v_2 \\ v_3 \\ \cdots \\ v_{N-1} \\ v_N \end{pmatrix} \tag{6.36}$$

The associated control inputs are now defined as

$$\begin{aligned} v_1 &= \tfrac{K}{\Delta x^2}\phi_0 + f(y_{1,1}) \\ v_2 &= f(y_{1,2}) \\ v_3 &= f(y_{1,3}) \\ &\cdots \\ v_{N-1} &= f(y_{1,N-1}) \\ v_N &= \tfrac{K}{\Delta x^2}\phi_{N+1} + f(y_{1,N}) \end{aligned} \tag{6.37}$$

By selecting measurements from a subset of points $x_j \ j\in[1, 2, \cdots, m]$, the associated observation (measurement) equation remains as in Eq. (6.34), i.e.

$$\begin{pmatrix} z_1 \\ z_2 \\ \cdots \\ z_m \end{pmatrix} = \begin{pmatrix} 1\,0\,0 \cdots 0\,0 \\ 0\,0\,0 \cdots 0\,0 \\ \cdots\cdots\cdots\cdots \\ 0\,0\,0 \cdots 1\,0 \\ 0\,0\,0 \cdots 0\,0 \end{pmatrix} \begin{pmatrix} y_{1,1} \\ y_{2,1} \\ y_{1,2} \\ y_{2,2} \\ \cdots \\ y_{1,N} \\ y_{2,N} \end{pmatrix} \tag{6.38}$$

For the linear description of the system in the form of Eq. (6.35) one can perform estimation using the standard Kalman Filter recursion. The discrete-time Kalman filter can be decomposed into two parts: (1) time update (prediction stage), and (2) measurement update (correction stage).

The discrete-time equivalents of matrices A, B, C in Eq. (6.36) and Eq. (6.38) are computed using common discretization methods. These are denoted as A_d, B_d, and C_d respectively. Then the Kalman Filter recursion becomes:

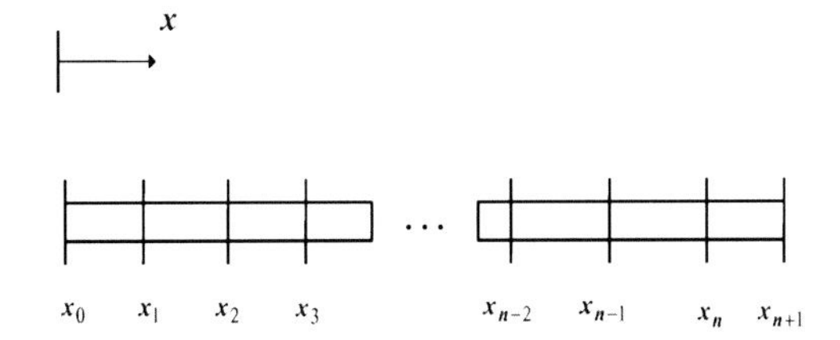

Fig. 6.2 Grid points for measuring $\phi(x,t)$

Measurement update:

$$\begin{aligned}K(k) &= P^{-}(k)C_d^T[C_d \cdot P^{-}(k)C_d^T + R]^{-1}\\ \hat{y}(k) &= \hat{y}^{-}(k) + K(k)[z(k) - C_d\hat{y}^{-}(k)]\\ P(k) &= P^{-}(k) - K(k)C_d P^{-}(k)\end{aligned} \tag{6.39}$$

Time update:

$$\begin{aligned}P^{-}(k+1) &= A_d(k)P(k)A_d^T(k) + Q(k)\\ \hat{y}^{-}(k+1) &= A_d(k)\hat{y}(k) + B_d(k)u(k)\end{aligned} \tag{6.40}$$

Therefore, by taking measurements of $\phi(x,t)$ at time instant t at a small number of measuring points, as shown in Fig. 6.2, $j = 1, \cdots, n_1$ it is possible to estimate the complete state vector, i.e. to get values of ϕ in a mesh of points that covers efficiently the variations of $\phi(x,t)$. By processing a sequence of output measurements of the system, one can obtain local estimates of the state vector $\hat{y}$. The measuring points can vary in time provided that the observability criterion for the state-space model of the PDE holds.

Remark. The proposed derivative-free nonlinear Kalman Filter is of improved precision because unlike other nonlinear filtering schemes, e.g. the Extended Kalman Filter it does not introduce cumulative numerical errors due to approximative linearization of the system's dynamics. Besides it is computationally more efficient (faster) because it does not require to calculate Jacobian matrices and partial derivatives.

6.6 Simulation Tests

6.6.1 Evaluation Experiments

The proposed filtering scheme was tested in estimation of the dynamics of a wave equation of the form of Eq. (6.24) under unknown boundary and initial conditions. It has been shown that it is possible to obtain voltage measurements from dendritic trees using, for example, laser-based scanning techniques [69, 84, 202]. In the simulation experiments the following wave-type dynamics in the neuron's membrane was considered

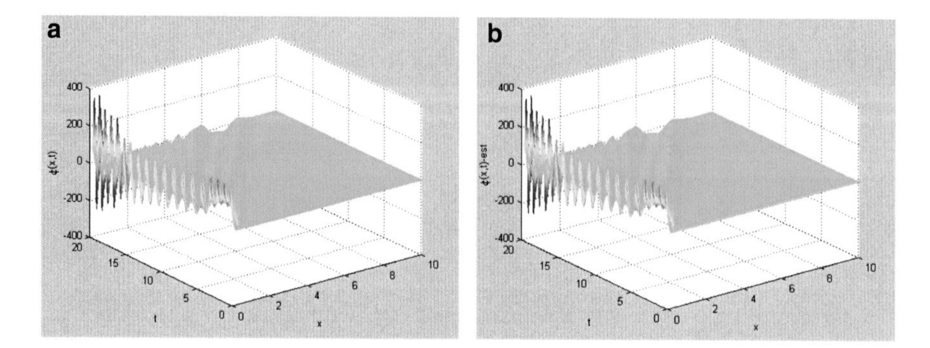

Fig. 6.3 Test case 1: (**a**) spatiotemporal variation of the wave function $\phi(x,t)$. (**b**) estimate $\hat{\phi}(x,t)$ of the wave function provided by the derivative-free nonlinear Kalman Filter

$$\frac{\partial^2\phi}{\partial t^2} = \frac{\partial}{\partial x}\{[c_0^2 + \rho\phi + q\phi^2]\frac{\partial\phi}{\partial x}\} \Rightarrow$$
$$\frac{\partial^2\phi}{\partial t^2} = (\rho + 2q\phi)(\frac{\partial\phi}{\partial x})^2 + [c_0^2 + \rho\phi + q\phi^2]\frac{\partial^2\phi}{\partial x^2} \tag{6.41}$$

By denoting $f(\phi) = (\rho + 2q\phi)(\frac{\partial\phi}{\partial x})^2$ and $K(x,t) = [c_0^2 + \rho\phi + q\phi^2]$, one finally obtains the following description for the wave-type dynamics of the membrane

$$\frac{\partial^2\phi}{\partial t^2} = K(x,t)\frac{\partial^2\phi}{\partial x^2} + f(\phi) \tag{6.42}$$

The estimator was based on the Derivative-free nonlinear Kalman Filter as analyzed in Sect. 6.5 and made use of a grid of $N = 50$ points, out of which $n = 25$ were measurement points. At each measurement point a value of the state variable $x_1 = \phi(x,t)$ was obtained. In Fig. 6.3 the estimate of the wave function provided by the distributed filtering algorithm is compared to the real wave function dynamics. Due to boundary conditions set at points $x = 0$ and $x = 10$ of the x-axis wave type dynamics $\psi(x,t)$ is generated and propagates in time. The nonlinear Kalman Filter estimates the wave-type dynamics without knowledge of the initial conditions of the PDE. As it can be noticed the derivative-free nonlinear Kalman Filter approximates closely the real wave function. Indicative results about the estimation obtained at local grid points is given in Figs. 6.4, 6.5, and 6.6.

The evaluation tests were repeated for different values in the parameters of the wave function coefficient $K(x,t)$. In Fig. 6.7 a new value of gain $K(x,t)$ is considered and the estimate of the wave function provided by the distributed filtering algorithm is compared again to the real wave function dynamics. Additional results about the accuracy of estimation obtained in local grid points, when using different coefficients c_0, ρ, and q in the computation of $K(x,t)$ are given in Figs. 6.8, 6.9, and 6.10. As it can be observed again, the derivative-free nonlinear Kalman Filter resulted in very accurate estimates of the wave-type dynamics.

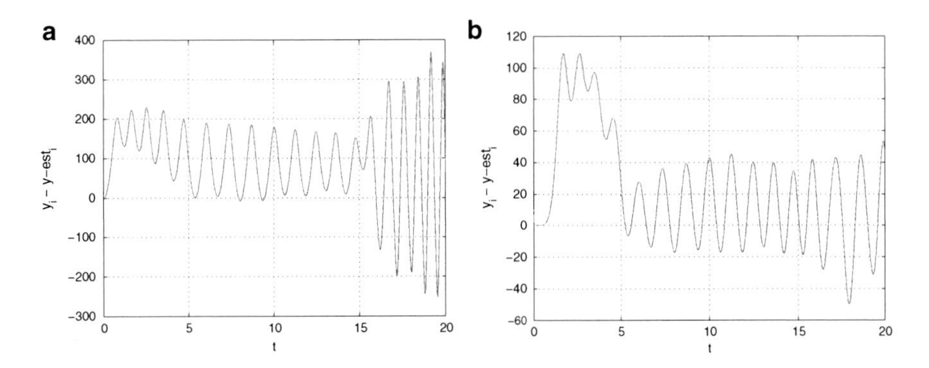

Fig. 6.4 Test case 1: estimate (*dashed green line*) and real value (*red continuous line*) (**a**) of $\phi(x_1, t)$ at grid point 1 (**b**) of $\phi(x_2, t)$ at grid point 2

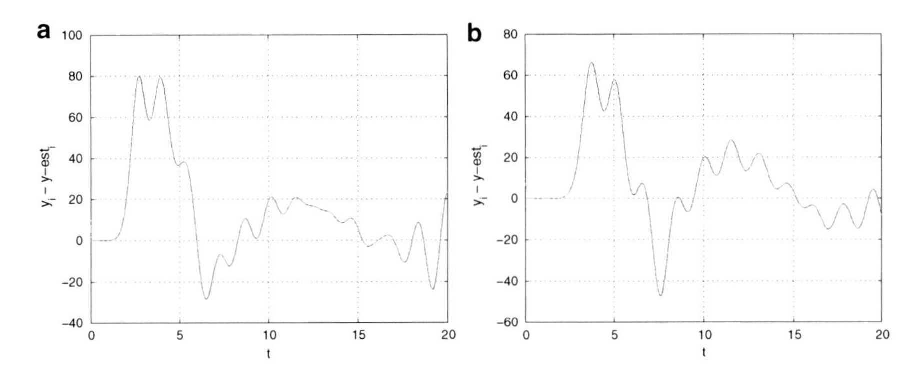

Fig. 6.5 Test case 1: estimate (*dashed green line*) and real value (*red continuous line*) (**a**) of $\phi(x_5, t)$ at grid point 5 (**b**) of $\phi(x_7, t)$ at grid point 7

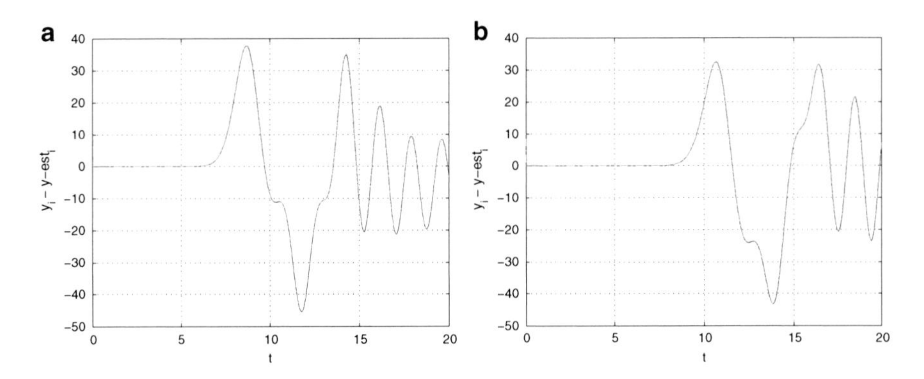

Fig. 6.6 Test case 1: estimate (*dashed green line*) and real value (*red continuous line*) (**a**) of $\phi(x_15, t)$ at grid point 15 (**b**) of $\phi(x_{20}, t)$ at grid point 20

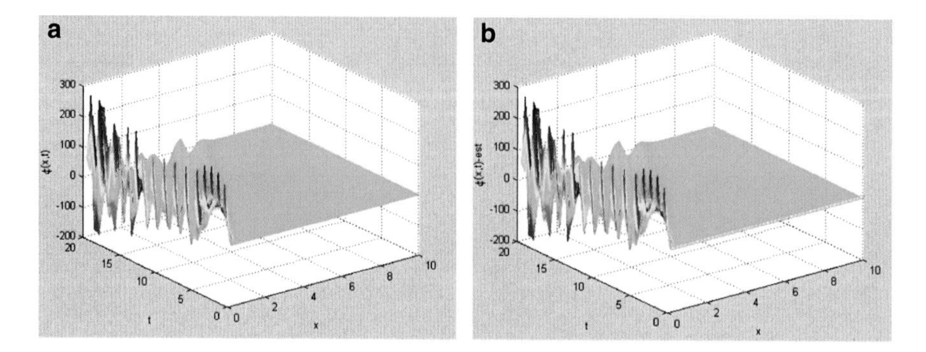

Fig. 6.7 Test case 2: (**a**) spatiotemporal variation of the wave function $\phi(x, t)$ (**b**) estimate $\hat{\phi}(x, t)$ of the wave function provided by the derivative-free nonlinear Kalman Filter

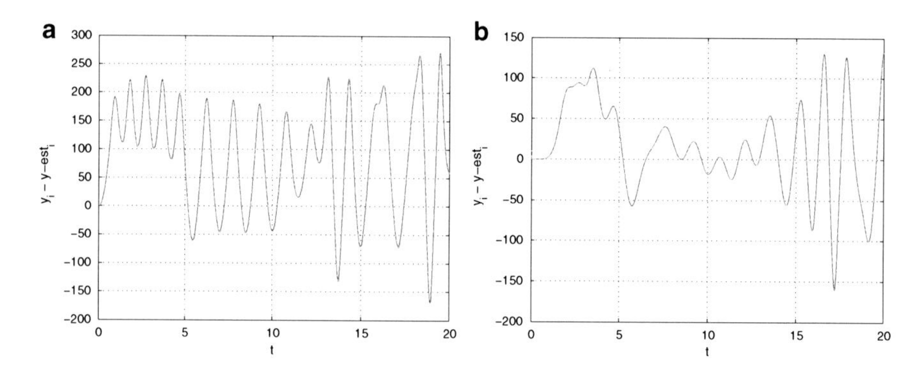

Fig. 6.8 Test case 2: estimate (*dashed green line*) and real value (*red continuous line*) (**a**) of $\phi(x_1, t)$ at grid point 1 (**b**) of $\phi(x_2, t)$ at grid point 2

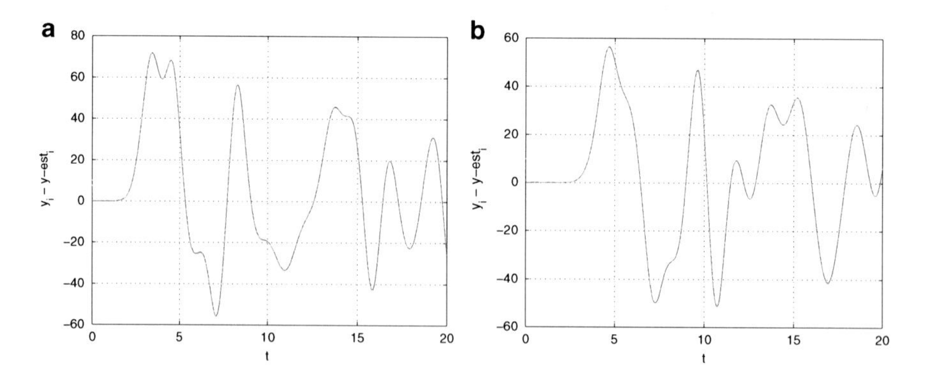

Fig. 6.9 Test case 2: estimate (*dashed green line*) and real value (*red continuous line*) (**a**) of $\phi(x_5, t)$ at grid point 5 (**b**) of $\phi(x_7, t)$ at grid point 7

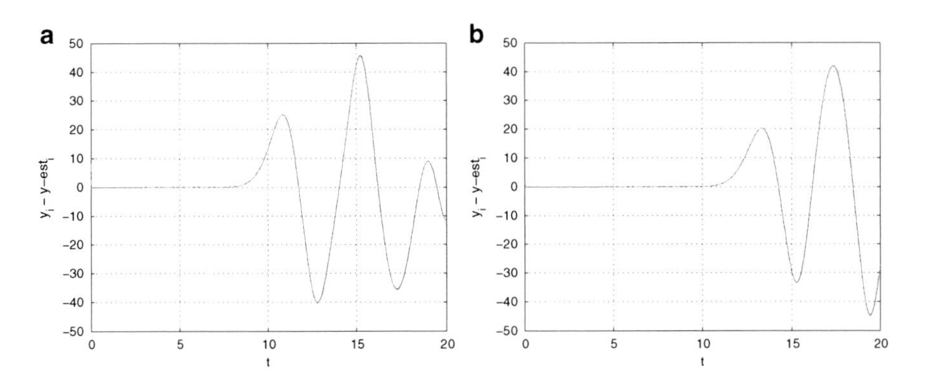

Fig. 6.10 Test case 2: estimate (*dashed green line*) and real value (*red continuous line*) (**a**) of $\phi(x_15, t)$ at grid point 15 (**b**) of $\phi(x_{20}, t)$ at grid point 20

6.6.2 Assessment of the Filter's Performance

6.6.2.1 Stability of the Filtering Method

The stability and convergence conditions for the proposed filtering algorithm are those of the standard Kalman Filter, which means that the filter converges if the system's model in its linearized canonical form satisfies obvervability (detectability) criteria. About the method for solution of a PDE through its decomposition into a set of local ODEs, numerical stability conditions have been provided and these are related to the step of spatio-temporal discretization of the initial PDE [139]. The method converges and remains numerically stable if the discretization steps are taken to be sufficiently small. For the basic wave-type PDE $\frac{\partial^2 y}{\partial t^2} = K\frac{\partial^2 y}{\partial x^2}$ the convergence condition is $2K\frac{Dt}{Dx^2} \leq 1$ where Dt and Dx are the discretization steps in space and time [139].

6.6.2.2 Implementation Stages and Advantages of the Filtering Method

The first stage of the filter's design is to analyze the PDE that describes the system's dynamics into a set of ordinary differential equations. Actually, at each iteration of the algorithm in time, a numerical solution of the PDE is computed at a spatial grid of N points and at each point of the grid the system's dynamics is described by a nonlinear ODE. At a second stage differential flatness theory and a diffeomorphism (change of coordinates) is used to transform the local nonlinear ODEs into linear ODEs which in turn can take the form of a linear state space equation in the canonical (Brunovsky) form. The Derivative-free nonlinear Kalman Filter consists of the standard Kalman Filter recursion on the linearized equivalent model of the valve and on computation of state and disturbance estimates using the diffeomorphism (relations about state variables transformation) provided by differential flatness theory.

Comparing to other nonlinear filtering approaches that could have been used, the proposed filtering approach exhibits significant advantages: (1) it is based on an exact linearization of the local ODEs thus avoiding numerical errors due to approximative linearization, (2) in terms of computation it is faster than other nonlinear filtering approaches, (3) it exhibits the optimality properties of the standard Kalman Filter recursion for estimation under measurement noise.

6.6.2.3 Methods for Real-Time Monitoring of Neuron's Membrane Dynamics

An established method for recording neuron's activity at low spatial resolution is based on the use of multi-electrode arrays. Monitoring of the neurons' activity at low resolution can be also obtained with the use of positron emission tomography. This technique involves usage of radioactive tracers which bind to receptors. The tracers emit γ rays which are detected by a scanning device and reflect the dynamics of the binding sites. It has been also shown that it is possible to obtain voltage measurements at high resolution from dendritic trees using fMRI techniques and optical imaging. fMRI is a special case of magnetic resonance imaging. It is based on the different magnetization features between oxygenated and non-oxygenated tissue. fMRI provides recording of the activity of the neuron's membrane with high spatial resolution (of the order of tens of μm) and high sampling rate.

Optical methods are the most suitable for measuring voltage variations in neuron membranes at high resolution. This can be succeeded, for example, through the monitoring of molecules called chromophores or voltage sensitive dyes that change their color according to voltage levels in the membrane. More recently, the use of quantum dots as voltage signaling nanoparticles has enabled to further improve the spatial resolution of optical imaging techniques. The resolution of the optical imaging methods is at the μm-scale and enables to obtain a sufficient number of measurements for filtering and state estimation purposes [123–138].

6.6.2.4 Practical Applications of the Proposed Filtering Method

The proposed filtering method for distributed parameter systems has a wide range of applications: (1) Estimation of the dynamics of the neuron's membrane with the use of a small number of measurements. Due to practical constraints the spatiotemporal variations of the membrane's voltage have to be performed using only a limited measurements set. Moreover, knowledge about initial and boundary conditions of the system is in most cases unknown. To estimate the dynamics of this type of distributed parameter system the use of the proposed filtering method that is based on the Derivative-free nonlinear Kalman Filter is of great importance, (2) Estimation of the parameters of the models that describe the dynamics of biological neurons. Actually, one has to do with a dual identification problem, that is estimation of the non-measurable elements of the state vector of the biological neuron and estimation of the unknown model parameters. The proposed Derivative-free nonlinear Kalman Filter can offer an efficient solution to this problem (3) Change detection about the parameters of the model that describes the neuron's dynamics. This is a fault diagnosis problem and can result in efficient method of medical diagnosis of neurological diseases. The filter can be parameterized to reproduce the dynamics of the healthy neuron. By comparing real measurements from the neuron against measurements obtained from the filter and by performing statistical processing for the associated differences (residuals) one can diagnose deviation of the neuron's functioning from the normal ranges. This can be particularly useful for diagnosis of neurological diseases related to neurons decay, at their early stages. Moreover, it is possible to perform fault isolation, which means to find which is the distorted parameter in the neuron's model that is responsible for deviation from the healthy functioning. This can be particularly useful for developing more efficient medical treatments of neurological diseases (4) More efficient modelling of brain functioning and of the nervous system. By detecting changes in the parameters of neurons' model one can find out how the response and adaptation of the neurons' functioning is associated with external stimuli and medication (pharmaceutical treatment).

6.7 Conclusions

This chapter has analyzed wave-type PDEs that appear in the transmission of neural signals and has proposed filtering for estimating the dynamics of the neurons' membrane. It has been shown that in specific neuron models the spatiotemporal variations of the membrane's voltage follow PDEs of the wave-type while in other models such variations are associated with the propagation of solitary waves through the membrane. To compute the dynamics of the membrane's PDE model without knowledge of initial conditions and through the processing of a small number of measurements, a new filtering method, under the name Derivative-free nonlinear Kalman Filtering, has been proposed.

The method is based into decomposition of the initial wave-type PDE that describes the dynamics of the distributed parameter system, into a set of nonlinear ordinary differential equations. Next, with the application of a change of coordinates (diffeomorphism) that is based on differential flatness theory, the local nonlinear differential equations are turned into linear ones. This enables to describe the wave dynamics with a state-space equation that is in the linear canonical (Brunovsky) form. For the linearized equivalent of the wave system dynamics it is possible to perform state estimation with the use of the standard Kalman Filter recursion. The efficiency of the derivative free nonlinear Kalman Filter has been confirmed through numerical simulation experiments.

Chapter 7
Stochastic Models of Biological Neuron Dynamics

Abstract The chapter examines neural networks in which the synaptic weights correspond to diffusing particles. Each diffusing particle (stochastic weight) is subject to the following forces: (1) a spring force (drift) which is the result of the harmonic potential and tries to drive the particle to an equilibrium and (2) a random force (noise) which is the result of the interaction with neighboring particles. This interaction can be in the form of collisions or repulsive forces. It is shown that the diffusive motion of the stochastic particles (weights' update) can be described by Fokker–Planck's, Ornstein–Uhlenbeck, or Langevin's equation which under specific assumptions are equivalent to Schrödinger's diffusion equation. It is proven that Langevin's equation is a generalization of the conventional gradient algorithms.

7.1 Outline

Conventional neural networks may prove insufficient for modelling memory and cognition, as suggested by effects in the functioning of the nervous system, which lie outside classical physics [19, 48, 59, 73, 75, 136, 168]. One finds ample support for this in an analysis of the sensory organs, the operation of which is quantized at levels varying from the reception of individual photons by the retina, to thousands of phonon quanta in the auditory system. Of further interest is the argument that synaptic signal transmission has a quantum character, although the debate on this issue has not been conclusive. For instance, it has been mentioned that superposition of quantum states takes place in the microtubules of the brain and that memory recall is the result of quantum measurement [73]. Moreover, it has been argued that human cognition must involve an element inaccessible to simulation on classical neural networks and this might be realized through a biological instantiation of quantum computation.

To evaluate the validity of the aforementioned arguments, neural structures with weights that follow the model of the quantum harmonic oscillator (QHO) will be studied in this chapter. Connectionist structures which are compatible with

G.G. Rigatos, *Advanced Models of Neural Networks*,
DOI 10.1007/978-3-662-43764-3_7,

the theory of quantum mechanics and demonstrate the particle-wave nature of information have been analyzed in [34, 120, 127, 133]. The use of neural networks compatible with quantum mechanics principles can be found in [6, 135, 137, 145], while the relation between random oscillations and diffusion equations has been studied in [66, 179]. Other studies on neural models with quantum mechanical properties can be found in [143, 149, 167, 198].

In this chapter it is assumed that the weights of neural networks are stochastic variables which correspond to diffusing particles, and interact to each other as the theory of Brownian motion (Wiener process) predicts. Brownian motion is the analogous of the QHO, i.e. of Schrödinger's equation under harmonic (parabolic) potential. However, the analytical or numerical solution of Schrödinger's equation is computationally intensive, since for different values of the potential $V(x)$ it is required to calculate the modes $\psi_k(x)$ in which the particle's wave-function $\psi(x)$ is decomposed. Moreover, the solution of Schrödinger's equation contains non-easily interpretable terms such as the complex number probability amplitudes which are associated with the modes $\psi_k(x)$, or the path integrals that constitute the particle's trajectory. On the other hand, instead of trying to solve Schrödinger's equation for various types of $V(x)$ one can study the time evolution of the particle through an equivalent diffusion equation, assuming probability density function depending on QHO's ground state, i.e. $\rho_0(x) = |\psi_0(x)|^2$ [56].

This chapter extends the results on neural structures compatible with quantum mechanics principles presented in [163, 165]. Moreover, the chapter extends the results on quantum neural structures, where the basic assumption was that the neural weights correspond to diffusing particles under Schrödinger's equation with zero or constant potential [149]. The diffusing particle (stochastic weight) is subject to the following forces: (1) a spring force (drift) which is the result of the harmonic potential and tries to drive the particle to an equilibrium and (2) a random force (noise) which is the result of the interaction with neighboring particles. This interaction can be in the form of collisions or repulsive forces. It is shown that the diffusive motion of the stochastic particles (weights' update) can be described by Langevin's equation which is a stochastic linear differential equation [56, 66]. Following the analysis of [22, 52, 93] it is proven that Langevin's equation is a generalization of the conventional gradient algorithms. Therefore neural structures with crisp numerical weights can be considered as a subset of NN with stochastic weights that follow the QHO model.

7.2 Wiener Process and Its Equivalence to Diffusion

7.2.1 *Neurons' Dynamics Under Noise*

Up to now, neuronal dynamics has been expressed in a deterministic manner. However, stochastic terms such as noise may also be present in the neurons' dynamic model. Reasons for the appearance of noise in neurons are: (1) randomness in the opening or closing of the ion channels, (2) randomness in the release of transmitters which in turn leads to random charging and discharging of the neuron's membrane [121, 196, 197]. The main effects of the noise in neurons are as follows: (1) firing of the neurons even under sub-threshold inputs, (2) increase of the pattern storage capacity.

A basic tool for studying the stochastic dynamics of neurons is a stochastic differential equation known as Langevin's equation:

$$dx = A(x,t)dt + B(x,t)dW(t) \tag{7.1}$$

For the numerical solution of Langevin's equation usually the following solution is used

$$x(n+1) = x(n) + hA(x(n), t_n) + B(x(n), t_n)\sqrt{h}\hat{N}(0,1) \tag{7.2}$$

where h is the discretization step, and $\hat{N}(0,1)$ is a vector of independent, identically distributed variables.

7.2.2 Wiener Walk and Wiener Process

The Wiener walk and the Wiener process are essential mathematical tools for describing the stochastic neuron dynamics. Wiener walk will be analyzed and the Wiener process will be derived as a limit case of the walk [37,90]. The Wiener walk describes a simple symmetric random walk. Assume $\xi_1, \cdots, \xi_n$ a finite sequence of independent random variables, each one of which takes the values ± 1 with the same probability. The random walk is the sequence

$$s_k = \xi_1 + \xi_2 + \cdots + \xi_k, \quad 0 \le k \le n \tag{7.3}$$

The n-step Wiener walk is considered in the time interval $[0, T]$, where the time step Δt is associated with the particle's displacement Δx

$$w(t_k) = \xi_1 \Delta x + \cdots + \xi_k \Delta x \tag{7.4}$$

The random function that is described by Eq. (7.4) is the Wiener walk [56]. A sample of the Wiener walk in depicted in Fig. 7.1a.

The Wiener walk is an important topic in the theory of stochastic processes, is also known as *Brownian motion* and it provides a model for the motion of a particle under the effect of a potential. The Wiener process is the limit of the Wiener walk for $n \to \infty$, and using the central limit theorem (CLT) it can be shown that the distribution of the Wiener process is Gaussian. Indeed, since the random variable $w(t_k)$ of Eq. (7.4) is the sum of an infinitely large number of increments

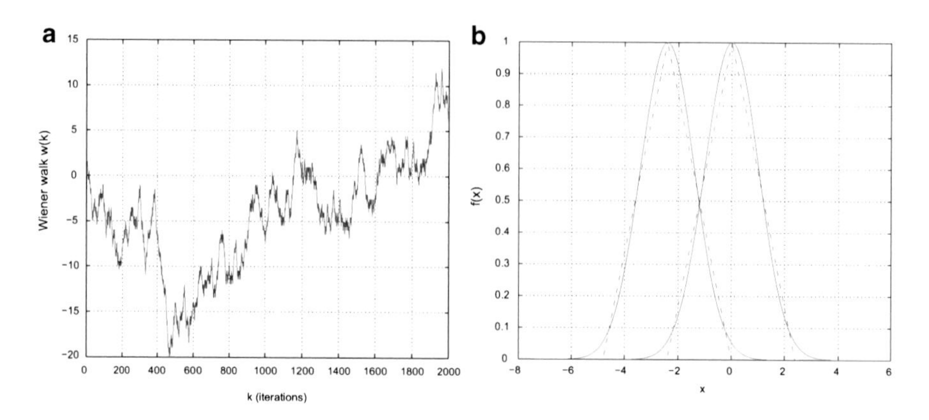

Fig. 7.1 (**a**) Wiener random walk (**b**) Solution of Schrödinger's equation: Shifted Gaussians that give the probability density function of a stationary diffusion process (*continuous curves*) and the approximation of the p.d.f. by symmetric triangular possibility distributions (*dashed lines*)

(or decrements), then according to the CLT it must follow a Gaussian distribution. Thus one obtains:

$$E\{w(t)\} = 0 \text{ while } E[w(t) - E\{w(t)\}]^2 = \sigma^2 t \tag{7.5}$$

7.2.3 *Outline of Wiener Process Properties*

Wiener's process is a stochastic process that satisfies the following conditions:

1. $w(0) = 0$
2. The distribution of the stochastic variable $w(t)$ is a Gaussian one with a p.d.f. $\rho(x,t)$ which satisfies the following relation

$$\frac{\partial\rho(x,t)}{\partial t} = \frac{\sigma^2}{2}\frac{\partial^2\rho(x,t)}{\partial x^2},\ \rho(x,0) = \delta(x) \tag{7.6}$$

   For a limited number of time instants $t_1 < t_2 < \cdots < t_n$ the stochastic variables $w(t_j) - w(t_{j-1})$ are independent.
3. It holds that $E\{w(t)\} = 0$ and $E\{[w(t) - w(s)]^2 = \sigma^2(t-s)\}$ for all $0 \le s \le t$.
4. $w(t)$ is a continuous process

The standard Wiener process evolves in time according to the following relation

$$w(t+h) = w(t) + \sqrt{h}N(0,1) \tag{7.7}$$

The Wiener process actually describes a diffusion phenomenon and the associated p.d.f. of stochastic variable is given by

$$\rho(x,t) = \frac{1}{\sqrt{2\pi t}}e^{\frac{-x^2}{2t}} \tag{7.8}$$

7.2.4 *The Wiener Process Corresponds to a Diffusion PDE*

It has been shown that the limit of the Wiener walk for $n \to \infty$ is the Wiener process, which corresponds to the partial differential equation of a diffusion [56]. For $t > 0$, function $\rho(x,t)$ is defined as the probability density function (p.d.f.) of the Wiener process, i.e.

$$E[f(w(t))] = \int_{-\infty}^{+\infty} f(x)\rho(x,t)dx \tag{7.9}$$

As explained in Eq. (7.6), the p.d.f. $\rho(x,t)$ is a Gaussian variable with mean value equal to 0 and variance equal to $\sigma^2 t$ which satisfies a diffusion p.d.e of the form

$$\frac{\partial \rho}{\partial t} = \frac{1}{2}\sigma^2 \frac{\partial^2 \rho}{\partial t^2} \tag{7.10}$$

which is the simplest diffusion equation (heat equation). The generalization of the Wiener process in an infinite dimensional space is the *Ornstein–Uhlenbeck* process, where the joint probability density function is also Gaussian [17, 56]. In that case there are n Brownian particles, and each one performs a Wiener walk, given by $w^i(t_k), i = 1, \cdots, n$ of Eq. (7.4).

7.2.5 *Stochastic Integrals*

For the stochastic variable $w(t)$ the following integral is computed

$$I = \int_{t_0}^{t} G(s)dw(s) \tag{7.11}$$

or

$$S_n = \sum_{j=1}^{n} G(\tau_j)[w(t_j) - w(t_{j-1})] \tag{7.12}$$

where $G(t)$ is a piecewise continuous function. To compute the stochastic integral the selection of sample points τ_j is important. If $\tau_j = t_{j-1}$, then one has the Itô calculus. If $\tau_j = (t_{j-1} + t_j)/2$, then one has the Stratonovic calculus.

7.2.6 *Ito's Stochastic Differential Equation*

From the relation about the standard Wiener process given in Eq. (7.7), and considering that the variance of noise is equal to 1, it holds that

$$E[(w(t+h) - w(t))]^2 = hE[N(0,1)^2] = h \tag{7.13}$$

Using Langevin's equation one has

$$dx = \alpha(x,t)dt + w(t) \tag{7.14}$$

Assume that function $y = f(x)$ where f is twice differentiable. It holds that

$$\begin{aligned} dy = & \quad f(x+dx) - f(x) = f'(x)dx + \tfrac{1}{2}f''(x)dx^2 + \cdots = \\ & f'(x)[\alpha(x,t)dt + b(x,t)dw(t)] + \tfrac{1}{2}f''(x)b^2(x,t)(dw(t))^2 + \cdots = \\ & [f'(x)\alpha(x,t) + \tfrac{1}{2}f''(x)b^2(x,t)]dt + f'(x)b(x,t)dw(t) + \cdots \end{aligned} \tag{7.15}$$

The relation

$$\begin{aligned} df[x(t)] = & \left\{ f'(x)\alpha(x,t) + \frac{1}{2}f''(x)b^2(x,t) \right\} dt \\ & + f'(x)b(x,t)dw(t) \end{aligned} \tag{7.16}$$

is Itô's formula and can be generalized to the multi-dimensional case.

7.3 Fokker–Planck's Partial Differential Equation

7.3.1 *The Fokker–Planck Equation*

Fokker–Planck's equation stands for a basic tool for the solution of stochastic differential equation. Instead of attempting an analytical solution of the stochastic differential equation one can attempt to find the equivalent solution of Fokker–Planck equation which stands for a partial differential equation.

It is assumed that $\rho(x,t)$ denotes the probability density function the stochastic variable X to take the value x at time instant t. One can obtain the Fokker–Planck equation after a series of transformations applied to Langevin's equation, thus confirming that the two relations are equivalent. Thus one starts from a Langevin-type equation

$$dx = f(x,t)dt + g(x,t)dt \tag{7.17}$$

Next the transformation $y = h(x)$ is introduced, where h is an arbitrary function that is twice differentiable. It holds that

$$dh(x) = h'(x)f(x,t)dt + h''g^2(x,t)/2dt + h'(x)gdw \tag{7.18}$$

By computing mean values one gets

$$\tfrac{d}{dt}E\{h(x)\} = E\{h'(x)f(x,t)\} + E\{h''(x)g^2(x,t)/2\} \tag{7.19}$$

Assume that $\rho(x,t)$ is the probability density function for the stochastic variable x. It is noted that

$$E\{U(x,t)\} = \int U(x,t)\rho(x,t)dx \tag{7.20}$$

Using Taylor series expansion one gets

$$\frac{d}{dt}\int h(x)\rho(x,t)dx = \int [h'(x)f(x,t) + h''g^2(x,t)/2]\rho(x,t)dx \tag{7.21}$$

which is also written as

$$\frac{d}{dt}\int h(x)\rho(x,t)dx = \int [-(f(x,t)\rho(x,t))' + (g^2(x,t)/2\rho(x,t))'']h(x)dx \tag{7.22}$$

and since the relation described in Eq. (7.22) should hold for all $h(x)$ one gets

$$\frac{\partial \rho}{\partial t} = \frac{\partial}{\partial x}[-f(x,t)\rho(x,t) + \frac{1}{2}\frac{\partial}{\partial x}[g^2(x,t)\rho(x,t)]] \tag{7.23}$$

This is Fokker–Planck equation.

7.3.2 *First Passage Time*

This is a parameter which permits to compute the firing rate of the neurons. In such a case, the aim is to compute the statistical distribution that is followed by the time instant T in which stochastic variable x exits the interval $[a,b]$ (Fig. 7.2).

The probability for x to belong in the interval $[a,b]$ at time instant t is denoted as $G(x,t)$. Thus if t is the time instant at which x exits the interval $[a,b]$, then $G(x,t) = \text{prob}(T \geq t)$.

In the case of neurons, the first passage time corresponds to the time instant where the membrane's potential exceeds a threshold and a spike is generated. It holds that $G(x,t)$ satisfies Fokker–Planck equation, that is

$$\frac{\partial G(x,t)}{\partial t} = f(x,t)\frac{\partial G(x,t)}{\partial x} + \frac{\sigma^2}{2}\frac{\partial^2 G(x,t)}{\partial x^2} \tag{7.24}$$

Regarding boundary conditions it holds that $G(x,0) = 1$ if $a \leq x < b$.

7.3.3 *Meaning of the First Passage Time*

In biological neuron models, such as the Morris–Lecar model, a spike is generated if variable $w(t)$ exceeds a threshold $\bar{w}$. One can write the complete Fokker–Planck equations for a wide domain, compute the outgoing flux (rate in time that the voltage exceeded certain boundaries) and thus compute the firing rate.

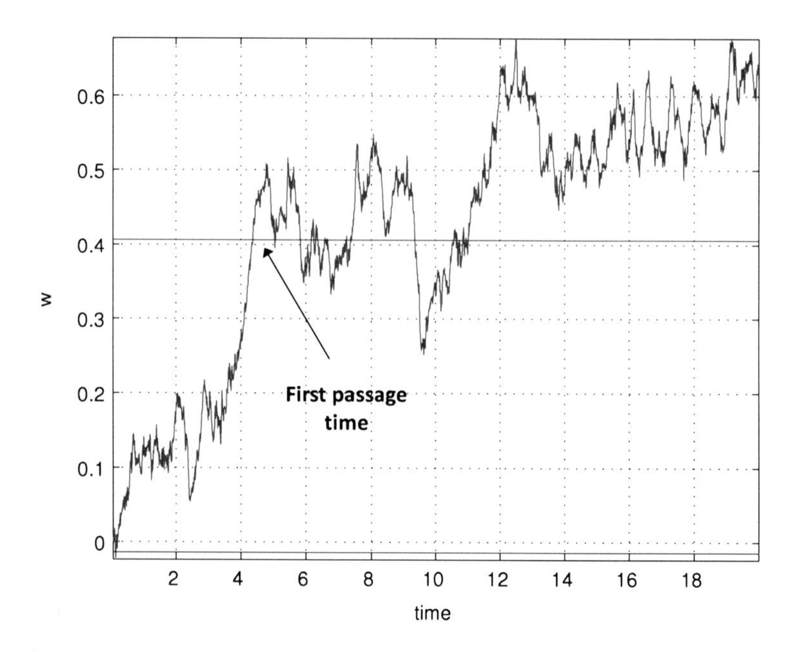

Fig. 7.2 First passage time for a wiener process

For example, in case of a 2D neuron model one has

$$\begin{aligned} dV &= f(V,w)dt + \sigma dw \\ dw &= g(V,w)dt \end{aligned} \tag{7.25}$$

and the associated Fokker–Planck equation is written as

$$\frac{\partial P}{\partial t} = \frac{\sigma^2}{2}\frac{\partial^2 P}{\partial V}[f(V,w)P]_V - [g(V,w)P]_w \tag{7.26}$$

Assume that the stationary solution of the Fokker–Planck equation is computed, denoted as $P(V,w)$. This variable defines the probability distribution for the values of V and w. Then one has firing of a spike if w exceeds threshold $\bar{w}$. The firing rate is computed as

$$F = \int_{V_1}^{V_2} J_w(V,\bar{w})dV \tag{7.27}$$

where $J_w(V,w) = g(V,w)P(V,w)$.

7.4 Stochastic Modelling of Ion Channels

By introducing stochastic dynamics to the neurons' model (i.e, to the Morris–Lecar model), one obtains equations of the Fokker–Planck or Langevin type. Assume that the number of ion channels in the membrane is N, α is the transition rate from open to close, β is the transition rate from closed to open. Moreover it is assumed that the number of open channels in the membrane is o.

Then $N-o$ is the number of closed channels at time instant t, $r_1=\alpha(N-o)$ is the total rate of transitions from open to closed, and $r_2=\beta o$ is the total number of transitions from closed to open. The aggregate rate for all events is $r=r_1+r_2$.

Denoting $X_1 \in (0,1)$ the probability of appearance of an event until time instant t_{new} (under exponential waiting time) one has

$$X_1 = e^{-rt_{new}} \Rightarrow t_{new} = -\frac{1}{r} ln(X_1) \tag{7.28}$$

Regarding the Morris–Lecar neuron model the following may hold: (1) opening of a calcium channel, (2) closing of a calcium channel, (3) opening of a potassium channel, (4) closing of a potassium channel. It has been shown that the rate of opening or closing of the channels is associated with the voltage of the membrane which is described by the relation

$$C_n \frac{dV}{dt} = I - g_l(V-E_l) - (g_k w/N_w)(V_{E_k}) - g_{Ca}(M/N_m)(V-E_{Ca}) \tag{7.29}$$

where W is the total number of open ion channels K^+, M is the total number of total ion channels Ca^{2+}. Equation (7.29) is written also as

$$\frac{dV}{dt} = (V_\infty - V)g \tag{7.30}$$

where V_∞ and g are functions of the parameters W and M.

When introducing stochastic models to describe the change of the membrane's voltage then either a Fokker–Planck or a Langevin equation follows, that is

$$dx = [\alpha(1-x) + \beta x]dt + \sigma_x dW(t) \tag{7.31}$$

where the standard deviation is approximated by $\sigma_x^2 = \frac{\alpha(1-x)+\beta x}{N}$ (standard deviation).

7.5 Fokker–Planck Equation and the Integrate-and-Fire Neuron Model

7.5.1 *The Integrate-and-Fire Neuron Model*

A simple model of the neuron's dynamics (apart from the elaborated model of the FitzHugh–Nagumo, Hodgkin–Huxley, or Morris–Lecar neuron) is that of the integrate and fire neuron, and this is given by

$$dV = (I - aV)dt + \sigma dW_t \tag{7.32}$$

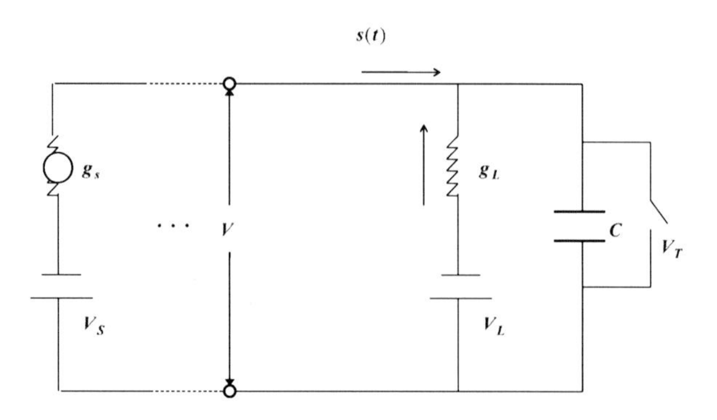

Fig. 7.3 The integrate-and-fire biological neuron model is represented as a resistance–capacitance circuit

which in discrete form is written as

$$V_{n+h} = V_n + h(I_n - aV_n) + \sigma\sqrt{h}\hat{N}(0,1) \tag{7.33}$$

By integrating Langevin's equation in time one obtains

$$x(t) = x(t_0) + \int_{t_0}^{t} a(x(s),s)ds + \int_{t_0}^{t} b(x(s),s)dW(s) \tag{7.34}$$

In the integrate-and-fire model the neuron dynamics is described as an electric circuit, as shown in Fig. 7.3 [45, 75]. This is in accordance with the simplified Hodgkin–Huxley model in which spatial variation of the voltage is not taken into account. The state variable is taken to be the membrane's potential V

$$C\dot{V} = \sum_i I_i(t) \tag{7.35}$$

where C is the neuron membrane capacitance, V is the neuron membrane voltage, and I_i is a current (flow of charge) associated with the membrane. The equivalent circuit representation of the integrate-and-fire neuron is a resistance–capacitance (RC) circuit.

In Fig. 7.4 it is shown how the parts of the biological neuron are modeled by the local RC circuits and how spikes are generated. The neuron's membrane is activated by input current $s(t)$. The membrane's capacitance C_m is in parallel to the membrane's resistance R_m (or equivalently to the membrane's conductance g_m) and is driven by the synaptic current $s(t)$. The synaptic current $s(t)$ is affected by the synapse's resistance R_{syn} and the synapse's capacitance C_{syn}. When the membrane's (capacitor C_m) voltage V exceeds threshold V_T a spike occurs and the voltage is reset to V_R. The integrate-and-fire model of the neuron dynamics can be extended to the case of multiple interconnected neurons, as shown in Fig. 7.5.

Currents $s(t)$ are considered to be due to the inflow and outflow of ions such as K^+, Cl^-, Na^+, or Ca^+. The membrane potential was shown to be associated with the concentration of electric charges. Currents $s(t)$ which appear in the previous

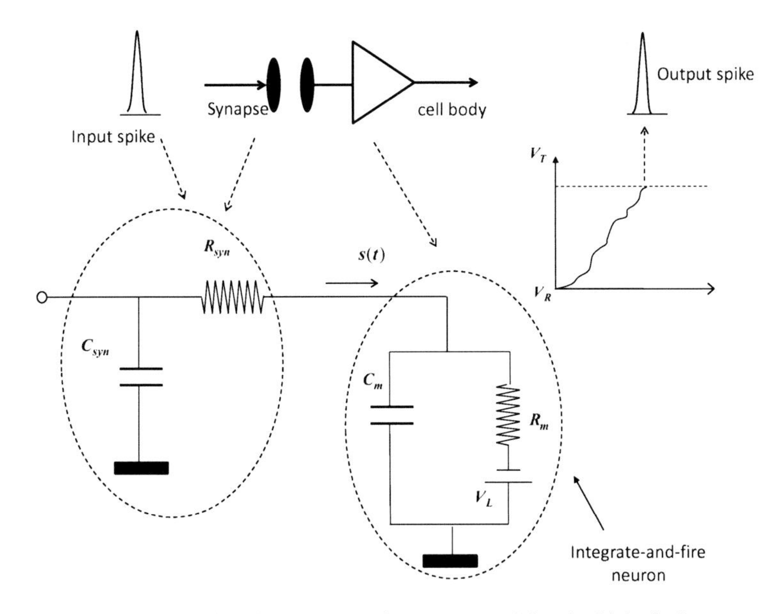

Fig. 7.4 Correspondence of the integrate-and-fire neuron model to the biological neuron

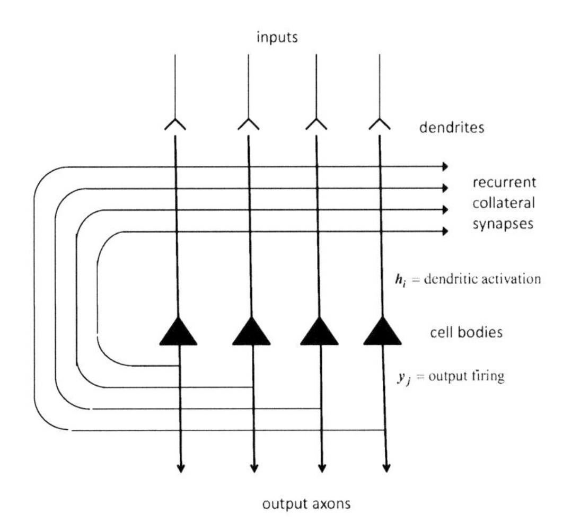

Fig. 7.5 Model of the biological associative memory

neuron model represent a continuous flow of charge. Considering both ionic currents and synaptic currents excitation Eq. (7.35) can be written as

$$C\dot{V} = g_L(V_L - V) + s(t) \tag{7.36}$$

where g_L is the membrane's (ion's channel) conductance, V_L is an equilibrium potential, and $s(t)$ is the synaptic current input. Spiking is modeled as a threshold process, i.e. once membrane potential V exceeds a threshold value V_T, a spike is assumed and the membrane potential is reset to V_R where $V_R \leq V_L \leq V_T$. An even more elaborated integrate-and-fire neuron model has been proposed, which takes the form

$$\dot{V} = \frac{1}{C}[g_L(V_L - V) + s(t)] + \alpha(V_R - V)\beta \tag{7.37}$$

The spiking period T varies according to

$$\begin{aligned} \dot{T} &= 1 - \alpha T H(V), \quad \beta = \exp\left(-\frac{T^2}{2\gamma^2}\right) \\ H(V) &= \begin{cases} 1 & V \geq V_T \\ 0 & V < V_T \end{cases} \end{aligned} \tag{7.38}$$

The integration of Eq. (7.36) provides also the spiking period of the neuron

$$\int_{V_R}^{V_T} \frac{C}{g_L(V_L - V) + s(t)} dV = \int dT = T \tag{7.39}$$

When $s(t)$ is deterministic then the neuron spikes with a specific period. On the other hand, when the input $s(t)$ of the neuron is a stochastic variable then the neuron spikes with a random period. In stochastic neurons there are stochastic flows of charges and stochastic currents $s(t)$ which cause spiking at a random rate. The models of Eqs. (7.35) and (7.37) can be also extended to include various types of ion-channel dynamics, and to model accordingly spike-rate adaptation and synaptic transmission [75].

7.5.2 *Stochastic Integrate-and-Fire Neuron Model and the Fokker–Planck Equation*

The dynamics of the previously described integrate-and-fire stochastic neuron can be associated with Fokker–Planck equation [75]. Now, input $s(t)$ to the neuron's model is a random variable (e.g., there is randomness in the flow of ions and randomness in the associated currents).

As a consequence, the spiking period in the stochastic neuron cannot be analytically defined [see Eq. (7.36)], but is sampled from a probability distribution [88]. Using the stochastic input $s(t)$ in the equations of neuron dynamics, i.e. Eqs.(7.35) and (7.37), the spatio-temporal variations of the neuron's voltage will no longer be associated with deterministic conditions, but will be given by a probability density function $\rho(x,t)$. In [75] it has been proposed to associate ions density function $\rho(x,t)$ with a Fokker–Planck equation (or an advection-diffusion equation) of the form

$$\frac{\partial \rho}{\partial t} = -\frac{\partial u(x)\rho}{\partial x} + \frac{\sigma^2}{2}\frac{\partial^2 \rho}{\partial x^2} \tag{7.40}$$

where $u(x)$ is an external potential function (drift function) and σ^2 is a diffusion constant. This is in accordance with the Fokker–Planck equation and can be used to describe the stochastic neuron dynamics [149, 156].

7.5.3 *Rate of Firing for Neural Models*

Most of the integrate-and-fire neural models activate a reset condition when the voltage of the membrane exceeds a specific threshold denoted as V_{spike}. Since voltage is reset when the threshold value is reached, from that point on the probability to detect this voltage level at the neuron's membrane becomes zero. What can be detected instead is the generated current.

A generic scalar model of the voltage of the neuron's membrane is used first

$$dV = f(V,t)dt + \sigma dw(t) \tag{7.41}$$

When the voltage $V(t)$ reaches the threshold value V_{spike} then it is reset to the initial value V_{reset}. If $f(V,t) = -\alpha V + I$, then one has the leaky integrate-and-fire model of the neuron. If $f(V,t) = \alpha V^2 + I$, then one has the quadratic integrate-and-fire neuron's model.

As explained in the previous subsections, to compute the mean firing rate one should compute the solution of the Fokker–Planck equation in steady state, that is

$$\frac{\partial G(x,t)}{\partial t} = f(x,t)\frac{\partial G(x,t)}{\partial x} + \frac{\sigma^2}{2}\frac{\partial^2 G(x,t)}{\partial x^2} \tag{7.42}$$

7.6 Stochasticity in Neural Dynamics and Relation to Quantum Mechanics

7.6.1 *Basics of Quantum Mechanics*

In quantum mechanics the state of an isolated quantum system Q is represented by a vector $|\psi(t)>$ in a Hilbert space. This vector satisfies Schrödinger's diffusion equation [34, 133]

$$i\hbar\frac{d}{dt}|\psi(t)>= H\psi(t) \tag{7.43}$$

where H is the Hamiltonian operator that gives the total energy of a particle (potential plus kinetic energy) $H = \frac{p^2}{2m} + V(x)$. Equation (7.43) denotes that the particle with momentum p and energy E is diffused in the wave with a probability density proportional to $|\psi(x,t)|^2$. Equivalently, it can be stated that $|\psi(x,t)|^2$ denotes the probability for the particle to be at position x at time t. The external potential V is defined as $V = -\frac{p^2}{2m} + E$ where E is the eigenvalue (discrete energy level) of the Hamiltonian H. For $V = 0$ or constant, the solution of Eq. (7.43) is a superposition of plane waves of the form

$$|\psi(x,t)> = e^{\frac{i(px-Et)}{\hbar}} \tag{7.44}$$

where i is the imaginary unit, x is the position of the particle, and $\hbar$ is Planck's constant. These results can be applied to the case of an harmonic potential if only the basic mode was taken into account.

The probability to find the particle between x and $x + dx$ at the time instant t is given by $P(x)dx = |\psi(x,t)|^2 dx$. The total probability should equal unity, i.e. $\int_{-\infty}^{\infty} |\psi(x,t)|^2 dx = 1$. The average position x of the particle is given by

$$< x >= \int_{-\infty}^{\infty} P(x)xdx = \int_{-\infty}^{\infty} (\psi^* x\psi)dx \tag{7.45}$$

where ψ^* is the conjugate value of ψ. Now, the wave function $\psi(x,t)$ can be analyzed in a set of orthonormal eigenfunctions in a Hilbert space

$$\psi(x,t) = \sum_{n=1}^{\infty} c_n \psi_n \tag{7.46}$$

The coefficients c_n are an indication of the probability to describe the particle's position x at time t by the eigenfunction ψ_n and thanks to the orthonormality of the ψ_n's, the c_n's are given by $c_n = \int_{-\infty}^{\infty} \psi_n{}^* \psi dx$. Moreover, the eigenvalues and eigenvectors of the quantum operator of position x can be defined as $x\psi_n = a_n \psi_n$, where ψ_n is the eigenvector and a_n is the associated eigenvalue. Using Eqs. (7.45) and (7.46) the average position of the particle is found to be

$$< x >= \sum_{n=1}^{\infty} ||c_n||^2 a_n \tag{7.47}$$

with $||c_n||^2$ denoting the probability that the particle's position be described by the eigenfunction ψ_n. When the position x is described by ψ_n the only measurement of x that can be taken is the associated eigenvalue a_n. This is the so-called filtering problem, i.e. when trying to measure a system that is initially in the state $\psi = \sum_{n=1}^{\infty} c_n \psi_n$ the measurement effect on state ψ is to change it to an eigenfunction ψ_n with measurable value only the associated eigenvalue a_n. The eigenvalue a_n is chosen with probability $P \propto ||c_n||^2$.

7.6.2 Schrödinger's Equation with Non-zero Potential and Its Equivalence to Diffusion with Drift

In Sect. 7.2 the equivalence between Wiener process and the diffusion process was demonstrated. Next, the direct relation between diffusion and quantum mechanics will be shown [56]. The basic equation of quantum mechanics is Schrödinger's equation, i.e.

$$i\frac{\partial\psi}{\partial t} = H\psi(x,t) \tag{7.48}$$

where $|\psi(x,t)|^2$ is the probability density function of finding the particle at position x at time instant t, and H is the system's Hamiltonian, i.e. the sum of its kinetic and potential energy, which is given by $H = p^2/2m + V$, with p being the momentum of the particle, m the mass and V an external potential. It holds that

$$\frac{p^2}{2m} = -\frac{1}{2}\frac{\hbar}{m}\frac{\partial^2}{\partial x^2} \tag{7.49}$$

thus the Hamiltonian can be also written as

$$H = -\frac{1}{2}\frac{\hbar}{m}\frac{\partial^2}{\partial x^2} + V. \tag{7.50}$$

The solution of Eq. (13.1) is given by [34]

$$\psi(x,t) = e^{-iHt}\psi(x,0) \tag{7.51}$$

A simple way to transform Schrödinger's equation into a diffusion equation is to substitute variable it with t. This transformation enables the passage from imaginary time to real time. However in the domain of non-relativistic quantum mechanics there is a closer connection between diffusion theory and quantum theory. In stochastic mechanics, the real time of quantum mechanics is also the real time of diffusion and in fact quantum mechanics is formulated as conservative diffusion [56]. This change of variable results in the diffusion equation

$$\frac{\partial\rho}{\partial t} = \left[\frac{1}{2}\frac{\sigma^2}{\partial^2}\partial x^2 - V(x)\right]\rho \tag{7.52}$$

Equation (7.52) can be also written as $\frac{\partial\rho}{\partial t} = -H\rho$, where H is the associated Hamiltonian and the solution is of the form $\rho(x,t) = e^{-tH}\rho(x)$, and variable σ^2 is a diffusion constant. The probability density function ρ satisfies also the *Fokker–Planck* partial differential equation

$$\frac{\partial\rho}{\partial t} = \left[\frac{1}{2}\sigma^2\frac{\partial^2}{\partial x^2} - \frac{\partial}{\partial x}u(x)\right]\rho \tag{7.53}$$

where $u(x)$ is the *drift function*, i.e. a function related to the derivative of the external potential V. In that case Eq. (7.53) can be rewritten as

$$\frac{\partial \rho}{\partial t} = -\hat{H}\rho \tag{7.54}$$

where $\hat{H} = -\left[\frac{1}{2}\sigma^2\frac{\partial^2}{\partial x^2} - \frac{\partial}{\partial x}u(x)\right]$, while the probability density function $\rho(x,t)$ is found to be

$$\rho(x,t) = e^{-\hat{H}}\rho(x) \tag{7.55}$$

It has to be noted that the Fokker–Planck equation is equivalent to Langevin's equation and can be also used for the calculation of the mean position of the diffused particle, as well as for the calculation of its variance [66].

Now, the solution $\rho(x)$ is a wave-function for which holds $\rho(x) = |\psi(x)|^2$ with $\psi(x) = \sum_{i=1}^{N} c_k\psi_k(x)$, where $\psi_k(x)$ are the associated eigenfunctions [34, 127]. It can be assumed that $\rho_0(x) = |\psi_0^2(x)|$, i.e. the p.d.f. includes only the basic mode, while higher order modes are truncated, and the drift function $u(x)$ of Eq. (7.53) is taken to be [56]

$$u(x) = \frac{1}{2}\sigma^2\frac{1}{\rho_0(x)}\frac{\partial\rho_0(x)}{\partial x} \tag{7.56}$$

Thus it is considered that the initial probability density function is $\rho(x) = \rho_0(x)$, which is independent of time, thus from Eq. (7.54) one has $\hat{H}\rho_0 = 0$, which means that the p.d.f. remains independent of time and the examined diffusion process is a stationary one, i.e. $\rho(x,t) = \rho_0(x)\ \forall t$. A form of the probability density function for the stationary diffusion is that of shifted, partially overlapping Gaussians, which is depicted in Fig. 7.1b. In place of Gaussian p.d.f., symmetric triangular possibility distributions have been also proposed [163]. The equation that describes the shifted Gaussians is (Fig. 7.1b)

$$\rho_0(x) = \frac{1}{2}C^2 e^{-\frac{\omega}{\sigma^2}(x-a)^2} + \frac{1}{2}C^2 e^{-\frac{\omega}{\sigma^2}(x+a)^2} \tag{7.57}$$

7.6.3 *Study of the QHO Model Through the Ornstein–Uhlenbeck Diffusion*

The Ornstein–Uhlenbeck diffusion is a model of the Brownian motion [17]. The particle tries to return to position $x(0)$ under the influence of a linear force vector, i.e. there is a spring force applied to the particle as a result of the potential $V(x)$. The corresponding phenomenon in quantum mechanics is that of the QHO [34, 66]. In the QHO the motion of the particle is affected by the parabolic (harmonic) potential

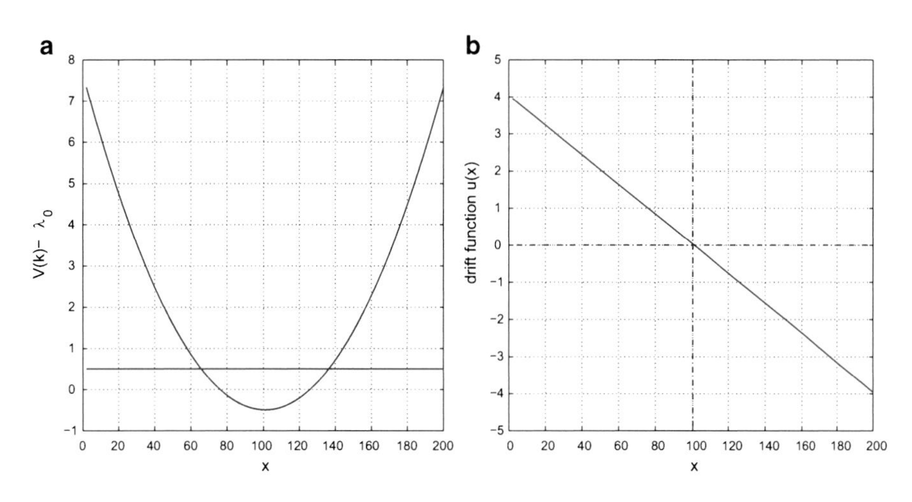

Fig. 7.6 (**a**) Diagram of $V(x) - \lambda_0$ where $V(x)$ is the harmonic potential of the QHO and λ_0 is the associated eigenvalue (**b**) Linear drift force applied to the diffused particle as a result of the harmonic potential $V(x)$

$$V(x) = \frac{1}{2}\frac{\omega^2}{\sigma^2}x^2 \tag{7.58}$$

It is known that the ground mode of the QHO of Eq. (7.48) is a Gaussian function [56, 164], i.e.

$$\psi_0(x) = Ce^{\frac{-\omega x^2}{2\sigma^2}} \tag{7.59}$$

while it can be proved easily that the associated eigenvalue is $\lambda_0 = \frac{1}{2}\omega$. A diagram of $V(x) - \lambda_0$ is given in Fig. 7.6a.

For the diffusion constant σ holds $\sigma^2 = \frac{\hbar}{m}$ where $\hbar$ is Planck's constant and finally gives $V(x) = \frac{1}{2}m\omega^2x^2$. Assuming the stationary p.d.f. of Eq. (7.57)

$$\rho(x) = \psi_0(x)^2 = C^2e^{\frac{-\omega x^2}{2\sigma^2}} \tag{7.60}$$

the force applied to the particle due to the harmonic potential $V(x)$ is given by Eq. (7.56), and is found to be

$$u(x) = \sigma^2\frac{1}{\psi_0(x)}\frac{\partial\psi_0(x)}{\partial x} \Rightarrow u(x) = -\omega x \tag{7.61}$$

which means that the drift is a spring force applied to the particle and which aims at leading it to an equilibrium position. The drift force is depicted in Fig. 7.6b.

7.6.4 *Particle's Motion Is a Generalization of Gradient Algorithms*

As analyzed, the Wiener process describes the Brownian motion of a particle. In this section, this motion is also stated in terms of equations of Langevin's equation. The stochastic differential equation for the position of the particle is [56]:

$$dx(t) = u(x(t))dt + dw(t) \tag{7.62}$$

where $u(x(t))$ is the so-called *drift function*, and is usually given in the form of a spring force, i.e. $u(x) = -kx$ which tries to bring the particle to the equilibrium $x = 0$ and is the result of a parabolic potential applied to the particle, i.e. $V(x) = kx^2$. The term $w(t)$ denotes a random force (due to interaction with other particles, e.g. collision) and follows a Wiener walk. For each continuous random path $w(t)$, a continuous random path $x(t)$ is also generated, which can be written in the form

$$x(t) = x(0) + \int_0^t u(x(s))ds + w(t) \tag{7.63}$$

The integration of Langevin's equation and certain assumptions about the noise $w(t)$, for instance white noise, dichotomic noise (also known as Ornstein–Uhlenbeck noise), etc., enable the calculation of the mean position of the particle $E\{x\}$ and also to find its variance $E\{x - E\{x\}\}^2$ [66].

Knowing that the QHO model imposes to the particle the spring force of Eq. (7.61), the kinematic model of the diffusing particle becomes

$$dx(t) = -\omega x(t)dt + dw(t) \tag{7.64}$$

with initial condition $x(0) = 0$. The first term in the right part is the drift, i.e. the spring forces that makes the particle return to $x(0)$, while the second term is the random diffusion term (noise).

The extension of Langevin's equation gives a model of an harmonic oscillator, driven by noise. Apart from the spring force, a friction force that depends on the friction coefficient γ and on the velocity of the particle is considered. The generalized model of motion of the particle can then be also written as [13, 66]:

$$\frac{d^2x}{dt^2} + 2\gamma\frac{dx}{dt} + \omega^2 x = \xi(t) \tag{7.65}$$

Thus an equation close to electrical or mechanical oscillators is obtained [163, 195]. Equation (7.64) can be also written as

$$\frac{dx(t)}{dt} = h(x(t)) + \eta(t) \Rightarrow dx(t) = h(x(t))dt + w(t) \tag{7.66}$$

where $h(x(t)) = \alpha\frac{\partial V(x)}{\partial t}$, with α being a learning gain, $V(x)$ being the harmonic potential, and $\eta(t)$ being a noise function. Equation (7.64) is a generalization of gradient algorithms based on the ordinary differential equation (ODE) concept, where the gradient algorithms are described as trajectories towards the equilibrium of an ODE [22, 52]. Indeed, conventional gradient algorithms with diminishing step are written as

$$dx(t) = h(x(t))dt \tag{7.67}$$

The comparison of Eqs. (7.64) and (7.67) verifies the previous argument. The update of the neural weights that follow the model of the QHO given by Eq. (7.64). The force that drives $x(t)$ to the equilibrium is the derivative of the harmonic potential, and there is also an external noisy force $w(t)$ which is the result of collisions or repulsive forces due to interaction with neighboring particles. Similarly, in the update of neural weights with a conventional gradient algorithm, the weight is driven towards an equilibrium under the effect of a potential's gradient (where the potential is usually taken to be a quadratic error cost function).

7.7 Conclusions

In this chapter, neural structures with weights which are stochastic variables and which follow the model of the QHO have been studied. The implications of this assumptions for neural computation were analyzed. The neural weights were taken to correspond to diffusing particles, which interact to each other as the theory of Wiener process (Brownian motion) predicts. The values of the weights are the positions of the particles and the probability density function $|\psi(x,t)|^2$ that describes their position is derived by Schrödinger's equation. Therefore the dynamics of the neural network is given by the time-dependent solution of Schrödinger's equation under a parabolic (harmonic) potential.

However, the time-dependent solution of Schrödinger's equation is difficult to be computed, either analytically or numerically. Moreover, this solution contains terms which are difficult to interpret, such as the complex number probability amplitudes associated with the modes $\psi_k(x)$ of the wave-function $\psi(x)$ (solution of Schrödinger's equation), or the path integrals that describe the particles' motion. Thus in place of Schrödinger's equation the solution of a stationary diffusion equation (Ornstein–Uhlenbeck diffusion) with drift was studied, assuming probability density function that depends on the QHO's ground state $\rho_0(x) = |\psi_0(x)|^2$.

It was shown that in neural structures with weights that follow the QHO model, the weights update is described by Langevin's stochastic differential equation. It was proved that conventional gradient algorithms are a subcase of Langevin's equation. It can be also stated that neural networks with crisp numerical weights and conventional gradient learning algorithms are a subset of neural structure with weights based on the QHO model. In that way the complementarity between classical and quantum physics was also validated in the field of neural computation.

Chapter 8
Synchronization of Stochastic Neural Oscillators Using Lyapunov Methods

Abstract A neural network with weights described by the position of interacting Brownian particles is considered. Each weight is taken to correspond to a Gaussian particle. Neural learning aims at leading a set of M weights (Brownian particles) with different initial values on the 2-D phase plane, to the desirable final position. A Lyapunov function describes the evolution of the phase diagram towards the equilibrium Convergence to the goal state is assured for each particle through the negative definiteness of the associated Lyapunov function. The update of the weight (trajectory in the phase plane) is affected by (1) a drift force due to the harmonic potential, and (2) the interaction with neighboring weights (particles). It is finally shown that the mean of the particles will converge to the equilibrium while using LaSalle's theorem it is shown that the individual particles will remain within a small area encircling the equilibrium.

8.1 Representation of the Neurons' Dynamics as Brownian Motion

It will be shown that the model of the interacting coupled neurons becomes equivalent to the model of interacting Brownian particles, and that each weight is associated with a Wiener process. In such a case, the neural network proposed by Hopfield can be described by a set of ordinary differential equations of the form

$$C_i \dot{x}_i(t) = -\frac{1}{R_i} x_i(t) + \sum_{j=1}^{n} T_{ij} g_j(x_j(t)) + I_i \quad 1 \leq i \leq n \tag{8.1}$$

Variable $x_i(t)$ represents the voltage at the membrane of the i-th neuron and I_i is the external current input to the i-th neuron. Each neuron is characterized by an input capacitance C_i and a transfer function $g_i(u)$ which represents connection to neighboring neurons. The connection matrix element T_{ij} has a value $1/R_{ij}$ when the output of the j-th neuron is connected to the input of the i-th neuron through a resistance R_{ij}, and a value $-1/R_{ij}$ when the inverting output of the j-th neuron

G.G. Rigatos, *Advanced Models of Neural Networks*,
DOI 10.1007/978-3-662-43764-3_8, © Springer-Verlag Berlin Heidelberg 2015

is connected to the input of the i-th neuron through a resistance R_{ij}. The nonlinear function that describes connection between neurons $g_i(u)$ is a sigmoidal one. By defining the parameters

$$b_i = \frac{1}{C_i R_i} \quad \alpha_{ij} = \frac{T_{ij}}{C_i} \quad c_i = \frac{I_i}{C_i} \tag{8.2}$$

Thus, Eq. (8.1) can be rewritten as

$$\dot{x}_i(t) = -b_i x_i(t) + \sum_{j=1}^{n} \alpha_{ij} g_j(x_j(t)) + c_i \quad 1 \leq i \leq n \tag{8.3}$$

which can be also written in matrix form as

$$\dot{x}(t) = -Bx(t) + Ag(x(t)) + C \tag{8.4}$$

where

$$\begin{array}{ccc} x(t) = (x_1(t), \cdots, x_n(t))^T & B = \text{diag}(b_1, \cdots, b_n) & A = (a_{ij})_{n \times n} \\ C = (c_1, \cdots, c_n)^T & g(x) = (g_1(x_1), \cdots, g_n(x_n))^T & \end{array} \tag{8.5}$$

One can consider the case

$$b_i = \sum_{j=1}^{n} |\alpha_{ij}| > 0, \; c_i \geq 0 \; 1 \leq i \leq n \tag{8.6}$$

Moreover, by assuming a symmetric network structure it holds

$$\alpha_{ij} = \alpha_{ji} \; 1 \leq i \quad j \leq n \tag{8.7}$$

which means that A is a symmetric matrix. It is also known that neural networks are subjected to noise. For example, if every external input I_i is perturbed in the way $I_i \rightarrow I_i + \epsilon_i w_i(t)$, where $w_i(t)$ is white noise, then the stochastically perturbed neural network is described by a set of stochastic differential equations (Wiener processes) of the form

$$dx(t) = [-Bx(t) + Ag(x(t)) + C]dt + \sigma_1 dw_1(t) \tag{8.8}$$

where $\sigma_1 = (\epsilon_1/C_1, \cdots, \epsilon_n/C_n)^T$. Moreover, if the connection matrix element T_{ij} is perturbed in the way $T_{ij} \rightarrow T_{ij} + \epsilon_{ij} \dot{w}_2(t)$, where $w_2(t)$ is another white noise independent of $w_1(t)$, then the stochastically perturbed neural network can be described as

$$dx(t) = [-Bx(t) + Ag(x(t)) + C]dt + \sigma_1 dw_1(t) + \sigma_2 g(x(t)) dw_2(t) \tag{8.9}$$

where $\sigma_2 = (\epsilon_{ij}/C_i)_{n \times n}$. In general one can describe the stochastic neural network by a set of stochastic differential equations of the form

$$dx(t) = [-Bx(t) + Ag(x(t)) + C]dt + \sigma dw(t) \tag{8.10}$$

In the above relation $w(t) = [w_1(t), \cdots, w_m(t)]^T$ is an m-dimensional Brownian motion defined on a probability space (Ω, F, P) with a natural filtration $\{F_{t \geq 0}\}$, i.e. $F_t = \sigma\{w(s) : 0 \leq s \leq t\}$ and $\sigma : R^n \rightarrow R^m$ i.e. $\sigma(x) = (\sigma_{ij}(x))_{n \times m}$ which is called the noise intensity matrix.

8.2 Interacting Diffusing Particles as a Model of Neural Networks

8.2.1 Weights' Equivalence to Brownian Particles

An equivalent concept of a neural network with weights described by interacting Brownian particles can be also found in [83, 87]. Here, a neural structure of M neurons is considered, e.g. an associative memory (Fig. 8.1). For the weight w, the error vector $[e, \dot{e}]$ is defined, where e denotes the distance of the weight from the desirable value w^* and $\dot{e}$ the rate of change of e. Thus, each weight can be mapped to the 2-D plane using the notation $[x, y] = [e, \dot{e}]$. Moreover, each weight is taken to correspond to a Brownian particle. The particles are considered to have mechanical properties, that is to be subjected to acceleration due to forces. The objective of learning is to lead a set of M weights (Brownian particles) with different initial values on the 2-D plane, to the desirable final position $[0, 0]$.

Each particle (weight) is affected by the rest of the $M - 1$ particles. The cost function that describes the motion of the i-th particle towards the equilibrium is denoted as $V(x^i) : R^n \rightarrow R$, with $V(x^i) = \frac{1}{2}{e^i}^2$ [147, 150–152]. Convergence to the goal state should be succeeded for each particle through the negative definiteness of the associated Lyapunov function, i.e. it should hold $\dot{V}^i(x^i) = \dot{e}^i(t)^T e^i(t) < 0$ [92].

As already mentioned, in the quantum harmonic oscillator (QHO) model of the neural weights the update of the weight (motion of the particle) is affected by (1) the drift force due to the harmonic potential, and (2) the interaction with neighboring weights (particles). The interaction between the i-th and the j-th particle is [64]:

$$g(x^i - x^j) = -(x^i - x^j)[g_a(||x^i - x^j||) - g_r(||x^i - x^j||)] \tag{8.11}$$

where $g_a()$ denotes the attraction term and is dominant for large values of $||x^i - x^j||$, while $g_r()$ denotes the repulsion term and is dominant for small values of $||x^i - x^j||$. Function $g_a()$ can be associated with an attraction potential, i.e.

$$\nabla_{x_i} V_a(||x^i - x^j||) = (x^i - x^j)g_a(||x^i - x^j||) \tag{8.12}$$

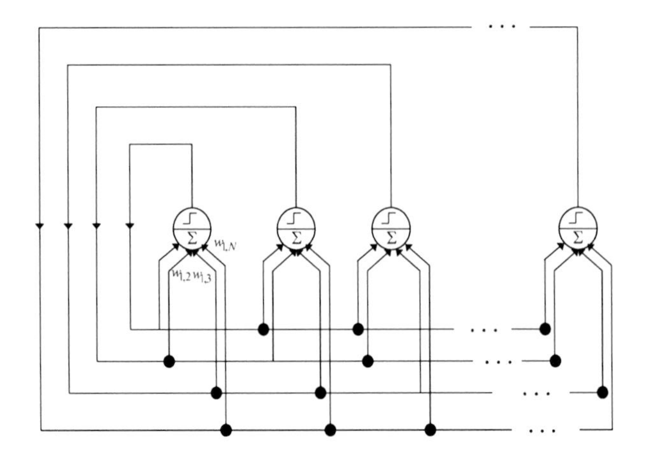

Fig. 8.1 The model of a Hopfield associative memory with feedback weights

Function $g_r()$ can be associated with a repulsion potential, i.e.

$$\nabla_{x_i} V_r(||x^i - x^j||) = (x^i - x^j)g_r(||x^i - x^j||) \tag{8.13}$$

A suitable function $g()$ that describes the interaction between the weights is given by [150]

$$g(x^i - x^j) = -(x^i - x^j)(a - be^{\frac{||x^i - x^j||^2}{\sigma^2}}) \tag{8.14}$$

where the parameters a, b, and c may also be subjected to stochastic variations and noise terms. It holds that $g_a(x^i - x^j) = -a$, i.e. attraction has a linear behavior (spring-mass system) $||x^i - x^j||g_a(x^i - x^j)$. Moreover,

$$g_r(x^i - x^j) = be^{\frac{-||x^i - x^j||^2}{\sigma^2}} \tag{8.15}$$

which means that $g_r(x^i - x^j)||x^i - x^j|| \leq b$ is bounded. For the i-th particle with unit mass $m^i = 1$ holds

$$\dot{x}^i = v^i, \;\; m^i \dot{v}^i = U^i \tag{8.16}$$

where the aggregate force is $U^i = f^i + F^i$. The term $f^i = -K_v v^i$ denotes friction, while the term F^i is the propulsion. Assuming zero acceleration $\dot{v}^i = 0$ one gets $F^i = K_v v^i$, which for $K_v = 1$ and $m^i = 1$ gives $F^i = v^i$. Thus an approximate kinematic model is

$$\dot{x}^i = F^i \tag{8.17}$$

The propulsion F^i is equal to the derivative of the total potential of each particle (drift term), i.e.

$$F^i = -\nabla_{x^i}\{V^i(x^i) + \tfrac{1}{2}\textstyle\sum_{i=1}^M \sum_{j=1,j\neq i}^M [V_a(||x^i - x^j|| + V_r(||x^i - x^j||)]\} \Rightarrow$$

$$F^i = -\nabla_{x^i}\{V^i(x^i)\} + \textstyle\sum_{j=1,j\neq i}^M [\nabla_{x^i} V_a(||x^i - x^j||) - \nabla_{x^i} V_r(||x^i - x^j||)] \Rightarrow$$

$$F^i = -\nabla_{x^i}\{V^i(x^i)\} + \textstyle\sum_{j=1,j\neq i}^M [-(x^i - x^j)g_a(||x^i - x^j||) - (x^i - x^j)g_r(||x^i - x^j||)] \Rightarrow$$

$$F^i = -\nabla_{x^i}\{V^i(x^i)\} - \textstyle\sum_{j=1,j\neq i}^M g(x^i - x^j)$$

Taking also into account that the force F^i which excites the particles' motion (weights variation in time) may be subjected to stochastic perturbations and noise, that is

$$F^i = -\nabla_{x^i}\{\nabla(x^i) - \textstyle\sum_{j=1,j\neq i}^M g(x^i - x^j) + \eta^i\} \tag{8.18}$$

one has finally that the particles' motion is a stochastic process. Next, substituting in Eq. (8.17) one gets Eq. (8.31), i.e.

$$\begin{array}{c} x^i(t+1) = x^i(t) + \gamma^i(t)[-\nabla_{x^i} V^i(x^i) + e^i(t+1)] - \sum_{j=1,j\neq i}^M g(x^i - x^j) \\ i = 1, 2, \cdots, M, \text{ with } \gamma^i(t) = 1, \end{array} \tag{8.19}$$

which verifies that the kinematic model of a multi-particle system is equivalent to a distributed gradient search algorithm.

8.2.2 *Stability Analysis for a Neural Model with Brownian Weights*

The stability of the system consisting of Brownian particles is determined by the behavior of its center (mean position of the particles x^i) and of the position of each particle with respect to this center. The center of the multi-particle system is given by

$$\begin{aligned} \bar{x} = E(x^i) = \tfrac{1}{M}\textstyle\sum_{i=1}^M x^i \Rightarrow \dot{\bar{x}} = \tfrac{1}{M}\sum_{i=1}^M \dot{x}^i \Rightarrow \dot{\bar{x}} \\ = \tfrac{1}{M}\textstyle\sum_{i=1}^M \left[-\nabla_{x^i} V^i(x^i) - \sum_{j=1,j\neq i}^M (g(x^i - x^j))\right] \end{aligned} \tag{8.20}$$

From Eq. (8.14) it can be seen that $g(x^i - x^j) = -g(x^j - x^i)$, i.e. $g()$ is an odd function. Therefore, it holds that

$$\frac{1}{M}\left(\sum_{j=1,j\neq i}^{M} g(x^i - x^j)\right) = 0 \tag{8.21}$$

$$\dot{\bar{x}} = \frac{1}{M}\sum_{i=1}^{M}[-\nabla_{x^i} V^i(x^i)] \tag{8.22}$$

Denoting the goal position by x^* , and the distance between the i-th particle and the mean position of the multi-particle system by $e^i(t) = x^i(t) - \bar{x}$ the objective of the learning algorithm for can be summarized as follows: (i) $\lim_{t\to\infty}\bar{x} = x^*$, i.e. the center of the multi-particle system converges to the goal position, (ii) $\lim_{t\to\infty}x^i = \bar{x}$, i.e. the i-th particle converges to the center of the multi-particle system, (iii) $\lim_{t\to\infty}\dot{\bar{x}} = 0$, i.e. the center of the multi-particle system stabilizes at the goal position. If conditions (i) and (ii) hold then, $\lim_{t\to\infty}x^i = x^*$. Furthermore, if condition (iii) also holds, then all particle will stabilize close to the goal position [64, 108, 150, 152].

To prove the stability of the multi-particle system the following simple Lyapunov function is considered for each particle:

$$V_i = \frac{1}{2}{e^i}^T e^i \Rightarrow V_i = \frac{1}{2}||e_i||^2 \tag{8.23}$$

Thus, one gets

$$\begin{aligned}
\dot{V}^i &= {e^i}^T \dot{e}^i \Rightarrow \dot{V}^i = (\dot{x}^i - \dot{\bar{x}})e^i \Rightarrow \\
\dot{V}^i &= \left[-\nabla_{x^i} V^i(x^i) - \textstyle\sum_{j=1,j\neq i}^{M} g(x^i - x^j) + \frac{1}{M}\sum_{j=1}^{M}\nabla_{x^j} V^j(x^j)\right] e^i.
\end{aligned}$$

Substituting $g(x^i - x^j)$ from Eq. (8.14) yields

$$\begin{aligned}
\dot{V}_i = &\left[-\nabla_{x^i} V^i(x^i) - \sum_{j=1,j\neq i}^{M} (x^i - x^j)a \right. \\
&\left. + \sum_{j=1,j\neq i}^{M} (x^i - x^j) g_r(||x^i - x^j||) + \frac{1}{M}\sum_{j=1}^{M}\nabla_{x^j} V^j(x^j)\right] e^i
\end{aligned}$$

which gives

$$\dot{V}_i = -a\left[\sum_{j=1,j\neq i}^{M}(x^i - x^j)\right]e^i$$
$$+\sum_{j=1,j\neq i}^{M} g_r(||x^i - x^j||)(x^i - x^j)^T e^i - \left[\nabla_{x^i}V^i(x^i) - \frac{1}{M}\sum_{j=1}^{M}\nabla_{x^j}V^j(x^j)\right]^T e^i.$$

It holds that

$$\sum_{j=1}^{M}(x^i - x^j) = Mx^i - M\tfrac{1}{M}\sum_{j=1}^{M}x^j = Mx^i - M\bar{x} = M(x^i - \bar{x}) = Me^i$$

therefore

$$\dot{V}_i = -aM||e^i||^2 + \sum_{j=1,j\neq i}^{M} g_r(||x^i - x^j||)(x^i - x^j)^T e^i$$
$$-\left[\nabla_{x^i}V^i(x^i) - \frac{1}{M}\sum_{j=1}^{M}\nabla_{x^j}V^j(x^j)\right]^T e^i \tag{8.24}$$

It assumed that for all x^i there is a constant $\bar{\sigma}$ such that

$$||\nabla_{x^i}V^i(x^i)|| \leq \bar{\sigma} \tag{8.25}$$

Equation (8.25) is reasonable since for a particle moving on a 2-D plane, the gradient of the cost function $\nabla_{x^i}V^i(x^i)$ is expected to be bounded. Moreover it is known that the following inequality holds:

$$\sum_{j=1,j\neq i}^{M} g_r(x^i - x^j)^T e^i \leq \sum_{j=1,j\neq i}^{M} be^i \leq \sum_{j=1,j\neq i}^{M} b||e^i||.$$

Thus the application of Eq. (8.24) gives:

$$\dot{V}^i \leq aM||e^i||^2 + \sum_{j=1,j\neq i}^{M} g_r(||x^i - x^j||)||x^i - x^j|| \cdot ||e^i||$$
$$+||\nabla_{x^i}V^i(x^i) - \tfrac{1}{M}\sum_{j=1}^{M}\nabla_{x^j}V^j(x^j)||||e^i||$$
$$\Rightarrow \dot{V}^i \leq aM||e^i||^2 + b(M-1)||e^i|| + 2\bar{\sigma}||e^i||$$

where it has been taken into account that

$$\sum_{j=1,j\neq i}^{M} g_r(||x^i - x^j||)^T||e^i|| \leq \sum_{j=1,j\neq i}^{M} b||e^i|| = b(M-1)||e^i||,$$

and from Eq. (8.25)

$$||\nabla_{x^i}V^i(x^i) - \tfrac{1}{M}\sum_{j=1}^{M}\nabla_{x^i}V^j(x^j)|| \leq ||\nabla_{x^i}V^i(x^i)||$$
$$+\tfrac{1}{M}||\sum_{j=1}^{M}\nabla_{x^i}V^j(x^j)|| \leq \bar{\sigma} + \tfrac{1}{M}M\bar{\sigma} \leq 2\bar{\sigma}.$$

Thus, one gets

$$\dot{V}^i \leq aM||e^i||\cdot\left[||e^i|| - \tfrac{b(M-1)}{aM} - 2\tfrac{\bar{\sigma}}{aM}\right] \tag{8.26}$$

The following bound ϵ is defined:

$$\epsilon = \frac{b(M-1)}{aM} + \frac{2\bar{\sigma}}{aM} = \frac{1}{aM}(b(M-1) + 2\bar{\sigma}) \tag{8.27}$$

Thus, when $||e^i|| > \epsilon$, $\dot{V}_i$ will become negative and consequently the error $e^i = x^i - \bar{x}$ will decrease. Therefore the error e^i will remain in an area of radius ϵ, i.e. the position x^i of the i-th particle will stay in the cycle with center $\bar{x}$ and radius ϵ.

8.2.2.1 Stability in the Case of a Quadratic Cost Function

The case of a convex quadratic cost function is examined, for instance

$$V^i(x^i) = \frac{A}{2}||x^i - x^*||^2 = \frac{A}{2}(x^i - x^*)^T(x^i - x^*) \tag{8.28}$$

where $x^* = 0$ and $\dot{x}^* = 0$ defines a minimum point (attractor) for which holds $V^i(x^i = x^*) = 0$. The weights error is expected to converge to x^*. The stochastic weights (particles) will follow different trajectories on the 2-D plane and will end at the attractor.

Using Eq.(8.28) yields $\nabla_{x^i} V^i(x^i) = A(x^i - x^*)$. Moreover, the assumption $\nabla_{x^i} V^i(x^i) \leq \bar{\sigma}$ can be used, since the gradient of the cost function remains bounded. The particles will gather round $\bar{x}$ and will stay in a radius ϵ given by Eq. (8.27). The motion of the mean position $\bar{x}$ of the particles is

$$\dot{\bar{x}} = -\tfrac{1}{M}\textstyle\sum_{i=1}^{M}\nabla_{x^i} V^i(x^i) \Rightarrow \dot{\bar{x}} = -\tfrac{A}{M}(x^i - x^*) \Rightarrow$$

$$\dot{\bar{x}} - \dot{x}^* = -\tfrac{A}{M}x^i + \tfrac{A}{M}x^* \Rightarrow \dot{\bar{x}} - \dot{x}^* = -A(\bar{x} - x^*).$$

The variable $e_\sigma = \bar{x} - x^*$ is defined, and consequently

$$\dot{e}_\sigma = -Ae_\sigma \Rightarrow \epsilon_\sigma(t) = c_1 e^{-At} + c_2, \text{ with } c_1 + c_2 = e_\sigma(0) \tag{8.29}$$

Equation (8.29) is an homogeneous differential equation, which for $A > 0$ results into

$$\lim_{t\to\infty} e_\sigma(t) = 0, \text{ thus } \lim_{t\to\infty}\bar{x}(t) = x^*$$

It is left to make more precise the position to which each particle converges.

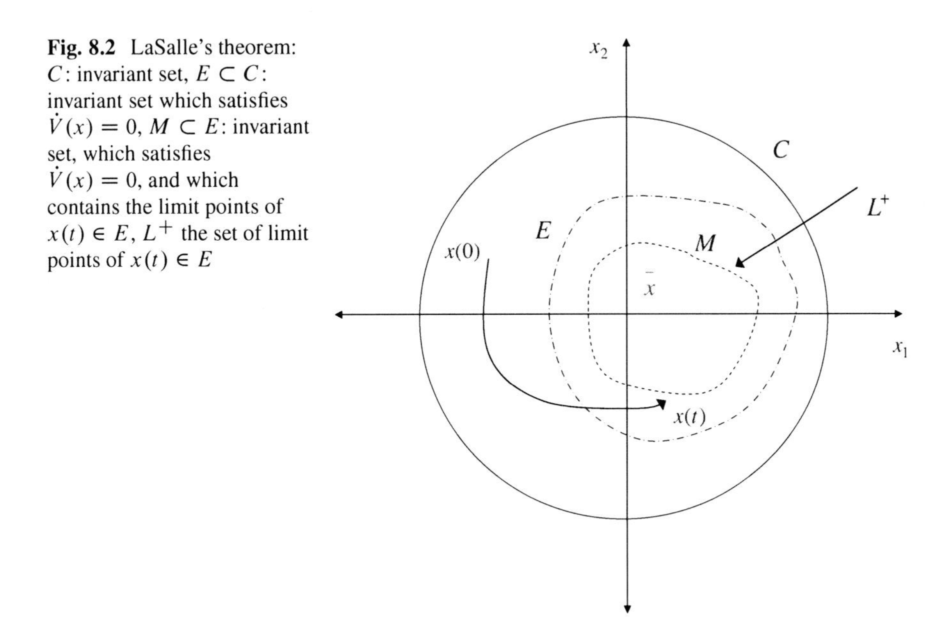

Fig. 8.2 LaSalle's theorem: C: invariant set, $E \subset C$: invariant set which satisfies $\dot{V}(x) = 0$, $M \subset E$: invariant set, which satisfies $\dot{V}(x) = 0$, and which contains the limit points of $x(t) \in E$, L^+ the set of limit points of $x(t) \in E$

8.2.2.2 Convergence Analysis Using La Salle's Theorem

It has been shown that $\lim_{t\to\infty}\bar{x}(t) = x^*$ and from Eq. (8.26) that each particle will stay in a cycle C of center $\bar{x}$ and radius ϵ given by Eq. (8.27). The Lyapunov function given by Eq. (8.23) is negative semi-definite, therefore asymptotic stability cannot be guaranteed. It remains to make precise the area of convergence of each particle in the cycle C of center $\bar{x}$ and radius ϵ. To this end, La Salle's theorem can be employed [64, 92].

La Salle's Theorem: Assume the autonomous system $\dot{x} = f(x)$ where $f : D \to R^n$. Assume $C \subset D$ a compact set which is positively invariant with respect to $\dot{x} = f(x)$, i.e. if $x(0) \in C \Rightarrow x(t) \in C \ \forall\ t$. Assume that $V(x) : D \to R$ is a continuous and differentiable Lyapunov function such that $\dot{V}(x) \leq 0$ for $x \in C$, i.e. $V(x)$ is negative semi-definite in C. Denote by E the set of all points in C such that $\dot{V}(x) = 0$. Denote by M the largest invariant set in E and its boundary by L^+, i.e. for $x(t) \in E : \lim_{t\to\infty}x(t) = L^+$, or in other words L^+ is the positive limit set of E (see Fig. 8.2b). Then every solution $x(t) \in C$ will converge to M as $t \to \infty$.

La Salle's theorem is applicable to the case of the multi-particle system and helps to describe more precisely the area round $\bar{x}$ to which the particle trajectories x^i will converge. A generalized Lyapunov function is introduced which is expected to verify the stability analysis based on Eq. (8.26). It holds that

$$V(x)=\sum_{i=1}^{M}V^i(x^i)+\tfrac{1}{2}\sum_{i=1}^{M}\sum_{j=1,j\neq i}^{M}\{V_a(||x^i-x^j||-V_r(||x^i-x^j||)\}\Rightarrow$$

$$V(x)=\sum_{i=1}^{M}V^i(x^i)+\tfrac{1}{2}\sum_{i=1}^{M}\sum_{j=1,j\neq i}^{M}\{a||x^i-x^j||-V_r(||x^i-x^j||)\text{ and}$$

$$\nabla_{x^i}V(x)=[\sum_{i=1}^{M}\nabla_{x^i}V^i(x^i)]+\tfrac{1}{2}\sum_{i=1}^{M}\sum_{j=1,j\neq i}^{M}\nabla_{x^i}\{a||x^i-x^j||-V_r(||x^i-x^j||)\}\Rightarrow$$

$$\nabla_{x^i}V(x)=[\sum_{i=1}^{M}\nabla_{x^i}V^i(x^i)]+\sum_{j=1,j\neq i}^{M}(x^i-x^j)\{g_a(||x^i-x^j||)-g_r(||x^i-x^j||)\}\Rightarrow$$

$$\nabla_{x^i}V(x)=[\sum_{i=1}^{M}\nabla_{x^i}V^i(x^i)]+\sum_{j=1,j\neq i}^{M}(x^i-x^j)\{a-g_r(||x^i-x^j||)\}$$

and using Eq. (8.19) with $\gamma^i(t)=1$ yields $\nabla_{x^i}V(x)=-\dot{x}^i$, and

$$\dot{V}(x)=\nabla_x V(x)^T\dot{x}=\sum_{i=1}^{M}\nabla_{x^i}V(x)^T\dot{x}^i\Rightarrow\dot{V}(x)=-\sum_{i=1}^{M}||\dot{x}^i||^2\leq 0 \qquad (8.30)$$

Therefore, in the case of a quadratic cost function it holds $V(x)>0$ and $\dot{V}(x)\leq 0$ and the set $C=\{x:V(x(t))\leq V(x(0))\}$ is compact and positively invariant. Thus, by applying La Salle's theorem one can show the convergence of $x(t)$ to the set $M\subset C,\ M=\{x:\dot{V}(x)=0\}\Rightarrow\ M=\{x:\dot{x}=0\}$.

8.3 Convergence of the Stochastic Weights to an Equilibrium

To visualize the convergence of the stochastic weights to an attractor, the interaction between the weights (Brownian particles) can be given in the form of a distributed gradient algorithm, as described in Eq. (8.19):

$$x^i(t+1)=x^i(t)+\gamma^i(t)\left[h(x^i(t))+\eta^i(t)+\sum_{j=1,j\neq i}^{M}g(x^i-x^j)\right],\ i=1,2,\cdots,M \qquad (8.31)$$

The term $h(x(t)^i)=-\nabla_{x^i}V^i(x^i)$ indicates a local gradient algorithm, i.e. motion in the direction of decrease of the cost function $V^i(x^i)=\frac{1}{2}e^i(t)^Te^i(t)$. The term $\gamma^i(t)$ is the algorithms step while the stochastic disturbance $e^i(t)$ enables the algorithm to escape from local minima. The term $\sum_{j=1,j\neq i}^{M}g(x^i-x^j)$ describes the interaction between the i-th and the rest $M-1$ Brownian particles (feedback from neighboring neurons as depicted in Figs. 10.1 and 7.5).

In the conducted simulation experiments the multi-particle set consisted of ten particles (weights) which were randomly initialized in the 2-D field $[x,y]=[e,\dot{e}]$.

The relative values of the parameters a and b that appear in the potential of Eq. (8.14) affected the trajectories of the individual particles. For $a>b$ the cohesion

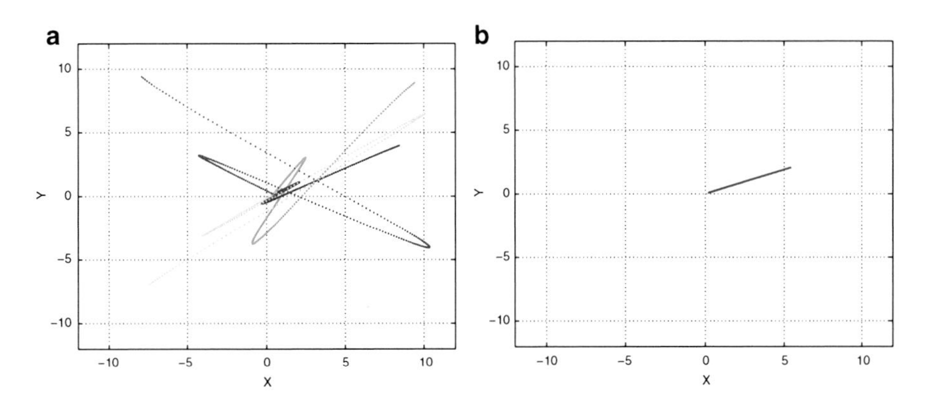

Fig. 8.3 (**a**) Convergence of the individual neural weights that follow the QHO model to an attractor (**b**) Convergence of the mean of the weights position to the attractor $[x^*, y^*] = [e^* = 0, \dot{e}^* = 0]$

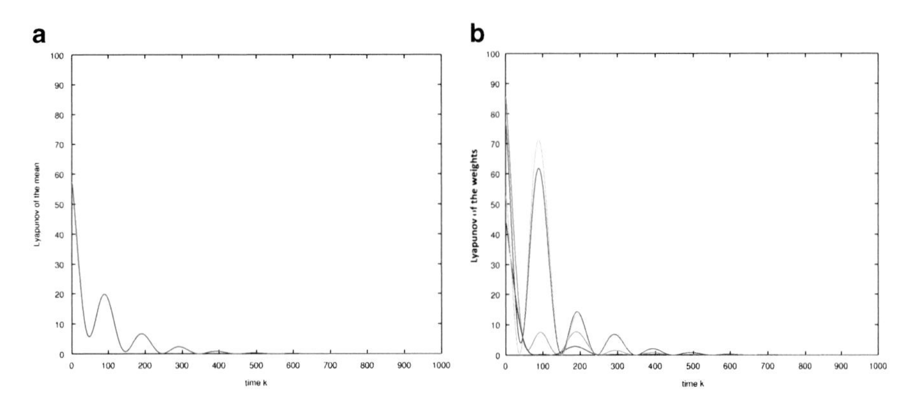

Fig. 8.4 (**a**) Lyapunov function of the individual stochastic weights (Brownian particles) in a 2D-attractors plane without prohibited regions (obstacle-free) and (**b**) Lyapunov function of the mean of the stochastic weights (multi-particle system) in a 2D-attractors plane without prohibited regions (obstacle-free)

of the particles was maintained and abrupt displacements of the particles were avoided (Fig. 8.3).

For the learning of the stochastic weights without constraints (i.e., motion of the Brownian particles in a 2D-attractors plane without prohibited areas), the evolution of the aggregate Lyapunov function is depicted in Fig. 8.4a. The evolution of the Lyapunov function corresponding to each stochastic weight (individual particle) is depicted in Fig. 8.4b.

8.4 Conclusions

Stochasticity in neuronal model was considered and neural weights with dynamics described by Brownian motion were studied. Convergence of learning of stochastic neural weights to attractors (equilibriums) was proved using Lyapunov stability analysis. The convergence of the weights is affected by a drift (elastic) term that is imposed by the potential (that is a function associated with the square of the weights' distance from the equilibrium) and a random term that is imposed by interaction with neighboring weights. Thus a proof has been given about the convergence (stability) of learning in models of stochastic neurons.

Chapter 9
Synchronization of Chaotic and Stochastic Neurons Using Differential Flatness Theory

Abstract This chapter presents a control method for neuron models that have the dynamics of chaotic oscillators. The proposed chaotic control method makes use of a linearized model of the chaotic oscillator which is obtained after transforming the oscillators dynamic model into a linear canonical form through the application of differential flatness theory. The chapter also analyzes a synchronizing control method (flatness-based control) for stochastic neuron models which are equivalent to particle systems and which can be modeled as coupled stochastic oscillators. It is explained that the kinematic model of the particles can be derived from the model of the quantum harmonic oscillator (QHO). It is shown that the kinematic model of the particles is a differentially flat system. It is also shown that after applying flatness-based control the mean of the particle system can be steered along a desirable path with infinite accuracy, while each individual particle can track the trajectory within acceptable accuracy levels.

9.1 Chaotic Neural Oscillators

9.1.1 Models of Chaotic Oscillators

Chaotic systems exhibit dynamics which are highly dependent on initial conditions. Since future dynamics are defined by initial conditions the evolution in time of chaotic systems appears to be random. The output or the state vector of a chaotic oscillator has been used to model the dynamics of biological neurons. For example, the Duffing oscillator has been used to model the EEG signal received that is from brain neuronal groups, as well as to describe the dynamics of variants of the FitzHugh–Nagumo neuron [63, 181, 184, 207, 214, 216]. Main types of chaotic oscillators are described in the sequel:

G.G. Rigatos, *Advanced Models of Neural Networks*,
DOI 10.1007/978-3-662-43764-3_9,

1. Duffing's chaotic system:

$$\begin{aligned} \dot{x}_1(t) &= x_2(t) \\ \dot{x}_2(t) &= 1.1x_1(t) - x_1^3(t) - 0.4x_2(t) + 1.8\cos(1.8t) \end{aligned} \tag{9.1}$$

2. The Genesio-Tesi chaotic system:

$$\begin{aligned} \dot{x}_1(t) &= x_2(t) \\ \dot{x}_2(t) &= x_3(t) \\ \dot{x}_3(t) &= -cx_1(t) - bx_2(t) - \alpha x_3(t) + x_1^2(t) \end{aligned} \tag{9.2}$$

   where a, b, and c are real constants. When at least one of the system's Lyapunov exponents is larger than zero the system is considered to be chaotic. For example, when $a = 1.2$, $b = 2.92$, and $c = 6$ the system behaves chaotically.
3. The Chen chaotic system

$$\begin{aligned} \dot{x}_1(t) &= -a(x_1(t) - x_2(t)) \\ \dot{x}_2(t) &= (c-a)x_1(t) + cx_2(t) - x_1(t)x_3(t) \\ \dot{x}_3(t) &= x_1(t)x_2(t) - bx_3(t) \end{aligned} \tag{9.3}$$

   where a, b, and c are positive parameters. When at least one of the system's Lyapunov exponents is larger than zero the system is considered to be chaotic. For example, when $a = 40$, $b = 3$, and $c = 3$ the system behaves chaotically.

The evolution in time of the phase diagram of the Duffing oscillator and of the Genesio-Tesi chaotic system are shown in Fig. 9.1. Other dynamical systems that exhibit chaotic behavior are Lorenz's system, Chua's system, and Rössler's system. It will be shown that for chaotic dynamical systems exact linearization is possible, through the application of a diffeomorphism that enables their description in new state space coordinates. As an example, the Duffing oscillator will be used, while it is straightforward to apply the method to a large number of chaotic oscillators.

9.1.2 *Differential Flatness of Chaotic Oscillators*

1. Differential flatness of Duffing's oscillator:
   By defining the flat output $y = x_1$ one has $x_1 = y$, $x_2 = \dot{y}$ and $u = 1.8\cos(1.8t) = \ddot{y} - 1.1y + y^3 + 0.4\dot{y}$. Thus all state variables and the control input of the Duffing oscillator are expressed as functions of the flat output and its derivatives, and differential flatness of the oscillator is proven. Moreover, by defining the new control input $v = \ddot{y} - 1.1y + y^3 + 0.4\dot{y}$ or $v = 1.1x_1 - x_1^3 - 0.4x_2 + u$, and the new state variables $y_1 = y$ and $y_2 = \dot{y}$

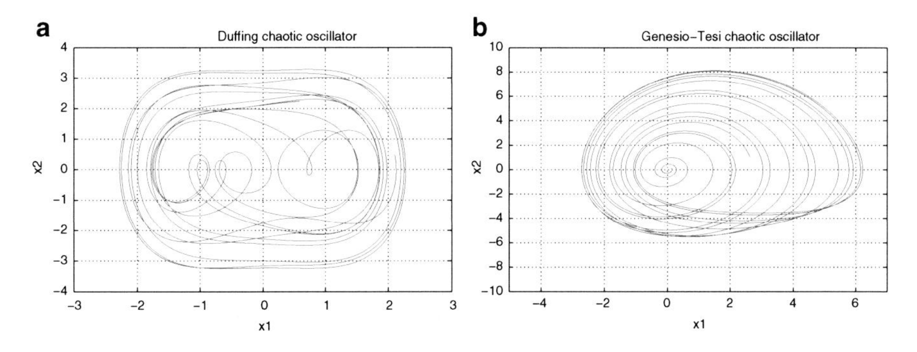

Fig. 9.1 Phase diagram of typical chaotic oscillators (**a**) Duffing's oscillator (**b**) Genesio-Tesi's oscillator

a description of the oscillator's dynamics in the linear canonical (Brunovsky's) form is obtained

$$\begin{pmatrix}\dot{y}_1\\ \dot{y}_2\end{pmatrix}=\begin{pmatrix}0 & 1\\ 0 & 0\end{pmatrix}\begin{pmatrix}y_1\\ y_2\end{pmatrix}+\begin{pmatrix}0\\ 1\end{pmatrix}v \tag{9.4}$$

2. Differential flatness of Genesio-Tesi's oscillator:

   By defining the flat output $y=x_1$ one has $x_1=y$, $x_2=\dot{y}$, $x_3=\ddot{y}$ and $y^{(3)}=\dot{x}_3=-cy-b\dot{y}-a\ddot{y}+y^2$. Thus all state variables of the Genesio-Tesi oscillator are expressed as functions of the flat output and its derivatives, and differential flatness of the oscillator is proven. Moreover, by defining as new control input $v=-cy-b\dot{y}-a\ddot{y}+y^2$, and the new state variables $y_1=y$, $y_2=\dot{y}$, and $y_3=\ddot{y}$ a description of the oscillator's dynamics in the linear canonical (Brunovsky's) form is obtained

$$\begin{pmatrix}\dot{y}_1\\ \dot{y}_2\\ \dot{y}_3\end{pmatrix}=\begin{pmatrix}0 & 1 & 0\\ 0 & 0 & 1\\ 0 & 0 & 0\end{pmatrix}\begin{pmatrix}y_1\\ y_2\\ y_3\end{pmatrix}+\begin{pmatrix}0\\ 0\\ 1\end{pmatrix}v \tag{9.5}$$

3. Differential flatness of Chen's oscillator:

   By defining the flat output $y=x_1$ one has $x_1=y$, $x_2=\frac{1}{a}(\dot{y}+ay)$, and $x_3=\frac{-\ddot{y}+(a-c)\dot{y}-(2ac-a^2)y}{ay}$ (for $y\neq 0$). From the third row of the associated state-space equations and after intermediate computations one obtains $y^{(3)}=-(a-c+1+a^bb)\ddot{y}-[-2(2ac-a^2)+a^2b(a+c)]\dot{y}-\frac{(a+c)}{y}\dot{y}^2-ay^2(\dot{y}+ay)-a^2b(2ac-a^2)y$. Thus all state variables of the Chen oscillator are expressed as functions of the flat output and its derivatives, and differential flatness of the oscillator is proven. Moreover, by defining as new control input $v=-(a-c+1+a^bb)\ddot{y}-[-2(2ac-a^2)+a^2b(a+c)]\dot{y}-\frac{(a+c)}{y}\dot{y}^2-ay^2(\dot{y}+ay)-a^2b(2ac-a^2)y$, and the new state variables $y_1=y$, $y_2=\dot{y}$, and $y_3=\ddot{y}$ a description of the oscillator's dynamics in the linear canonical (Brunovsky's) form is obtained

$$\begin{pmatrix} \dot{y}_1 \\ \dot{y}_2 \\ \dot{y}_3 \end{pmatrix} = \begin{pmatrix} 0 & 1 & 0 \\ 0 & 0 & 1 \\ 0 & 0 & 0 \end{pmatrix} \begin{pmatrix} y_1 \\ y_2 \\ y_3 \end{pmatrix} + \begin{pmatrix} 0 \\ 0 \\ 1 \end{pmatrix} v \tag{9.6}$$

It is noted that differential flatness can be proven for several other types of chaotic oscillators, such as Lorenz's system, Rössler's system, and Chua's system. By expressing all state variables and the control input of such systems as functions of the flat output and their derivatives it is possible to obtain again a description in the linear canonical (Brunovsky) form.

Having written the chaotic oscillator in the linearized canonical form

$$y^{(n)} = v \tag{9.7}$$

Then it suffices to apply a feedback control of the form

$$v = y_d^{(n)} v - k_1(y^{(n-1)} - y_d^{(n-1)}) - \cdots - k_{n-1}(\dot{y} - \dot{y}_d) - k_n(y - y_d) \tag{9.8}$$

to make the oscillator's output $y(t)$ convergence to the desirable setpoint $y_d(t)$. Considering that $v = f(y, \dot{y}, \cdots, \dot{y}^{(r)}) + u$ the control input that is finally applied to the system is $u = v - f(y, \dot{y}, \cdots, \dot{y}^{(r)})$. Using the control input u all state vector elements x_i of the chaotic oscillator will finally converge to their own setpoints.

9.2 Stabilization of Interacting Particles Which Are Modelled as Coupled Stochastic Oscillators

It will be shown that it is possible to succeed synchronizing control of models of stochastic neural oscillators, which are equivalent to interacting diffusing particles [35, 57, 124]. To this end one can consider either open-loop or closed-loop control approaches. Open-loop control methods are suitable for controlling micro and nano systems, since the control signal can be derived without need for on-line measurements [12, 28, 41, 102]. A different approach would be to apply closed-loop control using real-time measurements, taken at micro or nano-scale. The next sections of this chapter are concerned with open-loop control for particle systems. The proposed control approach is flatness-based control [58, 126, 153, 172].

A multi-particle system that consists of N particles is considered. It is assumed that the particles perform diffusive motion, and interact to each other as the theory of Brownian motion predicts. As explained, Brownian motion is the analogous of the quantum harmonic oscillator (QHO), i.e. of Schrödingers equation under harmonic (parabolic) potential [56]. Moreover, it has been shown that the diffusive motion of the particles (kinematic model of the particles) can be described by Langevin's equation which is a stochastic linear differential equation [56, 66].

Next, it will be also shown that the kinematic model of each individual particle is a differentially flat system and thus can be expressed using a flat output and its derivatives. It will be proven that flatness-based control can compensate for the effect of external potentials, and interaction forces, thus enabling the position of the multi-particle formation to follow the reference path. When flatness-based control is applied, the mean position of the formation of the N diffusing particles can be steered along any desirable position in the 2D plane, while the i-th particle can track this trajectory within acceptable accuracy levels.

9.3 Some Examples on Flatness-Based Control of Coupled Oscillators

Flatness-based control of N linear coupled oscillators has been analyzed in [172]. The generalized coordinates z_i are considered and n oscillators are taken. The oscillators can be coupled through an interaction term $f_i(z_1, z_2, \cdots, z_N)$ and through the common control input u. This means that the general oscillator model can be written as

$$\frac{d^2}{dt^2}z_i = -(\omega_i)^2 z_i + f_i(z_1, z_2, \cdots, z_N) + b_i u, \tag{9.9}$$

$i = 1, \cdots, N$. For $f_i(z_1, z_2, \cdots, z_N) = 0$ one obtains

$$\frac{d^2}{dt^2}z_i = -(\omega_i)^2 z_i + b_i u, \quad i = 1, \cdots, N \tag{9.10}$$

The terms $\omega_i > 0$ and $b_i \neq 0$ are constant parameters, while $T > 0$ and $D \neq 0$ are also defined. The objective is to find open-loop control $[0, T]$ with $t \to u(t)$ steering the system from an initial to a final state. In [172] it has been shown that such control can be obtained explicitly, according to the following procedure: using the Laplace transform of Eq. (9.10) and the notation $s = \frac{d}{dt}$ one has

$$(s^2 + (\omega_i)^2) z_i = b_i u, \quad i = 1, \cdots, N. \tag{9.11}$$

Then the system can be written in the form [110]:

$$\begin{array}{l} z_i = Q_i(s) y, \quad u = Q(s) y, \quad \text{with } y = \sum_{k=1}^{N} c_k z_k \\ Q_i(s) = \frac{b_i}{(\omega_i)^2} \prod_{k=1}^{N} (1 + (\frac{s}{\omega_k})^2) \text{ for } k \neq i, \\ Q(s) = \prod_{k=1}^{N} (1 + (\frac{s}{\omega_k})^2), \quad c_k = \frac{1}{Q_k(j\omega_k)} \in R \end{array} \tag{9.12}$$

The real coefficients q_k^i and q_k are defined as follows [172]:

$$\begin{array}{l} Q_i(s) = \sum_{k=0}^{N-1} q_k^i s^{2k} \\ Q(s) = \sum_{k=0}^{N} q_k s^{2k} \end{array} \tag{9.13}$$

This enables to express both the system's state variables $x(t)$ and the control input $u(t)$ as functions of the flat output $y(t)$, i.e.

$$x_i(t) = \sum_{k=0}^{n-1} q_k^i y^{(2k)}(t), \quad v(t) = \sum_{k=0}^{n} q_k y^{(2k)}(t) \tag{9.14}$$

Additionally, any piecewise continuous open-loop control $u(t)$, $t \in [0, T]$ steering from the steady-state $z_i(0) = 0$ to the steady-state $z_i(T) = \frac{b_i}{(\omega_i)^2} D$ can be written as

$$u(t) = \sum_{k=0}^{N} q_k y^{(2k)}(t), \tag{9.15}$$

for all functions $y(t)$ such that $y(0) = 0$, $y(T) = D, \forall\, i\, \{1, \cdots, 2n-1\}$, $y^{(i)}(0) = y^{(i)}(T) = 0$. The results can be extended to the case of an harmonic oscillator with damping. In that case Eq. (9.10) is replaced by

$$\frac{d^2 z_i}{dt^2} = -\omega_i{}^2 z_i - 2\xi_i \omega_i \dot{z}_i + b_i u, \quad i = 1, \cdots, n \tag{9.16}$$

where the damping coefficient is $\xi_i \geq 0$ [172]. Thus, one obtains

$$\begin{aligned} z_i &= Q_i(s)y, \quad u = Q(s)y \\ Q_i(s) &= \frac{b_i}{(\omega_i)^2} \prod_{k=1}^{n} \left(1 + 2\xi_k \left(\frac{s}{\omega_k}\right) + \left(\frac{s}{\omega_k^2}\right)\right) \quad k \neq i \\ Q(s) &= \prod_{k=1}^{n} \left(1 + 2\xi_k \left(\frac{s}{\omega_k}\right) + \left(\frac{s}{\omega_k}\right)^2\right) \end{aligned} \tag{9.17}$$

which proves again that the system's parameters (state variables) and the control input can be written as functions of the flat output y and its derivatives. In that case the flat output is of the form

$$y = \sum_{k=1}^{n} c_k z_k + d_k s z_k \tag{9.18}$$

where $s = \frac{d}{dt}$ and the coefficients c_k and d_k can be computed explicitly. According to [110] explicit descriptions of the system parameters via an arbitrary function y (flat output) and its derivatives are possible for any controllable linear system of finite dimension (controllability is equivalent to flatness).

The desirable setpoint for the flat output is

$$y_d = \sum_{k=1}^{N} c_k z_k^d + d_k \cdot s \cdot z_k^d \tag{9.19}$$

Using this in Eq. (9.15) enables to compute the open-loop control law that steers the system's state variables along the desirable trajectories.

9.4 Flatness-Based Control for the Multi-Particle System

First, the motion of particles is considered, i.e. it is assumed that no external control affects the particle's motion. The particles move on the 2D-plane only under the influence of an harmonic potential. The kinematic model of the particles is as described in Sect. 8.2.

$$\begin{aligned} \dot{x}^i &= F^i \\ F^i &= -\nabla_{x^i}\{V^i(x^i) + \tfrac{1}{2}\sum_{i=1}^M \sum_{j=1,j\neq i}^M [V_a(||x^i - x^j|| + V_r(||x^i - x^j||)]\} \end{aligned} \tag{9.20}$$

The interaction between the i-th and the j-th particle is

$$g(x^i - x^j) = -(x^i - x^j)[g_a(||x^i - x^j||) - g_r(||x^i - x^j||)] \tag{9.21}$$

where $g_a()$ denotes the attraction term and is dominant for large values of $||x^i - x^j||$, while $g_r()$ denotes the repulsion term and is dominant for small values of $||x^i - x^j||$. From Eq. (9.21) it can be seen that $g(x^i - x^j) = -g(x^j - x^i)$, i.e. $g()$ is an odd function. Therefore, for the center of the multi-particle system holds that

$$\begin{aligned} &\tfrac{1}{M}(\sum_{j=1,j\neq i}^M g(x^i - x^j)) = 0 \text{ and} \\ &\dot{\bar{x}} = \tfrac{1}{M}\sum_{i=1}^M [-\nabla_{x^i} V^i(x^i)] \end{aligned} \tag{9.22}$$

According to the Lyapunov stability analysis and application of LaSalle's theorem given in Chap. 8, it has been shown that the mean of the multi-particle system will converge exactly to the goal position $[x^*, y^*] = [0, 0]$ while each individual particle will remain in a small bounded area that encircles the goal position [64, 92, 150, 152].

Next, flatness-based control for the multi-particle system will be analyzed. The equivalent kinematic model of the i-th particle is given in Eq. (9.20) and can be written in the form:

$$\dot{x}^i = -\omega x^i + u^i + \eta^i \tag{9.23}$$

where $-\omega x^i$ is the drift term due to the harmonic potential, u^i is the external control, and η^i is a disturbance term due to interaction with the rest $N - 1$ particles, or due to the existence of noise. Then it can be easily shown that the system of Eq. (9.23) is differentially flat, while an appropriate flat output can be chosen to be $y = x^i$. Indeed all system variables, i.e. the elements of the state vector and the control input can be written as functions of the flat output y, and thus the model that describes the i-th particle is differentially flat.

An open-loop control input that makes the i-th particle track the reference trajectory y_r^i is given by

$$u^i = \omega x_r^i + \dot{x}_r^i + u_c^i, \tag{9.24}$$

where x_i^r is the reference trajectory for the i-th particle, and $\dot{x}_i^r$ is the derivative of the i-th desirable trajectory. Moreover $u_c^i = -\eta^i$ stands for an additional control term which compensates for the effect of the noise η_i on the i-th particle. Thus, if the disturbance η_i that affects the ith-particle is adequately approximated it suffices to set $u_c^i = -\eta_i$. The application of the control law of Eq. (9.24) to the model of Eq. (9.23) results in the error dynamics

$$\begin{gathered}\dot{x}^i = \dot{x}_r^i - \omega x^i + \omega x_r^i + \eta^i - u_c^i \Rightarrow \\ \dot{x}^i - \dot{x}_r^i + \omega(x_i - x_i^r) = \eta^i + u_c \Rightarrow \\ \dot{e}^i + \omega e^i = \eta_i + u_c.\end{gathered} \tag{9.25}$$

Thus, if $u_c = -\eta_i$, then $\lim_{t\to\infty} = 0$.

Next, the case of the N interacting particles will be examined. The control law that makes the mean of the multi-particle system follow a desirable trajectory $E\{x_r^i\}$ can be derived. The kinematic model of the mean of the multi-particle system is given by

$$E\{\dot{x}^i\} = -\omega E\{x^i\} + E\{u^i\} + E\{\eta^i\} \tag{9.26}$$

$i = 1, \cdots, N$, where $E\{x^i\}$ is the mean value of the particles' position, $E\{\dot{x}^i\}$ is the mean velocity of the multi-particle system, $E\{\eta^i\}$ is the average of the disturbance signal, and $E\{u^i\}$ is the control input that is expected to steer the mean of the multi-particle formation along a desirable path. The open-loop controller is selected as:

$$E\{u^i\} = \omega E\{x^i\}_r + E\{\dot{x^i}\}_r - E\{\eta^i\} \tag{9.27}$$

where $E\{x^i\}_r$ is the desirable trajectory of the mean. From Eq. (9.27) it can be seen that the particle's control consists of two parts: (1) a part that compensates for the interaction between the particles and, (2) a part that compensates for the forces which are due to the harmonic external potential (slowing down of the particles' motion).

Assuming that for the mean of the particles' system holds $E\{\eta^i\} = 0$, then the control law of Eq. (9.27) results in the error dynamics for the mean position of the particles

$$E\{\dot{e}^i\} + \omega E\{e^i\} = 0, \tag{9.28}$$

which assures that the mean position of the particles will track the desirable trajectory, i.e.

$$\lim_{t\to\infty} E\{e^i\} = 0. \tag{9.29}$$

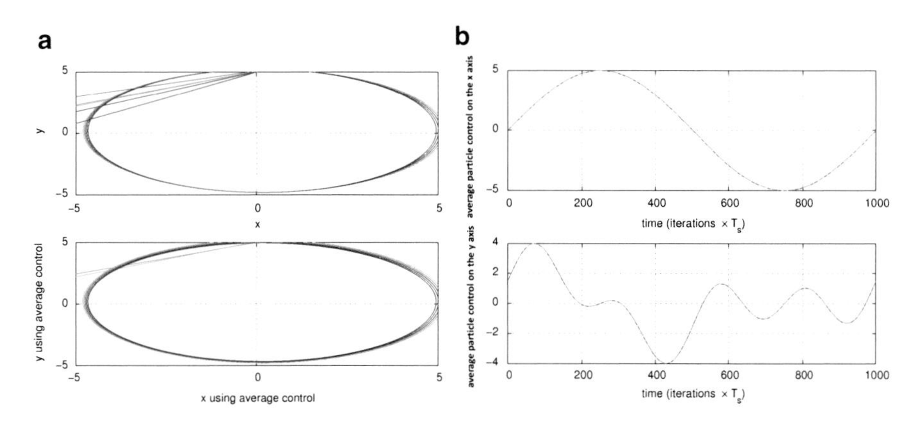

Fig. 9.2 (**a**) Particles following reference trajectory (i), (**b**) associated control input

For non-constant acceleration, the generalized model of the harmonic oscillator will be used and the results of Sect. 9.3 will be applied. Thus, the common control input $u(t)$ can be selected as $u = Q(s)y$, where y is the flat output, defined as $y = \sum_{k=1}^{n} c_k z_k + d_k s z_k$.

9.5 Simulation Tests

The particles (interacting weights) were initialized at arbitrary positions on the 2D-plane. Trajectory tracking under flatness-based control is examined. The following reference trajectories have been tested:

$$\begin{aligned} x_r(t) &= 5{\cdot}\sin(\tfrac{2\pi t}{T}) \\ y_r(t) &= 5{\cdot}\cos(\tfrac{2\pi t}{T}) \end{aligned} \tag{9.30}$$

$$\begin{aligned} x_r(t) &= 5{\cdot}\sin(\tfrac{2\pi t}{T}) \\ y_r(t) &= 1.5{\cdot}\cos(\tfrac{2\pi t}{T}) + 1.5{\cdot}\sin(\tfrac{4\pi t}{T}) + 1.5{\cdot}\sin(\tfrac{8\pi t}{T}) \end{aligned} \tag{9.31}$$

Open-loop flatness-based control has been applied, which means that the control signal consisted of the reference trajectory and its derivatives, while measurements about the position and the velocity of the particles have not been used. It can be observed that under flatness-based control the mean of the particle system can be steered along a desirable path with infinite accuracy, while each individual particle can track the trajectory within acceptable accuracy levels.

The top left diagram in Figs. 9.2 and 9.3 shows the trajectories followed by the particles when a separate control law was designed for each particle, while the bottom diagram shows the trajectories followed by the particles when the average control input was applied to each one of them. The right plot in Figs. 9.2 and 9.3

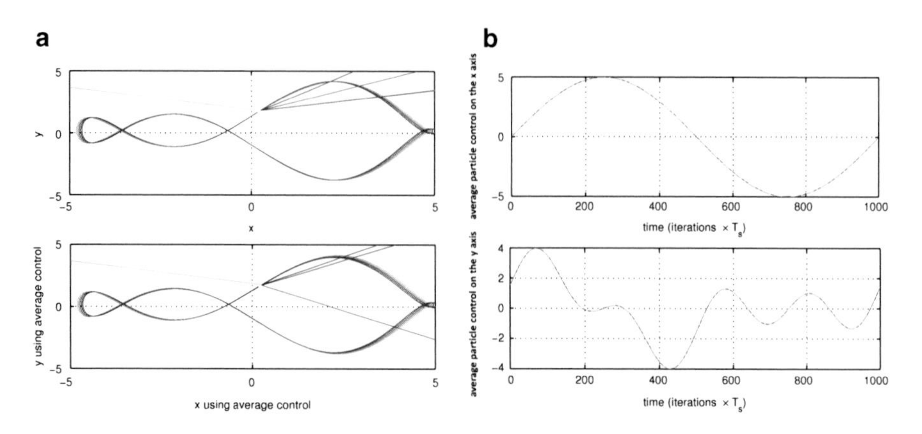

Fig. 9.3 (**a**) Particles following reference trajectory (ii), (**b**) associated control input

shows the control input along the x (top) and y axis (bottom) (continuous line: average control, dashed line: control designed individually for the i-th particle).

9.6 Conclusions

The chapter has presented first a control method for neuron models that have the dynamics of chaotic oscillators. The proposed chaotic control method makes use of a linearized model of the chaotic oscillator which is obtained after transforming the oscillators dynamic model into a linear canonical form through the application of differential flatness theory. Moreover, flatness-based control was applied to motion control of a multi-particle system. A set of M particles (stochastic weights) in a 2D-plane was considered and it was assumed that the interaction between the particles was given by forces as the ones that appear in Brownian motion. The multi-particle formation corresponds to an ensemble of coupled harmonic oscillators. The kinematic model of the particles comprised a drift term that was due to the harmonic potential, a disturbance term that was due to interaction with the rest of the particles and a term that stands for the control input. Next, by applying flatness-based control it was shown that the mean of the multi-particle formation can follow the reference trajectory with infinite accuracy, while each individual particle can be steered along the desirable trajectory within satisfactory accuracy ranges.

Chapter 10
Attractors in Associative Memories with Stochastic Weights

Abstract Neural associative memories are considered in which the elements of the weight matrix are taken to be stochastic variables. The probability density function of each weight is given by the solution of Schrödinger's diffusion equation. The weights of the proposed associative memories are updated with the use of a learning algorithm that satisfies quantum mechanics postulates. This learning rule is proven to satisfy two basic postulates of quantum mechanics: (a) existence in superimposing states, (b) evolution between the superimposing states with the use of unitary operators. Taking the elements of the weight matrix of the associative memory to be stochastic variables means that the initial weight matrix can be decomposed into a superposition of associative memories. This is equivalent to mapping the fundamental memories (attractors) of the associative memory into the vector spaces which are spanned by the eigenvectors of the superimposing matrices and which are related to each other via unitary rotations. In this way, it can be shown that the storage capacity of the associative memories with stochastic weights increases exponentially with respect to the storage capacity of conventional associative memories.

10.1 Weights Learning in Associative Memories Is a Wiener Process

10.1.1 The Weights of Associative Memories Are Equivalent to Brownian Particles

In Sect. 8.2 a neural network with weights that follow the QHO model was presented and stability of learning was analyzed. Interacting Brownian particles stood in place of the neural weights. Now, it will be shown that this model is a generalization of known associative memories, and since this model is compatible with quantum mechanics postulates it can be considered as a quantum associative

G.G. Rigatos, *Advanced Models of Neural Networks*,
DOI 10.1007/978-3-662-43764-3_10,

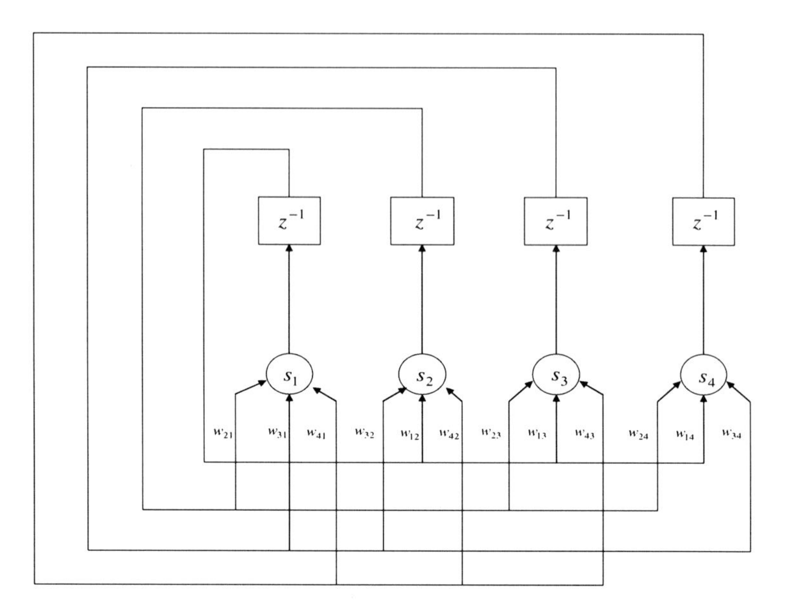

Fig. 10.1 A connectionist associative memory

memory. Moreover, it will be shown that the considered associative memory with stochastic weights exhibits an exponential increase in the number of attractors, comparing to conventional associative memories [83, 91, 156].

Suppose that a set of p N-dimensional vectors (*fundamental memories*) are to be stored in a Hopfield associative memory of N neurons, as the one depicted in Fig. 10.1. The output of neuron j is given by $s_j = sgn(\mathrm{v}_j),\ j = 1, 2, \cdots, N$ and the input v_j is given by $\mathrm{v}_j = \sum_{i=1}^{N} w_{ji} s_i - \theta_j\ j = 1, 2, \cdots, N$, where s_i are the outputs of the neurons connected to the j-th neuron, θ_j is the threshold, and w_{ji}'s are the associated weights [78, 82].

The weights of the associative memory are selected using Hebb's postulate of learning, i.e. $w_{ji} = \frac{1}{N}\sum_{k=1}^{p} x_k^j x_k^i,\ \ j \neq i$ and $w_{ji} = 0, j = i$, where $i, j = 1, 2, \cdots, N$. The above can be summarized in the network's correlation weight matrix which is given by

$$W = \frac{1}{N}\sum_{k=1}^{p} {x_k}^T x_k \tag{10.1}$$

From Eq. (10.1) it can be easily seen that in the case of binary fundamental memories the weights learning is given by Hebb's rule

$$w_{ji}(n+1) = w_{ji}(n) + sgn(x_{k+1}^j x_{k+1}^i) \tag{10.2}$$

where k denotes the k-th fundamental memory. The increment $sgn(x^j_{k+1}x^i_{k+1})$ may have the values $+1$ or -1. Thus, the weight w_{ji} is increased or decreased by 1 each time a new fundamental memory is presented to the network. The maximum number of fundamental memories p that can be retrieved successfully from an associative memory of N neurons (error probability $< 1\,\%$) is given by $p_{\max} = \frac{N}{2lnN}$ [7]. Thus, for $N = 16$ the number of fundamental memories should not be much larger than 2.

The equivalence between Eqs. (7.4) and (10.2) is obvious and becomes more clear if in place of $sgn(x^j_{k+1}x^i_{k+1})$, the variable $\xi_{k+1} = \pm 1$ is used, and if ξ_{k+1} is also multiplied with the (step) increment Δx. Thus the learning of the weights of an associative memory is a Wiener process. Consequently, the weights can be considered as Brownian particles and their mean value and variance can be calculated using the Central Limit Theorem (CLT) as in Sect. 7.2.2.

10.1.2 Mean Value and Variance of the Brownian Weights

The eigenvalues and the eigenvectors of the weight matrix W give all the information needed to interpret the storage capacity of a neural associative memory [7]. The eigenvalues λ_i $i = 1, 2, \cdots, N$ of the weight matrix W are calculated from the solution of $f(\lambda) = |W - \lambda I| = 0$. The eigenvectors q of matrix W satisfy $Wx = \lambda q \Rightarrow (W - \lambda I)q = 0$. Then according to the *spectral theorem* one gets [78]:

$$W = Q\Lambda Q^T \Rightarrow W = \sum_{i=1}^{M} \lambda_i \bar{q}_i \bar{q}_i^T \tag{10.3}$$

where λ_i is the i-th eigenvalue of matrix W and q_i is the associated $N \times 1$ eigenvector. Since the weight matrix W is symmetric, all the eigenvalues are real and the eigenvectors will be orthogonal [78]. The input vector x which is presented to the Hopfield network can be written as a linear combination of the eigenvectors q_i, i.e. $x = \sum_{i=1}^{M} \gamma_i q_i$. Due to the orthogonality of the eigenvectors the convergence of the Hopfield network to its fundamental memories (attractors) $\bar{x}$ gives $\bar{x} = \sum_{i=1}^{N} \lambda_i \gamma_i q_i$. The memory vectors $\bar{x}$ are given by the sum $\sum_{i=1}^{N} \lambda_i \gamma_i q_i$, which means that the eigenvectors of the symmetric matrix W consist of an orthogonal basis of the subspace which is spanned by the memory vectors $\bar{x}$.

It can be shown that the memory vectors become collinear to the eigenvectors of matrix W, thus constituting an orthogonal basis of the space V if the following two conditions are satisfied: (a) the number of neurons N of the Hopfield network is large (high dimensional spaces), and (b) the memory vectors are chosen randomly.

The following lemmas give sufficient conditions for the fundamental memory vectors (attractors) to coincide with the eigenvectors of the weight matrix W:

Lemma 1. *If the fundamental memory vectors (attractors) of the associative memory are chosen to be orthogonal, then they are collinear to the eigenvectors of matrix W [165].*

Proof. The fundamental memory vectors $\bar{x}$ are taken to be orthogonal to each other, i.e. $\bar{x}_i \bar{x}_j^T = \delta(i-j)$, where $i, j = 1, 2, \cdots, N$. The weight matrix W is given by Eq. (10.1).Thus, the following holds

$$W\bar{x}_k^T = \frac{1}{N}\{\sum_{i=1}^{p} \bar{x}_i^T \bar{x}_i\}\bar{x}_k^T = \frac{1}{N}\{\sum_{i=1}^{p} \bar{x}_i^T(\bar{x}_i \bar{x}_k^T)\} \Rightarrow W\bar{x}_k^T = \frac{1}{N}\bar{x}_k^T \quad (10.4)$$

From Eq. (10.4) it can be concluded that if the memory vectors are orthogonal then they are collinear to the eigenvectors of the matrix W.

Lemma 2. *If the memory vectors of an associative memory are chosen randomly and the number of neurons N is large, then there is high probability for them to be orthogonal [165].*

Proof. The normalized internal product of the memory vectors x_i and x_k is considered

$$\frac{1}{N}x_i x_k^T = \frac{1}{N}\sum_{j=1}^{N} x_i^j x_k^j = \sum_{j=1}^{N} \frac{x_i^j x_k^j}{N} = \sum_{j=1}^{N} Y_j \quad (10.5)$$

For large N and x_i^j, x_k^j randomly chosen from the discrete set $\{-1, 1\}$ it holds that the mathematical expectation of Y_j, denoted by $E(Y_j)$, is $E(Y_j) = 0$ and $E(Y_j - \bar{Y}_j)^2 = \frac{1}{N}\sum_{j=1}^{N}(Y_j - \bar{Y}_j)^2 = \frac{1}{N}\sum_{j=1}^{N} Y_j^2 = \frac{1}{N}\frac{1}{N^2}\sum_{j=1}^{N}(x_i^j x_k^j)^2$. Assuming binary patterns i.e. $x_i^j \in \{-1, 1\}$ then $(x_i^j x_k^j)^2 = 1$, i.e. $\sum_{j=1}^{N}(x_i^j x_k^j)^2 = N$, thus

$$E(Y_j - \bar{Y}_j)^2 = \frac{1}{N}\frac{1}{N^2}N \Rightarrow E(Y_j - \bar{Y}_j)^2 = \frac{1}{N^2} \quad (10.6)$$

Therefore $E(Y_j) = 0$ and $E(Y_j - \bar{Y}_j)^2 = \frac{1}{N^2}$. The Central Limit Theorem is applied here. This states:

"Consider $\{Y_k\}$" a sequence of mutually independent random variables $\{Y_k\}$ which follow a common distribution. It is assumed that Y_k has mean μ and variance σ^2, and let $Y = Y_1 + Y_2 + \cdots + Y_N = \sum_{i=1}^{N} Y_i$. Then as N approaches infinity the probability distribution of the sum random variable Y approaches a Gaussian distribution"

$$\frac{(Y - N\mu_i)}{\sqrt{N}\sigma} \cap N(0, 1) \quad (10.7)$$

According to CLT the probability distribution of the sum random variable $\frac{1}{N}\sum_{j=1}^{N} x_i^j x_k^j = \sum_{j=1}^{N} Y_j$ follows a Gaussian distribution of center $\mu = N \cdot 0$ and variance $\sigma^2 = N\frac{1}{N^2} = \frac{1}{N}$. Therefore for large number of neurons N, i.e.

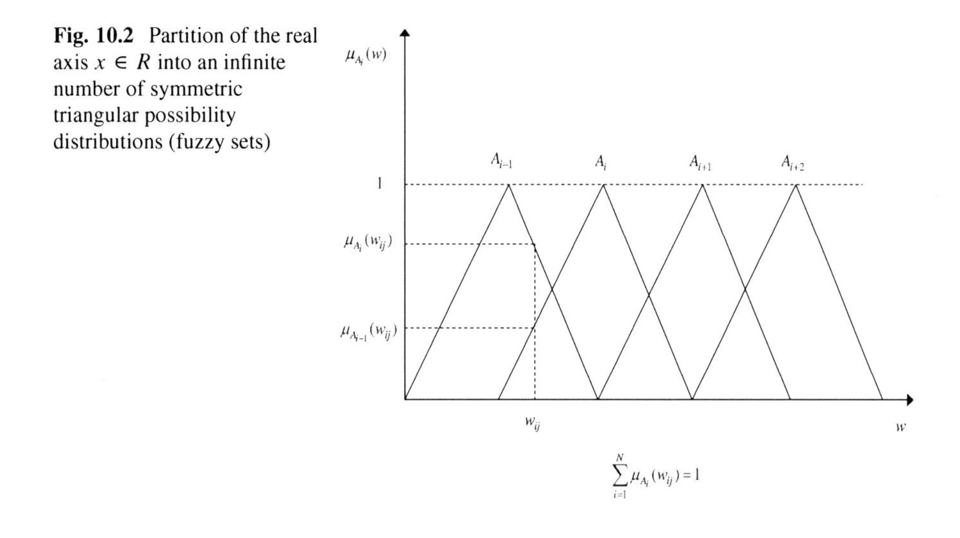

Fig. 10.2 Partition of the real axis $x \in R$ into an infinite number of symmetric triangular possibility distributions (fuzzy sets)

for high dimensional spaces $\frac{1}{N} \rightarrow 0$ and the vectors x_i and x_k will be practically orthogonal.

Thus, taking into account the orthogonality of the fundamental memories $\bar{x}_k$ and Lemma 1, it can be concluded that memory patterns in high dimensional spaces practically coincide with the eigenvectors of the weight matrix W.

10.1.3 Learning Through Unitary Quantum Mechanical Operators

As already mentioned, the update of the weights w_{ij} of the associative memories is performed according to Eq. (10.2), which implies that the value of w_{ij} is increased by or decreased as indicated by $sgn(x^i x^j)$. If the weight w_{ij} is assumed to be a stochastic variable, which is described by the probability (possibility) distribution of Fig. 7.1b, then the term $sgn(x_k^i x_k^j)$ of Hebbian learning can be replaced by a stochastic increment.

A way to substantiate this stochastic increment is to describe variable w_{ij} in terms of a possibility distribution (fuzzy sets). To this end, the real axis x, where the w_{ij} takes its values, is partitioned in triangular possibility distributions (fuzzy sets) $A_1, A_2, \cdots, A_{n-1}, A_n$ (see Fig. 10.2). These approximate sufficiently the Gaussians depicted in Fig. 7.1b. Then, the *increase* of the fuzzy (stochastic) weight is performed through the possibility transition matrix R_n^i which results into $A_n = R_n^i \circ A_{n-1}$, with $\circ$ being the max–min operator. Similarly, the *decrease* of the fuzzy weight is performed through the possibility transition matrix R_n^d, which results into $A_{n-1} = R_n^i \circ A_n$ [163, 195].

It has been shown that update of the weights based on the *increase* and *decrease* operators satisfies two basic postulates of quantum mechanics, i.e.: (a) existence of the stochastic weights w_{ij} in a superposition of states, (b) evolution of the weights with the use of unitary operators, i.e. $R_n^i \circ R_n^d = R_n^d \circ R_n^i = I$ [149, 165]. Moreover, it has been shown that the aforementioned operators are equivalent to quantum addition and quantum subtraction, respectively [163, 165].

10.2 Attractors in QHO-Based Associative Memories

10.2.1 Decomposition of the Weight Matrix into a Superposition of Matrices

Taking the weights w_{ij} of the weight matrix W to be stochastic variables with p.d.f. (or possibility distribution) as the one depicted in Fig. 7.1b means that W can be decomposed into a superposition of associative memories (Fig. 10.3). The equivalence of using probabilistic or possibilistic (fuzzy) variables in the description of uncertainty has been analyzed in [51].

Following the stochastic (fuzzy) representation of the neural weights, the overall associative memory W equals a weighted averaging of the individual weight matrices $\bar{W}_i$, i.e. $W = \sum_{i=1}^{m} \mu_i \bar{W}_i$, where the nonnegative weights μ_i indicate the contribution of each local associative memory $\bar{W}_i$ to the aggregate outcome [165].

Without loss of generality, a 3×3 weight matrix of a neural associative memory is considered. It is also assumed that the weights w_{ij} are stochastic (fuzzy) variables and that quantum learning has been used for the weights update. The weights satisfy the condition $\sum_{i=1}^{N} \mu_{A_i}(w_{ij}) = 1$ (strong fuzzy partition). Thus, the following combinations of membership values of the elements of the matrices $\bar{W}_i$ are possible:

$$\begin{array}{llll}
\bar{W}_1 : & \mu_{12}\ \mu_{13}\ \mu_{23} & \bar{W}_5 : & 1-\mu_{12}\ \mu_{13}\ \mu_{23} \\
\bar{W}_2 : & \mu_{12}\ \mu_{13}\ 1-\mu_{23} & \bar{W}_6 : & 1-\mu_{12}\ \mu_{13}\ 1-\mu_{23} \\
\bar{W}_3 : & \mu_{12}\ 1-\mu_{13}\ \mu_{23} & \bar{W}_7 : & 1-\mu_{12}\ 1-\mu_{13}\ \mu_{23} \\
\bar{W}_4 : & \mu_{12}\ 1-\mu_{13}\ 1-\mu_{23} & \bar{W}_8 : & 1-\mu_{12}\ 1-\mu_{13}\ 1-\mu_{23}
\end{array}$$

The decomposition of matrix W into a set of superimposing matrices $\bar{W}_i$ gives

$$\bar{W}_1 = \left\{ \begin{pmatrix} l* & \mu_{12} & \mu_{13} \\ \mu_{12} & * & \mu_{23} \\ \mu_{13} & \mu_{23} & * \end{pmatrix}, \begin{pmatrix} 0 & a_{12}^{A_i} & a_{13}^{A_i} \\ a_{12}^{A_i} & 0 & a_{23}^{A_i} \\ a_{13}^{A_i} & a_{23}^{A_i} & 0 \end{pmatrix} \right\}$$

$$\bar{W}_2 = \left\{ \begin{pmatrix} l* & \mu_{12} & \mu_{13} \\ \mu_{12} & * & 1-\mu_{23} \\ \mu_{13} & 1-\mu_{23} & * \end{pmatrix}, \begin{pmatrix} 0 & a_{12}^{A_i} & a_{13}^{A_i} \\ a_{12}^{A_i} & 0 & a_{23}^{A_i+1} \\ a_{13}^{A_i} & a_{23}^{A_i+1} & 0 \end{pmatrix} \right\}$$

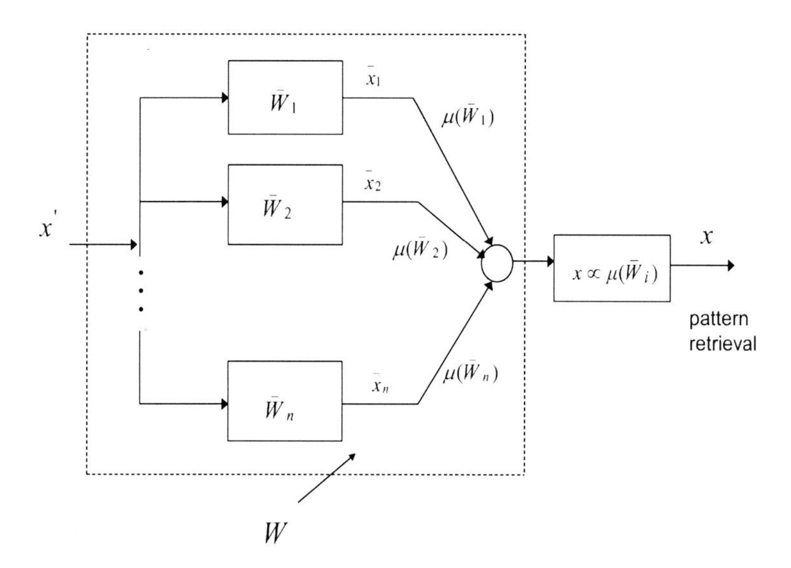

Fig. 10.3 Decomposition of an associative memory with stochastic weights into a superposition of weight matrices, each one having its own attractors

$$\bar{W}_3 = \left\{ \begin{pmatrix} l* & \mu_{12} & 1-\mu_{13} \\ \mu_{12} & * & \mu_{23} \\ 1-\mu_{13} & \mu_{23} & * \end{pmatrix}, \begin{pmatrix} 0 & a_{12}^{A_i} & a_{13}^{A_i+1} \\ a_{12}^{A_i} & 0 & a_{23}^{A_i+1} \\ a_{13}^{A_i} & a_{23}^{A_i+1} & 0 \end{pmatrix} \right\}$$

$$\bar{W}_4 = \left\{ \begin{pmatrix} l* & \mu_{12} & 1-\mu_{13} \\ \mu_{12} & * & 1-\mu_{23} \\ 1-\mu_{13} & 1-\mu_{23} & * \end{pmatrix}, \begin{pmatrix} 0 & a_{12}^{A_i} & a_{13}^{A_i+1} \\ a_{12}^{A_i} & 0 & a_{23}^{A_i+1} \\ a_{13}^{A_i+1} & a_{23}^{A_i+1} & 0 \end{pmatrix} \right\}$$

$$\bar{W}_5 = \left\{ \begin{pmatrix} l* & 1-\mu_{12} & \mu_{13} \\ 1-\mu_{12} & * & \mu_{23} \\ \mu_{13} & \mu_{23} & * \end{pmatrix}, \begin{pmatrix} 0 & a_{12}^{A_i+1} & a_{13}^{A_i} \\ a_{12}^{A_i+1} & 0 & a_{23}^{A_i} \\ a_{13}^{A_i} & a_{23}^{A_i} & 0 \end{pmatrix} \right\}$$

$$\bar{W}_6 = \left\{ \begin{pmatrix} l* & 1-\mu_{12} & \mu_{13} \\ 1-\mu_{12} & * & 1-\mu_{23} \\ \mu_{13} & 1-\mu_{23} & * \end{pmatrix}, \begin{pmatrix} 0 & a_{12}^{A_i+1} & a_{13}^{A_i} \\ a_{12}^{A_i+1} & 0 & a_{23}^{A_i} \\ a_{13}^{A_i} & a_{23}^{A_i+1} & 0 \end{pmatrix} \right\}$$

$$\bar{W}_7 = \left\{ \begin{pmatrix} l* & 1-\mu_{12} & 1-\mu_{13} \\ 1-\mu_{12} & * & \mu_{23} \\ 1-\mu_{13} & \mu_{23} & * \end{pmatrix}, \begin{pmatrix} 0 & a_{12}^{A_i} & a_{13}^{A_i+1} \\ a_{12}^{A_i+1} & 0 & a_{23}^{A_i} \\ a_{13}^{A_i+1} & a_{23}^{A_i} & 0 \end{pmatrix} \right\}$$

$$\bar{W}_8 = \left\{ \begin{pmatrix} l* & 1-\mu_{12} & 1-\mu_{13} \\ 1-\mu_{12} & * & 1-\mu_{23} \\ 1-\mu_{13} & 1-\mu_{23} & * \end{pmatrix}, \begin{pmatrix} 0 & a_{12}^{A_i+1} & a_{13}^{A_i+1} \\ a_{12}^{A_i+1} & 0 & a_{23}^{A_i+1} \\ a_{13}^{A_i+1} & a_{23}^{A_i+1} & 0 \end{pmatrix} \right\}$$

where A_i and A_{i+1} are the two adjacent fuzzy sets to which the weight w_{ij} belongs, with centers c^{A_i} and $c^{A_{i+1}}$, respectively (Fig. 10.2). The diagonal elements of the matrices $\bar{W}_i$ are taken to be 0 (no self-feedback in neurons is considered), and $*$ denotes that the membership value of the element w_{ii}, $i = 1, \cdots, 3$ is indifferent.

The matrices which have as elements the membership values of the weights w_{ij} are denoted by M_i and the associated $||L_1||$ are calculated. Each L_1 norm is divided by the number of the non-diagonal elements of the matrices M_i. This results to

$$W = \frac{\mu_{12}+\mu_{13}+\mu_{23}}{3}\begin{pmatrix} 0 & a_{12}^{A_i} & a_{13}^{A_i} \\ a_{12}^{A_i} & 0 & a_{23}^{A_i} \\ a_{13}^{A_i} & a_{23}^{A_i} & 0 \end{pmatrix} + \frac{\mu_{12}+\mu_{13}-\mu_{23}+1}{3}\begin{pmatrix} 0 & a_{12}^{A_i} & a_{13}^{A_i} \\ a_{12}^{A_i} & 0 & a_{23}^{A_i+1} \\ a_{13}^{A_i} & a_{23}^{A_i+1} & 0 \end{pmatrix}$$
$$+\frac{\mu_{12}-\mu_{13}+\mu_{23}+1}{3}\begin{pmatrix} 0 & a_{12}^{A_i} & a_{13}^{A_i+1} \\ a_{12}^{A_i} & 0 & a_{23}^{A_i} \\ a_{13}^{A_i+1} & a_{23}^{A_i} & 0 \end{pmatrix} + \frac{\mu_{12}-\mu_{13}-\mu_{23}+2}{3}\begin{pmatrix} 0 & a_{12}^{A_i} & a_{13}^{A_i+1} \\ a_{12}^{A_i} & 0 & a_{23}^{A_i+1} \\ a_{13}^{A_i+1} & a_{23}^{A_i+1} & 0 \end{pmatrix}$$
$$+\frac{-\mu_{12}+\mu_{13}+\mu_{23}+1}{3}\begin{pmatrix} 0 & a_{12}^{A_i+1} & a_{13}^{A_i} \\ a_{12}^{A_i+1} & 0 & a_{23}^{A_i} \\ a_{13}^{A_i} & a_{23}^{A_i} & 0 \end{pmatrix} + \frac{-\mu_{12}+\mu_{13}-\mu_{23}+2}{3}\begin{pmatrix} 0 & a_{12}^{A_i+1} & a_{13}^{A_i} \\ a_{12}^{A_i+1} & 0 & a_{23}^{A_i+1} \\ a_{13}^{A_i} & a_{23}^{A_i+1} & 0 \end{pmatrix}$$
$$+\frac{-\mu_{12}-\mu_{13}+\mu_{23}+2}{3}\begin{pmatrix} 0 & a_{12}^{A_i+1} & a_{13}^{A_i+1} \\ a_{12}^{A_i+1} & 0 & a_{23}^{A_i} \\ a_{13}^{A_i+1} & a_{23}^{A_i} & 0 \end{pmatrix} + \frac{-\mu_{12}-\mu_{13}-\mu_{23}+3}{3}\begin{pmatrix} 0 & a_{12}^{A_i+1} & a_{13}^{A_i+1} \\ a_{12}^{A_i+1} & 0 & a_{23}^{A_i+1} \\ a_{13}^{A_i+1} & a_{23}^{A_i+1} & 0 \end{pmatrix}$$

The following lemma holds [165]:

Lemma 3. *The* $||L_1||$ *of the matrices* M_i *(i.e.* $\sum_{i=1}^{N}\sum_{j=1}^{N}|\mu_{ij}|$*) divided by the number of the non-diagonal elements , i.e.* $N(N-1)$ *and by* 2^{N-1}*, where* N *is the number of neurons, equals unity.*

$$\frac{1}{N(N-1)2^{N-1}}\sum_{i=1}^{N}\sum_{j=1}^{N}|\mu_{ij}| = 1 \tag{10.8}$$

Proof. There are 2^{N-1} couples of matrices M_i. Due to the strong fuzzy partition there are always two matrices M_i and M_j with complementary elements, i.e. $\mu(w_{ij})$ and $1-\mu(w_{ij})$. Therefore the sum of the corresponding L_1 norms $||M_i||+||M_j||$ normalized by the number of the non-zero elements (i.e., $N(N-1)$) equals unity. Since there are 2^{N-1} couples of L_1 norm sums $||M_i||+||M_j||$ it holds $\frac{1}{2^{N-1}}\sum_{i=1}^{2^N}||M_i|| = 1$. This normalization procedure can be used to derive the membership values of the weight matrices $\bar{W}_i$.

Remark 1. The decomposition of the matrix W into a group of matrices $\bar{W}_i$ reveals the existence of non-observable attractors. According to Lemmas 1 and 2 these attractors coincide with the eigenvectors v_i of the matrices $\bar{W}_i$. Thus the patterns that can be recalled from an associative memory are more than the ones associated with the initial matrix W. For an associative memory of N neurons, the possible patterns become $N \times 2^N$.

10.2.2 Evolution Between the Eigenvector Spaces via Unitary Rotations

It will be shown that the transition between the vector spaces which are associated with matrices $\bar{W}_i$'s is described by unitary rotations. This is stated in the following theorem [165]:

Theorem 2. *The rotations between the spaces which are spanned by the eigenvectors of the weight matrices $\bar{W}_i$ are unitary operators.*

Proof. Let x_i, y_i, z_i and x_j, y_j, z_j be the unit vectors of the bases which span the spaces associated with the matrices $\bar{W}_i$ and $\bar{W}_j$, respectively. Then a memory vector p can be described in both spaces as: $p = (p_{x_i}, p_{y_i}, p_{z_i})^T$ and $p = (p_{x_j}, p_{y_j}, p_{z_j})^T$. Transition from the reference system $\bar{W}_i \rightarrow \{x_i, y_i, z_i\}$ to the reference system $\bar{W}_j \rightarrow \{x_j, y_j, z_j\}$ is expressed by the rotation matrix R, i.e. $p_{\bar{W}_i} = R \cdot p_{\bar{W}_j}$. The inverse transition is expressed by the rotation matrix Q, i.e. $p_{\bar{W}_j} = Q p_{\bar{W}_i}$.

Furthermore it is true that

$$\begin{pmatrix} p_{x_i} \\ p_{y_i} \\ p_{z_i} \end{pmatrix} = \begin{pmatrix} x_i x_j & x_i y_j & x_i z_j \\ y_i x_j & y_i y_j & y_i z_j \\ z_i x_j & z_i y_j & z_i z_j \end{pmatrix} \begin{pmatrix} p_{x_j} \\ p_{y_j} \\ p_{z_j} \end{pmatrix} \tag{10.9}$$

Thus, using Eq. (10.9) the rotation matrices R and Q are given by

$$R = \begin{pmatrix} x_i x_j & x_i y_j & x_i z_j \\ y_i x_j & y_i y_j & y_i z_j \\ z_i x_j & z_i y_j & z_i z_j \end{pmatrix}, \; Q = \begin{pmatrix} x_j x_i & x_j y_i & x_j z_i \\ y_j x_i & y_j y_i & y_j z_i \\ z_j x_i & z_j y_i & z_j z_i \end{pmatrix} \tag{10.10}$$

It holds that

$$\begin{aligned} & p_{\bar{W}_i} = R \cdot p_{\bar{W}_j} \text{ and } p_{\bar{W}_j} = Q p_{\bar{W}_i} \Rightarrow \\ & p_{\bar{W}_i} = R \cdot Q \cdot p_{\bar{W}_i} \text{ and } p_{\bar{W}_j} = Q \cdot R \cdot p_{\bar{W}_j} \Rightarrow \\ & Q \cdot R = R \cdot Q = I \Rightarrow Q = R^{-1} \end{aligned} \tag{10.11}$$

Moreover, since "dot products" are commutative, from Eq. (10.10) one obtains $Q = R^T$. Therefore it holds $Q = R^{-1} = R^T$, and the transition from the reference system $\bar{W}_i$ to the reference system $\bar{W}_j$ is described by unitary operators, i.e. $QR = R^T R = R^{-1} R = I$.

10.2.3 Applications of the Stochastic Associative Memory Model

The proposed stochastic associative memory model has also practical significance:

1. Stochastic neuron models enable the study and understanding of probabilistic decision making [31].
2. Stochastic neuron models can explain instabilities in short-term memory and attentional systems. These become particularly apparent when the basins of attraction become shallow or deep (in terms of an energy profile) and have been related to some of the symptoms of neurological diseases. The approach thus enables predictions to be made about the effects of pharmacological agents [43,44,75,114,115,171].
3. Stochastic neuron models can improve neural computation towards the direction of quantum computation [137,165].

10.3 Attractors in Associative Memories with Stochastic Weights

It will be examined how the concept of weights that follow the QHO model affects the number of attractors in associative memories. The storage and recall of patterns in quantum associative memories will be tested through a numerical example and simulation tests.

1. Superposition of weight matrices:

Assume that the fundamental memory patterns are the following binary vectors $s_1 = [1,1,1]$, $s_2 = [1,1,-1]$, $s_3 = [1,-1,1]$, which are linearly independent but not orthogonal. Orthogonality should be expected in high dimensional vector spaces if the elements of the memory vectors are chosen randomly. In this example, to obtain orthogonality of the memory vectors, Gramm–Schmidt orthogonalization is used according to

$$u_k = s_k - \sum_{j=1}^{k-1} \frac{u_j s_k}{u_j u_j^T} u_j \tag{10.12}$$

This gives the orthogonal vectors $u_1 = [1,1,1]$, $u_2 = [\frac{2}{3},\frac{2}{3},\frac{-4}{3}]$,$u_3 = [1,-1,0]$. The weight matrix which results from Hebbian learning asymptotically becomes a Wiener process. Matrix W is $W = \frac{1}{3}[u_1^T u_1 + u_2^T u_2 + u_3^T u_3]$, i.e.

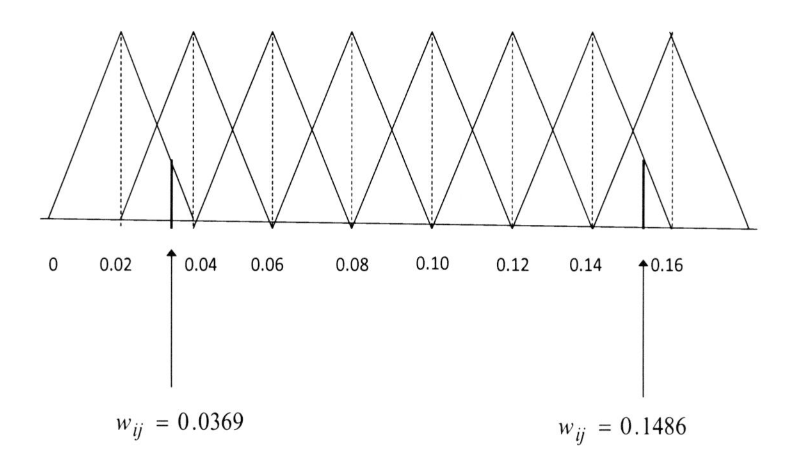

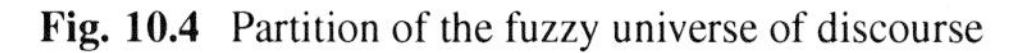

Fig. 10.4 Partition of the fuzzy universe of discourse

$$W = \begin{pmatrix} 0.8141 & 0.1481 & 0.0369 \\ 0.1481 & 0.8141 & 0.0369 \\ 0.0369 & 0.0369 & 0.9247 \end{pmatrix} \tag{10.13}$$

It can be easily shown that $Wu_1 = u_1$, $Wu_2 = u_2$, $Wu_3 = u_3$, i.e. u_1, u_2, u_3 are stable states (attractors) of the network. The eigenvalues of matrix W are $\lambda_1 = 0.667$, $\lambda_2 = 0.888$, and $\lambda_3 = 1.0$. The associated eigenvectors of W are $v_1 = [0.7071, -0.7071, 0]^T$, $v_2 = [-0.4066, -0.4066, 0.8181]^T$, and $v_3 = [0.5785, 0.5785, 0.5750]^T$. It can be observed that v_1 is collinear to u_3, v_2 is collinear to u_2, and v_3 is collinear to u_1. That was expected from Lemma 1, in Sect. 10.1.

Next, the elements of the weight matrix W are considered to be stochastic variables, with p.d.f. (possibility distribution) as the one depicted in Fig. 7.1b and thus matrix W, given by Eq. (10.13), can be decomposed into a superposition of weight matrices $\bar{W}_i$. Assume that only the non-diagonal elements of W are considered and that the possibility distribution of the stochastic variables w_{ij} is depicted in Fig. 10.4.

Then, the weight matrix W is decomposed into a superposition of weight matrices $\bar{W}_i, i = 1, \cdots, 8$:

$$W = \left\{ \begin{pmatrix} * & 0.405 & 0.155 \\ 0.405 & * & 0.155 \\ 0.155 & 0.155 & * \end{pmatrix}, \begin{pmatrix} 0 & 0.14 & 0.02 \\ 0.14 & 0 & 0.02 \\ 0.02 & 0.02 & 0 \end{pmatrix} \right\}$$

$$+ \left\{ \begin{pmatrix} * & 0.405 & 0.155 \\ 0.405 & * & 0.845 \\ 0.155 & 0.845 & * \end{pmatrix}, \begin{pmatrix} 0 & 0.14 & 0.02 \\ 0.14 & 0 & 0.04 \\ 0.02 & 0.04 & 0 \end{pmatrix} \right\}$$

$$+\left\{\begin{pmatrix} * & 0.405 & 0.845 \\ 0.405 & * & 0.155 \\ 0.845 & 0.155 & * \end{pmatrix}, \begin{pmatrix} 0 & 0.14 & 0.04 \\ 0.14 & 0 & 0.02 \\ 0.04 & 0.02 & 0 \end{pmatrix}\right\}$$

$$+\left\{\begin{pmatrix} * & 0.405 & 0.845 \\ 0.405 & * & 0.845 \\ 0.845 & 0.845 & * \end{pmatrix}, \begin{pmatrix} 0 & 0.14 & 0.04 \\ 0.14 & 0 & 0.04 \\ 0.04 & 0.04 & 0 \end{pmatrix}\right\}$$

$$+\left\{\begin{pmatrix} * & 0.595 & 0.155 \\ 0.595 & * & 0.155 \\ 0.155 & 0.155 & * \end{pmatrix}, \begin{pmatrix} 0 & 0.16 & 0.02 \\ 0.16 & 0 & 0.02 \\ 0.02 & 0.02 & 0 \end{pmatrix}\right\}$$

$$+\left\{\begin{pmatrix} * & 0.595 & 0.155 \\ 0.595 & * & 0.845 \\ 0.155 & 0.845 & * \end{pmatrix}, \begin{pmatrix} 0 & 0.16 & 0.02 \\ 0.16 & 0 & 0.04 \\ 0.02 & 0.04 & 0 \end{pmatrix}\right\}$$

$$+\left\{\begin{pmatrix} * & 0.595 & 0.845 \\ 0.595 & * & 0.155 \\ 0.845 & 0.155 & * \end{pmatrix}, \begin{pmatrix} 0 & 0.16 & 0.04 \\ 0.16 & 0 & 0.02 \\ 0.04 & 0.02 & 0 \end{pmatrix}\right\}$$

$$+\left\{\begin{pmatrix} * & 0.595 & 0.845 \\ 0.595 & * & 0.845 \\ 0.845 & 0.845 & * \end{pmatrix}, \begin{pmatrix} 0 & 0.16 & 0.04 \\ 0.16 & 0 & 0.04 \\ 0.04 & 0.04 & 0 \end{pmatrix}\right\}$$

Using Lemma 3, the membership μ_i of each matrix $\bar{W}_i$ is taken to be the norm $||L_1||$ of the matrix with elements the membership values of the weights w_{ij}, i.e. $\frac{1}{N(N-1)2^{N-1}}(\sum_{i=1}^{N}\sum_{j=1}^{N}|\mu(\bar{w}_{ij})|)$. According to Sect. 10.2.1 this gives $W = \mu_1 W_1 + \mu_2 W_2 + \cdots + \mu_8 W_8$, where the membership μ_i are: $\mu_1 = 0.0596$, $\mu_2 = 0.1171$, $\mu_3 = 0.1171$, $\mu_4 = 0.1746$, $\mu_5 = 0.0754$, $\mu_6 = 0.1329$, $\mu_7 = 0.1329$, and $\mu_8 = 0.1904$. By calculating the eigenvectors of matrices $\bar{W}_i$, the associated memory patterns can be found. These are non-observable attractors different from the attractors u_1, u_2, and u_3 of the initial weight matrix W. Thus, the number of memory patterns is increased by a factor $2^N = 8$.

The eigenstructure analysis of matrix $\bar{W}_1$ gives: $\lambda_1 = -0.14$, $\lambda_2 = 0.1455$, $\lambda_3 = -0.0055$, with associated eigenvectors $v_1^{\bar{W}_1} = [0.7071, -0.7071, 0]^T$, $v_2^{\bar{W}_1} = [0.6941, 0.6941, 0.1908]^T$, and $v_3^{\bar{W}_1} = [0.1349, 0.1349, -0.9816]^T$.

The eigenstructure analysis of matrix $\bar{W}_2$ gives: $\lambda_1 = -0.1415$, $\lambda_2 = -0.0104$, $\lambda_3 = 0.1519$, with associated eigenvectors $v_1^{\bar{W}_2} = [0.6921, -0.7143, 0.1041]^T$, $v_2^{\bar{W}_2} = [-0.2648, -0.1176, 0.9572]^T$, and $v_3^{\bar{W}_2} = [0.6715, 0.6900, 0.2701]^T$.

The eigenstructure analysis of matrix $\bar{W}_3$ gives: $\lambda_1 = -0.1415$, $\lambda_2 = -0.0104$, $\lambda_3 = 0.1519$, with associated eigenvectors are $v_1^{\bar{W}_3} = [0.7143, -0.6921, -0.1041]^T$, $v_2^{\bar{W}_3} = [-0.1170, -0.2648, 0.9572]^T$, and $v_3^{\bar{W}_3} = [0.6715, 0.6900, 0.2701]^T$.

The eigenstructure analysis of matrix $\bar{W}_4$ gives: $\lambda_1 = -0.14$, $\lambda_2 = 0.16$, $\lambda_3 = -0.02$, with associated eigenvectors $v_1^{\bar{W}_4} = [0.7071,\ -0.7071,\ 0]^T$, $v_2^{\bar{W}_4} = [0.6667,\ 0.6667,\ 0.3333]^T$, and $v_3^{\bar{W}_4} = [0.2357,\ 0.2357,\ -0.9428]^T$.

The eigenstructure analysis of matrix $\bar{W}_5$ gives: $\lambda_1 = -0.16$, $\lambda_2 = 0.1649$, $\lambda_3 = -0.0049$, with associated eigenvectors $v_1^{\bar{W}_5} = [0.7071,\ -0.7071,\ 0]^T$, $v_2^{\bar{W}_5} = [0.6969,\ 0.6969,\ 0.1691]^T$, and $v_3^{\bar{W}_5} = [0.1196,\ 0.1196,\ -0.9856]^T$.

The eigenstructure analysis of matrix $\bar{W}_6$ gives: $\lambda_1 = -0.1613$, $\lambda_2 = -0.0093$, $\lambda_3 = 0.1706$, with associated eigenvectors $v_1^{\bar{W}_6} = [0.6957,\ -0.7126,\ 0.0905]^T$, $v_2^{\bar{W}_6} = [-0.2353,\ -0.1071,\ 0.9660]^T$, and $v_3^{\bar{W}_6} = [0.6787,\ 0.6933,\ 0.2421]^T$.

The eigenstructure analysis of matrix $\bar{W}_7$ gives: $\lambda_1 = -0.1613$, $\lambda_2 = -0.0093$, $\lambda_3 = -0.1706$, with associated eigenvectors $v_1^{\bar{W}_7} = [-0.7126,\ 0.6957,\ 0.0905]^T$, $v_2^{\bar{W}_7} = [-0.1071,\ -0.2353,\ 0.9660]^T$, and $v_3^{\bar{W}_7} = [-0.6993,\ -0.6787,\ -0.2421]^T$.

The eigenstructure analysis of matrix $\bar{W}_8$ gives: $\lambda_1 = -0.1600$, $\lambda_2 = 0.1780$, $\lambda_3 = -0.0180$, with associated eigenvectors $v_1^{\bar{W}_8} = [0.7071,\ -0.7071,\ 0]^T$, $v_2^{\bar{W}_8} = [0.6739,\ 0.6739,\ 0.3029]^T$, and $v_3^{\bar{W}_8} = [0.2142,\ 0.2142,\ -0.9530]^T$.

Remark 2. In neural structures with weights that follow Schrödinger's equation with zero or constant potential (see the analysis in [165]), the probability to recall the pattern $v_k^{\bar{W}_i}$, $i = 1, \cdots, 8$, $k = 1, \cdots, 3$ is proportional to the membership μ_i of the matrix $\bar{W}_i$, i.e. $P \propto \mu(\bar{W}_i)$. The superimposing matrices $\bar{W}_i$'s describe a distributed associative memory [96].

Remark 3. The difference of the neural structures that follow the quantum harmonic oscillator (QHO) model comparing to neural structures that follow Schrödinger's equation with zero or constant potential is that convergence to an attractor is controlled by the drift force imposed by the harmonic potential [124]. Convergence to an attractor, as the results of Chap. 8 indicate, can be steered through the drift force, which in turn is tuned by the parameters a and b of Eq. (8.14).

2. *Unitarity of the rotation operators*

Here the analysis of second part of Sect. 10.2 will be verified. Matrix R of Eq. (10.10) which performs a rotation from the basis defined by the eigenvectors $v_1^{(\bar{W}_1)}, v_2^{(\bar{W}_1)}, v_3^{(\bar{W}_1)}$, to the basis defined by the vectors $v_1^{(\bar{W}_2)}, v_2^{(\bar{W}_2)}, v_3^{(\bar{W}_2)}$ is calculated as follows: $v_1^{(\bar{W}_1)} = [0.7071,\ -0.7071,\ 0]^T$, $v_2^{(\bar{W}_1)} = [0.6941,\ 0.6941,\ 0.1908]^T$ and $v_3^{\bar{W}_1} = [0.1349,\ 0.1349,\ -0.9816]^T$, while $v_1^{(\bar{W}_2)} = [0.6921,\ -0.7143,\ 0.1041]^T$, $v_2^{(\bar{W}_2)} = [-0.2648, 0.1176,\ 0.9972]^T$, and $v_3^{(\bar{W}_2)} = [0.6715,\ 0.6900,\ 0.2701]^T$.

The rotation matrix R is given by

$$R = \begin{pmatrix} v_1^{(\bar{W}_1)} v_1^{(\bar{W}_2)} & v_1^{(\bar{W}_1)} v_2^{(\bar{W}_2)} & v_1^{(\bar{W}_1)} v_3^{(\bar{W}_2)} \\ v_2^{(\bar{W}_1)} v_1^{(\bar{W}_2)} & v_2^{(\bar{W}_1)} v_2^{(\bar{W}_2)} & v_2^{(\bar{W}_1)} v_3^{(\bar{W}_2)} \\ v_3^{(\bar{W}_1)} v_1^{(\bar{W}_2)} & v_3^{(\bar{W}_1)} v_2^{(\bar{W}_2)} & v_3^{(\bar{W}_1)} v_3^{(\bar{W}_2)} \end{pmatrix} = \begin{pmatrix} 0.9945 & -0.1041 & -0.0131 \\ 0.0045 & -0.0752 & 0.9966 \\ -0.1052 & -1.0304 & -0.0815 \end{pmatrix} \tag{10.14}$$

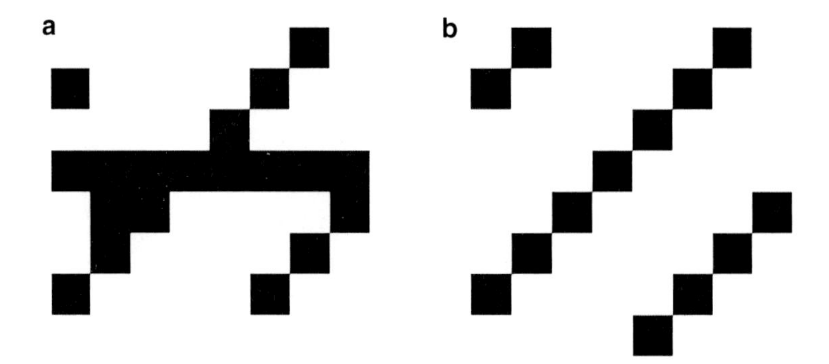

Fig. 10.5 (**a**) input pattern 1 (**b**) recalled pattern 1

It holds

$$RR^T = \begin{pmatrix} 1.000 & -0.0008 & 0.0037 \\ -0.0008 & 0.9989 & -0.0042 \\ 0.0037 & -0.0042 & 1.0794 \end{pmatrix} \Rightarrow RR^T \simeq \begin{pmatrix} 1 & 0 & 0 \\ 0 & 1 & 0 \\ 0 & 0 & 1 \end{pmatrix} \tag{10.15}$$

Thus, matrix R represents a unitary rotation from the vector space $v_1^{(\bar{W}_1)}, v_2^{(\bar{W}_1)}, v_3^{(\bar{W}_1)}$ to the vector space $v_1^{(\bar{W}_2)}, v_2^{(\bar{W}_2)}, v_3^{(\bar{W}_2)}$.

3. *Simulation tests*

Simulation tests have been carried out for the recognition of images stored in an associative memory. Two binary images were considered (see Figs. 10.5b and 10.7b). Both images consisted of 8×8 pixels. The images were selected, so as to reduce cross pattern interference [180].

The correlation weight matrix $W \in R^{64\times 64}$ was found. The images of Figs. 10.5a, 10.6a, 10.7a, 10.8a were given as input to the weight matrix W. The recalled patterns are given in Fig. 10.5b, 10.6b, 10.7b, 10.8b, respectively. This simulation test demonstrates the known error-correcting capability of neural associative memories. When the randomly distorted patterns were presented to the associative memory then after a number of iterations the associative memory converged to the stored images.

Next, the weight matrix W was viewed as an associative memory with stochastic weights. The superimposed weight matrices $\bar{W}_i$, $i = 1, 2, \cdots, 2^{64}$ were considered and the associated rotation matrices R_i's were calculated. The decomposition of W into the superposition of weight matrices $\bar{W}_i$ followed the analysis presented in Sect. 10.2.1. The calculation of the rotation matrices R_i followed the analysis presented in Sect. 10.2.2. The patterns of Figs. 10.9a and 10.10a were generated by applying the stored images of Figs. 10.5b and 10.7b to two different rotation matrices R_i, respectively. Then by applying the patterns of Figs. 10.9a and 10.10a as input to the matrices R_i^T recall of the initial patterns could be observed. This simulation test demonstrates that the rotation matrices R_i result in 2^{64} different perceptions of the initial memory patterns.

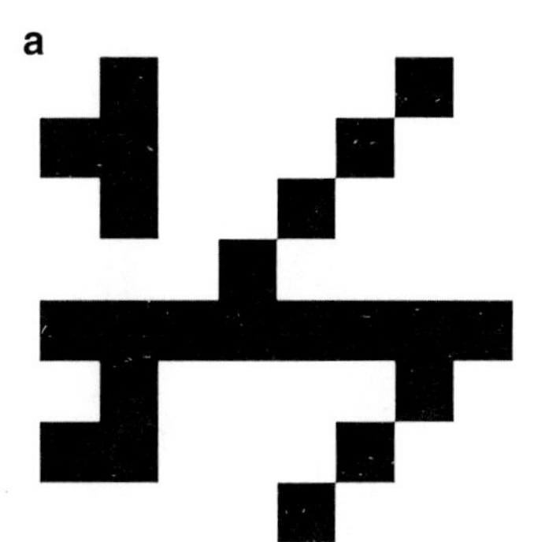

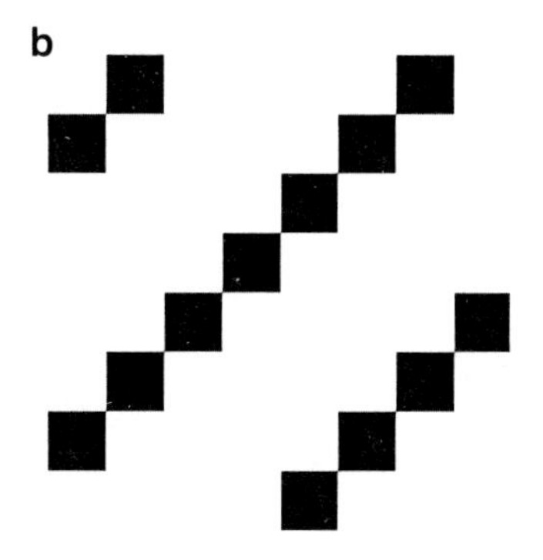

Fig. 10.6 (**a**) input pattern 2 (**b**) recalled pattern 2

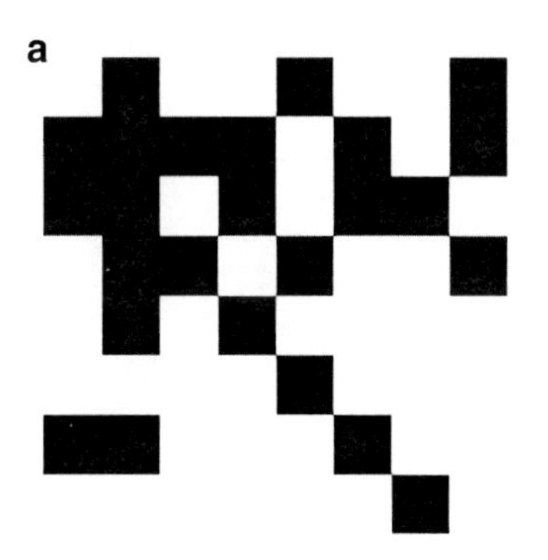

Fig. 10.7 (**a**) input pattern 3 (**b**) recalled pattern 3

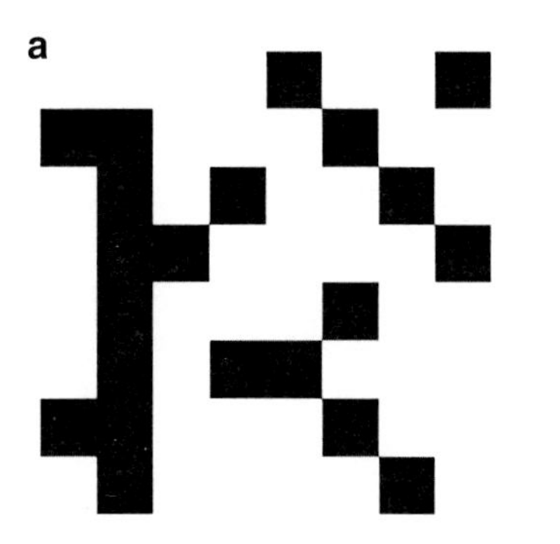

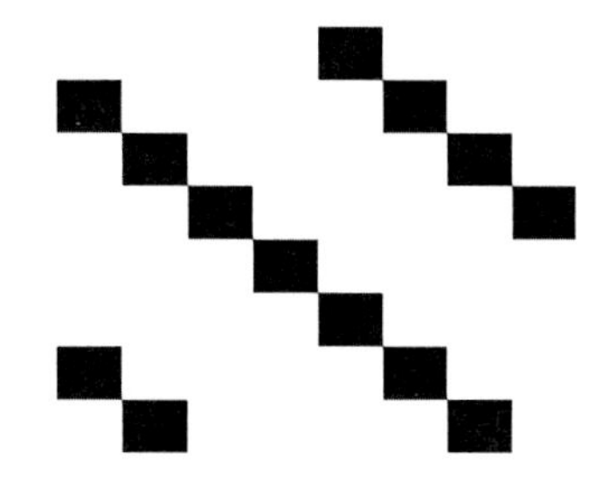

Fig. 10.8 (**a**) input pattern 4 (**b**) recalled pattern 4

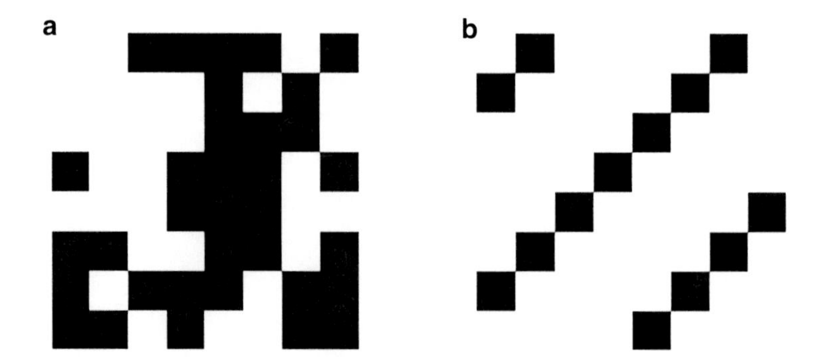

Fig. 10.9 Eigenspace of matrix W_i (**a**) Rotated pattern 1 (**b**) Retrieved pattern 1

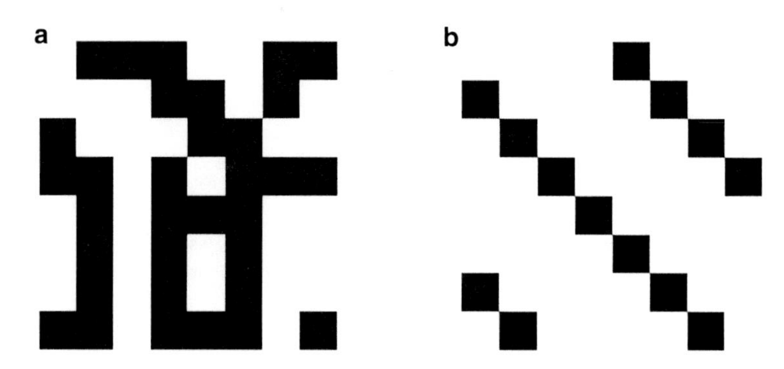

Fig. 10.10 Eigenspace of matrix W_i (**a**) Rotated pattern 2 (**b**) Retrieved pattern 2

10.4 Conclusions

Neural associative memories have been considered in which the elements of the weight matrix were taken to be stochastic variables. The probability density function of each weight was given by the solution of Schrödinger's equation under harmonic potential (quantum harmonic oscillator). Weights that follow the QHO model give to associative memories significant properties: (a) the learning of the stochastic weights is a Wiener process, (b) the update of weights can be described by unitary operators as the principles of quantum mechanics predict, (c) the number of attractors increases exponentially comparing to conventional associative memories. The above theoretical findings were further analyzed through simulation tests. The convergence of the QHO-driven weights to an equilibrium was observed and the existence of non-observable attractors in QHO-based associative memories was demonstrated.

Chapter 11
Spectral Analysis of Neural Models with Stochastic Weights

Abstract Spectral analysis of neural networks with stochastic weights (stemming from the solution of Schrödinger's diffusion equation) has shown that: (a) The Gaussian basis functions of the weights express the distribution of the energy with respect to the weights' value. The smaller the spread of the basis functions is, the larger becomes the spectral (energy) content that can be captured therein. Narrow spread of the basis functions results in wide range of frequencies of the Fourier transformed pulse, (b) The stochastic weights satisfy an equation which is analogous to the principle of uncertainty.

11.1 Overview

The energy spectrum of the stochastic weights that follow the quantum harmonic oscillator (QHO) model is studied. To this end, previous results on wavelets' energy spectrum are used [2,42,149,168]. Spectral analysis of the stochastic weights shows that: (a) The Gaussian membership functions of the weights express the distribution of energy with respect to the weights' value. The smaller the spread of the basis functions is, the larger becomes the spectral (energy) content that can be captured therein (b) The stochastic weights satisfy an equation which is analogous to the principle of uncertainty. Moreover, simulation and numerical results were provided to support the argument that the representation of the basis functions in the space and in the frequency domain cannot be, in both cases, too sharply localized.

G.G. Rigatos, *Advanced Models of Neural Networks*,
DOI 10.1007/978-3-662-43764-3_11, © Springer-Verlag Berlin Heidelberg 2015

11.2 Wavelet Basis Functions

11.2.1 Wavelet Frames

The continuous time wavelet is defined at scale a and b as

$$\psi_{a,b}(x) = \frac{1}{\sqrt{a}}\psi\left(\frac{x-b}{\alpha}\right) \tag{11.1}$$

It will be shown that a continuous time signal $f(x)$ can be expressed as a series expansion of discrete wavelet basis functions. The discrete wavelet has the form [2, 187, 191, 225]

$$\psi_{m,n}(x) = \frac{1}{\sqrt{\alpha_0^m}}\psi\left(\frac{x-nb_0\alpha_0^m}{\alpha_0^m}\right) \tag{11.2}$$

The wavelet transform of a continuous signal $f(x)$ using discrete wavelets of the form of Eq. (11.2) is given by

$$T_{m,n} = \int_{-\infty}^{+\infty} f(x)\frac{1}{\sqrt{\alpha_0^m}}\psi\left(\frac{x-nb_0\alpha_0^m}{\alpha_0^m}\right)dx \tag{11.3}$$

which can be also expressed as the inner product $T_{m,n} =< f(x), \psi_{m,n} >$. For the discrete wavelet transform, the values $T_{m,n}$ are known as wavelet coefficients. To determine how good the representation of a signal is in the wavelet space one can use the theory of wavelet frames. The family of wavelet functions that constitute a frame are such that the energy of the resulting wavelet coefficients lies within a certain bounded range of the energy of the original signal

$$A{\cdot}E \leq \sum_{m=-\infty}^{+\infty}\sum_{n=-\infty}^{+\infty} |T_{m,n}|^2 \leq B{\cdot}E \tag{11.4}$$

where $T_{m,n}$ are the discrete wavelet coefficients, A and B are the frame bounds, and E is the energy of the signal given by $E = \int_{-\infty}^{+\infty}|f(x)|^2dt = ||f(x)||^2$. The values of the frame bounds depend on the parameters α_0 and b_0 chosen for the analysis and the wavelet function used. If $A = B$, the frame is known as tight and has a simple reconstruction formula given by the finite series

$$f(x) = \frac{1}{A}\sum_{m=-\infty}^{+\infty}\sum_{n=-\infty}^{+\infty} T_{m,n}\psi_{m,n}(x) \tag{11.5}$$

A tight frame with $A = B > 1$ is redundant, with A being a measure of the redundancy. When $A = B = 1$ the wavelet family defined by the frame forms an orthonormal basis. Even if $A \neq B$ a reconstruction formula of $f(x)$ can be obtained in the form:

$$f'(x) = \frac{2}{A+B} \sum_{m=-\infty}^{+\infty} \sum_{n=-\infty}^{+\infty} T_{m,n} \psi_{m,n}(x) \tag{11.6}$$

where $f'(x)$ is the reconstruction which differs from the original signal $f(x)$ by an error which depends on the values of the frame bounds. The error becomes acceptably small for practical purposes when the ratio B/A is near unity. The closer this ratio is to unity, the tighter the frame.

11.2.2 Dyadic Grid Scaling and Orthonormal Wavelet Transforms

The dyadic grid is perhaps the simplest and most efficient discretization for practical purposes and lends itself to the construction of an orthonormal wavelet basis. Substituting $\alpha_0 = 2$ and $b_0 = 1$ into Eq. (11.2) the dyadic grid wavelet can be written as

$$\psi_{m,n} = \frac{1}{\sqrt{2^m}} \psi \left(\frac{x - n2^m}{2^m} \right) \tag{11.7}$$

or more compactly

$$\psi_{m,n}(t) = 2^{-\frac{m}{2}} \psi(2^{-m}x - n) \tag{11.8}$$

Discrete dyadic grid wavelets are commonly chosen to be orthonormal. These wavelets are both orthogonal to each other and normalized to have unit energy. This is expressed as

$$\int_{-\infty}^{+\infty} \psi_{m,n}(x) \psi_{m',n'}(x) dx = \begin{cases} 1 \text{ if } m = m', \text{ and } n = n' \\ 0 \text{ otherwise} \end{cases} \tag{11.9}$$

Thus, the products of each wavelet with all others in the same dyadic system are zero. This also means that the information stored in a wavelet coefficient $T_{m,n}$ is not repeated elsewhere and allows for the complete regeneration of the original signal without redundancy. In addition to being orthogonal, orthonormal wavelets are normalized to have unit energy. This can be seen from Eq. (11.9), as using $m = m'$ and $n = n'$ the integral gives the energy of the wavelet function equal to unity. Orthonormal wavelets have frame bounds $A = B = 1$ and the corresponding wavelet family is an orthonormal basis. An orthonormal basis has components which, in addition to being able to completely define the signal, are perpendicular to each other.

Using the dyadic grid wavelet of Eq. (11.7) the discrete wavelet transform is defined as

$$T_{m,n} = \int_{-\infty}^{+\infty} x(t)\psi_{m,n}(x)dt \tag{11.10}$$

By choosing an orthonormal wavelet basis $\psi_{m,n}(x)$ one can reconstruct the original signal $f(x)$ in terms of the wavelet coefficients $T_{m,n}$ using the inverse discrete wavelet transform:

$$f(x) = \sum_{m=-\infty}^{+\infty} \sum_{n=-\infty}^{+\infty} T_{m,n}\psi_{m,n}(x) \tag{11.11}$$

requiring the summation over all integers m and n. In addition, the energy of the signal can be expressed as

$$\int_{-\infty}^{+\infty} |f(x)|^2 dx = \sum_{m=-\infty}^{+\infty} \sum_{n=-\infty}^{+\infty} |T_{m,n}|^2 \tag{11.12}$$

11.2.3 The Scaling Function and the Multi-resolution Representation

Orthonormal dyadic discrete wavelets are associated with scaling functions and their dilation equations. The scaling function is associated with the smoothing of the signal and has the same form as the wavelet

$$\phi_{m,n}(x) = 2^{-m/2}\phi(2^{-m}x - n) \tag{11.13}$$

The scaling functions have the property

$$\int_{-\infty}^{+\infty} \phi_{0,0}(x)dx = 1 \tag{11.14}$$

where $\phi_{0,0}(x) = \phi(x)$ is sometimes referred as the father scaling function or father (mother) wavelet. The scaling function is orthogonal to translations of itself, but not to dilations of itself. The scaling function can be convolved with the signal to produce approximation coefficients as follows:

$$S_{m,n} = \int_{-\infty}^{+\infty} f(x)\phi_{m,n}(x)dx \tag{11.15}$$

One can represent a signal $f(x)$ using a combined series expansion using both the approximation coefficients and the wavelet (detail) coefficients as follows:

$$f(x) = \sum_{n=-\infty}^{+\infty} S_{m_0,n}\phi_{m_0,n} + \sum_{m=-\infty}^{m_0}\sum_{n=-\infty}^{+\infty} T_{m,n}\psi_{m,n}(x) \tag{11.16}$$

It can be seen from this equation that the original continuous signal is expressed as a combination of an approximation of itself, at arbitrary scale index m_0 added to a succession of signal details form scales m_0 down to negative infinity. The signal detail at scale m is defined as

$$d_m(x) = \sum_{n=-\infty}^{+\infty} T_{m,n}\psi_{m,n}(x) \tag{11.17}$$

and hence one can write Eq. (11.16)

$$f(x) = f_{m_0}(t) + \sum_{m=-\infty}^{m_0} d_m(x) \tag{11.18}$$

From this equation it can be shown that

$$f_{m-1}(x) = f_m(x) + d_m(x) \tag{11.19}$$

which shows that if one adds the signal detail at an arbitrary scale (index m) to the approximation at that scale he gets the signal approximation at an increased resolution (at a smaller scale index $m-1$). This is the so-called multi-resolution representation.

11.2.4 Examples of Orthonormal Wavelets

The scaling equation (or dilation equation) describes the scaling function $\phi(x)$ in terms of contracted and shifted versions of itself as follows [2, 119]:

$$\phi(x) = \sum_k c_k \phi(2x-k) \tag{11.20}$$

where $\phi(2x-k)$ is a contracted version of $\phi(t)$ shifted along the time axis by an integer step k and factored by an associated scaling coefficient c_k. The coefficient of the scaling equation should satisfy the condition

$$\sum_k c_k = 2 \tag{11.21}$$

$$\sum_k c_k c_{k+2k'} = \begin{cases} 2 \text{ if } k' = 0 \\ 0 \text{ otherwise} \end{cases} \tag{11.22}$$

This also shows that the sum of the squares of the scaling coefficients is equal to 2. The same coefficients are used in reverse with alternate signs to produce the associated wavelet equation

$$\psi(x) = \sum_k (-1)^k c_{1-k} \phi(2x - k) \tag{11.23}$$

This construction ensures that the wavelets and their corresponding scaling functions are orthogonal. For wavelets of compact support, which have a finite number of scaling coefficients N_k the following wavelet function is defined

$$\psi(x) = \sum_k (-1)^k c_{N_k - 1 - k} \phi(2x - k) \tag{11.24}$$

This ordering of scaling coefficients used in the wavelet equation allows for our wavelets and their corresponding scaling equations to have support over the same interval $[0, N_{k-1}]$. Often the reconfigured coefficients used for the wavelet function are written more compactly as

$$b_k = (-1)^k c_{N_k - 1 - k} \tag{11.25}$$

where the sum of all coefficients b_k is zero. Using this reordering of the coefficients Eq. (11.24) can be written as

$$\psi(x) = \sum_{k=0}^{N_k - 1} b_k \phi(2x - k) \tag{11.26}$$

From the previous equations and examining the wavelet at scale index $m + 1$ one can see that for arbitrary integer values of m the following holds

$$2^{-(m+1)/2} \phi\left(\frac{1}{2^{m+1}} - n\right) = 2^{-m/2} 2^{-1/2} \sum_k c_k \phi\left(\frac{2t}{2 \times 2^m} - 2n - k\right) \tag{11.27}$$

which may be written more compactly as

$$\phi_{m+1,n}(x) = \frac{1}{\sqrt{2}} \sum_k c_k \phi_{m,2n+k}(x) \tag{11.28}$$

That is the scaling function at an arbitrary scale is composed of a sequence of shifted functions at the next smaller scale each factored by their respective scaling coefficients. Similarly, for the wavelet function one obtains

$$\psi_{m+1,n}(x) = \frac{1}{\sqrt{2}} \sum_k b_k \phi_{m,2n+k}(x) \tag{11.29}$$

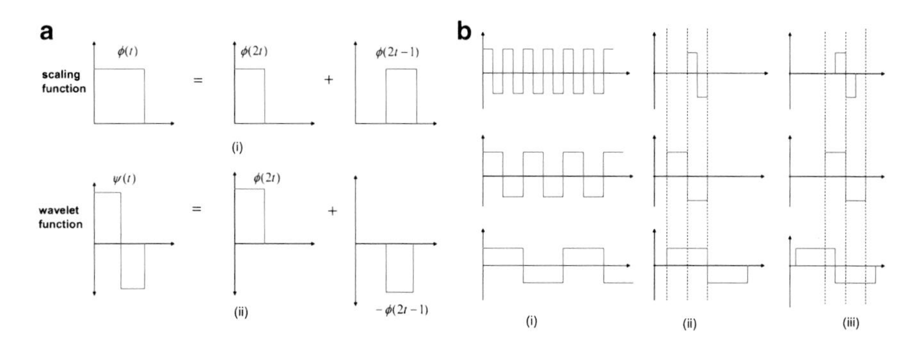

Fig. 11.1 (**a**) (i) The Haar scaling function in terms of shifted and dilated versions of itself, (ii) The Haar wavelet in terms of shifted and dilated versions of its scaling function, (**b**) (i) Three consecutive scales shown from the Haar wavelet family specified on a dyadic grid, e.g. from the bottom $\psi_{m,n}(x)$, $\psi_{m+1,n}(x)$, $\psi_{m+2,n}(x)$, (ii) Three Haar wavelets at three consecutive scales on a dyadic grid, iii) Three Haar wavelets at different scales. This time the Haar wavelets are not defined on a dyadic grid and are hence not orthogonal to each other

11.2.5 The Haar Wavelet

The Haar wavelet is the simplest example of an orthonormal wavelet. Its scaling equation contains only two non-zero scaling coefficients and is given by

$$\phi(x) = \phi(2x) + \phi(2x - 1) \tag{11.30}$$

that is, its scaling coefficients are $c_0 = c_1 = 1$. These values can be obtained from Eqs. (11.21) and (11.22). The solution of the Haar scaling equation is the single block pulse defined as

$$\phi(x) = \begin{cases} 1 & 0 \le x < 1 \\ 0 & \text{elsewhere} \end{cases} \tag{11.31}$$

Using this scaling function, the Haar wavelet equation is

$$\psi(x) = \phi(2x) - \phi(2x - 1) \tag{11.32}$$

The Haar wavelet is finally found to be

$$\psi(x) = \begin{cases} 1 & 0 \le x < \frac{1}{2} \\ -1 & \frac{1}{2} \le x < 1 \\ 0 & \text{elsewhere} \end{cases} \tag{11.33}$$

The mother wavelet for the Haar wavelet system $\psi(x) = \psi_{0,0}(x)$ is formed from two dilated unit block pulses placed next to each other on the time axis, with one of them inverted. From the mother wavelet one can construct the Haar system of wavelets on a dyadic grid $\psi_{m,n}(x)$ (Fig. 11.1).

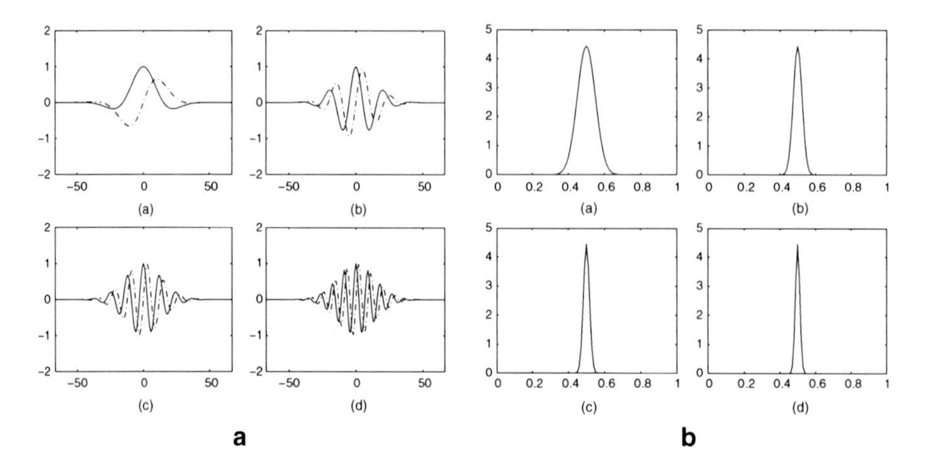

Fig. 11.2 (i) Real (continuous) and imaginary (*dashed*) part of Morlet wavelets for various frequencies: (**a**) $f_0 = 0.1$ Hz (**b**) $f_0 = 0.3$ Hz (**c**) $f_0 = 0.5$ Hz, (**d**) $f_0 = 0.7$ Hz (ii) Energy spectrum of the Morlet wavelet of Fig. 11.2c for central frequency $f_0 = 0.5$ Hz and different variances: (**a**) σ^2, (**b**) $4\sigma^2$, (**c**) $8\sigma^2$, (**d**) $16\sigma^2$

11.3 Spectral Analysis of the Stochastic Weights

Spectral analysis of associative memories with weights described by interacting Brownian particles (quantum associative memories) will be carried out following previous studies on wavelets power spectra [2, 42, 149, 168]. Spectral analysis in quantum associative memories shows that the weights w_{ij} satisfy the principle of uncertainty.

11.3.1 Spectral Analysis of Wavelets

11.3.1.1 The Complex Wavelet

The study of the energy spectrum of wavelets will be used as the basis of the spectral analysis of quantum associative memories. The Morlet wavelet is the most commonly used complex wavelet and in a simple form is given by

$$\psi(x) = \pi^{-\frac{1}{4}}\left(e^{i2\pi f_0 x} - e^{\frac{-(2\pi f_0)^2}{2}}\right)e^{\frac{-x^2}{2}} \tag{11.34}$$

This wavelet is simply a complex wave within a Gaussian envelope. The complex sinusoidal waveform is contained in the term $e^{i2\pi f_0 x} = \cos(2\pi f_0 x) + i\,\sin(2\pi f_0 x)$. The real and the imaginary part of the Morlet wavelet for various central frequencies are depicted in Fig. 11.2a.

It can be seen that the real and the imaginary part of the wavelet differ in phase by a quarter period. The $\pi^{-\frac{1}{4}}$ term is a normalization factor which ensures that the wavelet has unit energy.

11.3.1.2 Spectral Decomposition and Heisenberg Boxes

The Fourier transform of the Morlet wavelet is given by

$$\hat{\psi}(f) = \pi^{\frac{1}{4}}\sqrt{2}e^{\frac{1}{2}(2\pi f - 2\pi f_0)^2} \tag{11.35}$$

which has the form of a Gaussian function displaced along the frequency axis by f_0. The energy spectrum (the squared magnitude of the Fourier transform) is given by [2]

$$|\hat{\psi}(f)|^2 = 2\pi^{\frac{1}{2}}e^{-(2\pi f - 2\pi f_0)^2} \tag{11.36}$$

which is a Gaussian centered at f_0. The integral of (11.36) gives the energy of the Morlet wavelet. The energy spectrum of the Morlet wavelet depicted in diagram (c) of Fig. 11.2(i), for different values of the variance σ^2 is given in Fig. 11.2b.

The central frequency f_0 is the frequency of the complex sinusoid and its value determines the number of significant sinusoidal waveforms contained within the envelope. The dilated and translated Morlet wavelet $\psi(\frac{x-b}{a})$ is given by

$$\psi\left(\frac{x-b}{a}\right) = \pi^{-\frac{1}{4}}e^{i2\pi f_0\left(\frac{x-b}{a}\right)}e^{-\frac{1}{2}\left(\frac{x-b}{a}\right)^2} \tag{11.37}$$

The Heisenberg boxes in the x-frequency plane for a wavelet at different scales are shown in Fig. 11.3a. To evaluate frequency composition a sample of a long region of the signal is required. If instead, a small region of the signal is measured with accuracy, then it becomes very difficult to determine the frequency content of the signal in that region. That is, the more accurate the temporal measurement (smaller σ_x) is, the less accurate the spectral measurement (larger σ_f) becomes, and vice-versa [119].

The Morlet central frequency f_0 sets the location of the Heisenberg box in the x-frequency plane. If the x-length of the wavelets remains the same, then no matter the change of the central frequency f_0 the associated Heisenberg boxes will have the same dimensions. This is depicted in Fig. 11.3b.

Finally, in Fig. 11.4a are shown the Heisenberg boxes in the x-frequency plane for a number of wavelets with three different spectral frequencies (low, medium, and high). The confining Gaussian windows have the same dimensions along the x axis. Therefore, altering the central frequency of the wavelet shifts the Heisenberg box up and down the x-frequency plane without altering its dimensions.

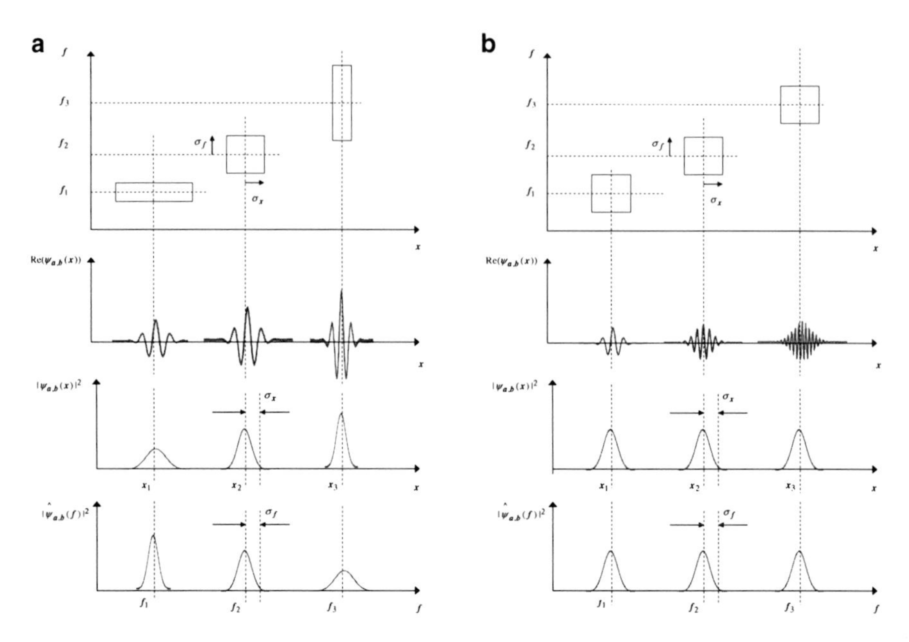

Fig. 11.3 (**a**) Heisenberg boxes in the x-frequency plane for a wavelet at various scales f_1, f_2, f_3 and of different length σ_x along the x-axis. σ_x is the standard deviation of the squared wavelet function $\psi(x)^2$; σ_f is the standard deviation of the spectrum around the mean spectral components f_1, f_2, and f_3 (**b**) Heisenberg boxes in the x-frequency plane for wavelet at various scales f_1, f_2, f_3 and of the same length σ_x along the x-axis. Changes of the central frequency f_0 do not affect the size of the Heisenberg boxes

11.3.2 Energy Spectrum of the Stochastic Weights

The following theorem holds [148, 149]:

Theorem 1. *The Gaussian basis functions of the weights w_{ij} of a quantum associative memory express the distribution of energy with respect to the value of w_{ij}. The smaller the spread σ of the basis functions is, the larger becomes the spectral (energy) content that can be captured therein.*

Proof. The Fourier transform of $g(x) = I_0 e^{-ax^2}$ is $G(f) = I_0\sqrt{\frac{\pi}{a}}e^{-\frac{f^2}{4a}}$. Consequently for $< w_{ij} >$ described by shifted overlapping Gaussians, i.e.

$$< w_{ij} >= \sum_{k=1}^{\infty} e^{-\frac{1}{2}\frac{(w_{ij}-kc)^2}{\sigma^2}} a_k \tag{11.38}$$

it holds that the associated Fourier transform is:

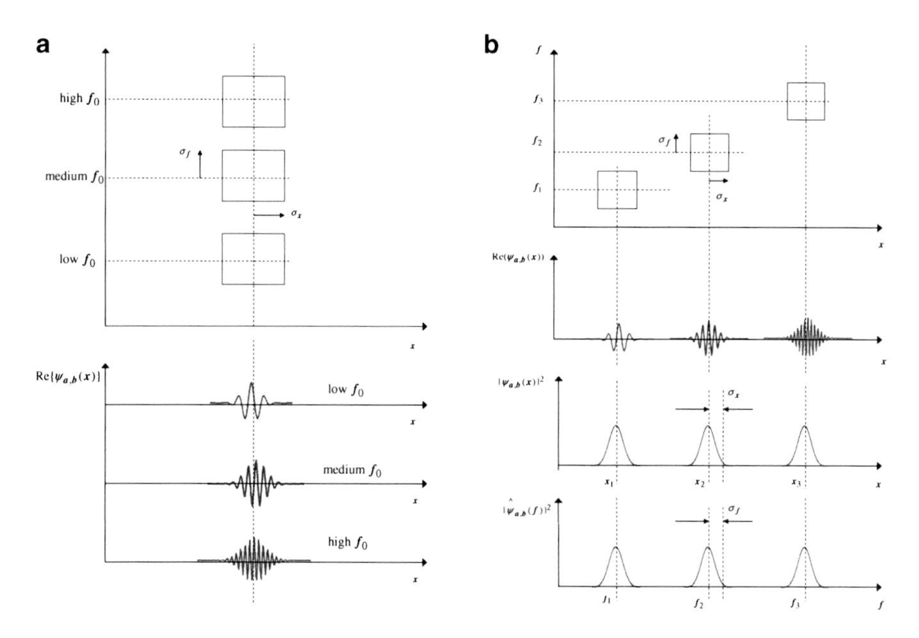

Fig. 11.4 (**a**) Heisenberg boxes in the x-frequency plane for a number of superimposing wavelets of the same length σ_x along the x-axis and at three different spectral frequencies set to low, medium, and high value. (**b**) Heisenberg boxes in the x-frequency plane for a number of Gaussian basis functions which have the same variance and which are shifted on the x axis

$$G\left\{\sum_{k=1}^{\infty} e^{-\frac{1}{2}\frac{(w_{ij}-kc)^2}{\sigma^2}} a_k\right\} = \sum_{k=1}^{\infty}(k\cdot c)e^{ifkc}\sqrt{2\pi}e^{-\frac{1}{2}\sigma^2 f^2}$$

where $\alpha_k = k\cdot c$, with c being the distance between the centers of two adjacent fuzzy basis functions. If only the real part of $G(f)$ is kept, then one obtains

$$G\left\{\sum_{k=1}^{\infty} e^{-\frac{1}{2}\frac{(w_{ij}-kc)^2}{\sigma^2}} a_k\right\} = \sum_{k=1}^{\infty}(kc)\cos(fkc)\sqrt{2\pi}e^{-\frac{1}{2}\sigma^2 f^2} \qquad (11.39)$$

If the weight w_{ij} is decomposed in the x domain into shifted Gaussians of the same variance σ^2 then, in the frequency domain, w_{ij} is analyzed in a superposition of filters, which have a shape similar to the one given in Fig. 11.4b.

Rayleigh's theorem states that the energy of a signal $f(x)$ $x \in [-\infty, +\infty]$ is given by $\int_{-\infty}^{+\infty} f^2(x)dx$. Equivalently, using the Fourier transform $F(s)$ of $f(x)$, the energy is given by $\int_{-\infty}^{+\infty} F^2(s)ds$. The energy distribution of a particle is proportional to the probability $|\psi(w_{ij})|^2$ of finding the particle between w_{ij} and $w_{ij} + \Delta w_{ij}$. Therefore, the energy of the particle will be given by the integral of the squared Fourier transform $|\hat{\psi}(f)|^2$. The energy spectrum of the weights of the quantum associative memories is depicted in Fig. 11.4b.

11.3.3 Stochastic weights and the Principle of Uncertainty

The significance of Eq. (11.39) is that the product between the information in the space domain $g(w_{ij})$ and the information in the frequency domain $G(s)$ cannot be smaller than a constant. In the case of a Gaussian function $g(w_{ij}) = e^{-\frac{{w_{ij}}^2}{2\sigma^2}}$ with Fourier transform $G(s) = e^{-\frac{1}{2}s^2\sigma^2}$ it can be observed that: (a) if σ is small, then $g(w_{ij})$ has a pick at $w_{ij} = 0$ while $G(s)$ tends to become flat, (b) if σ is large, then $g(w_{ij})$ is flat at $w_{ij} = 0$ while $G(s)$ makes a peak at $s = 0$.

This becomes more clear if the dispersion of function $g(w_{ij}) = e^{-\frac{w_{ij}^2}{2\sigma^2}}$ round $w_{ij} = 0$ is used [139]. The dispersion of $g(w_{ij})$ and of its Fourier transform $G(s)$ becomes

$$
\begin{aligned}
D(g) &= \frac{\int_{-\infty}^{+\infty} w_{ij}^2 e^{-\frac{1}{2}\frac{w_{ij}^2}{\sigma^2}} dw_{ij}}{\int_{-\infty}^{+\infty} e^{-2\frac{w_{ij}^2}{\sigma^2}} dw_{ij}} = \frac{1}{2}\sigma^2 \\
D(G) &= \frac{\int_{-\infty}^{+\infty} s^2 e^{-\frac{1}{2}s^2\sigma^2} ds}{\int_{-\infty}^{+\infty} e^{-\frac{1}{2}s^2\sigma^2} ds} = \frac{1}{2\sigma^2}
\end{aligned}
\tag{11.40}
$$

which results into the uncertainty principle for the weights of quantum associative memories

$$
D(g)D(G) = \frac{1}{4}. \tag{11.41}
$$

Equation (11.41) means that the accuracy in the calculation of the weight w_{ij} is associated with the accuracy in the calculation of its spectral content. When the spread of the Gaussians of the stochastic weights is large (small) then their spectral content is poor (rich). Equation (11.41) is an analogous of the quantum mechanics uncertainty principle, i.e. $\Delta x \Delta p \geq \hbar$, where Δx is the uncertainty in the measurement of particle's position, Δp is the uncertainty in the measurement of the particle's momentum, and $\hbar$ is Planck's constant.

It should be noted that (11.41) expresses a general property of the Fourier transform and that similar relations can be found in classical physics. For instance, in electromagnetism it is known that it is not possible to measure with arbitrary precision, at the same time instant, the variation of a wave function both in the time and frequency domain. What is really quantum in the previous analysis is the association of a wave function with a particle (stochastic weight) and the assumption that the wave length and the momentum of the particle satisfy the relation $p = \hbar k$ with $|k| = \frac{2\pi}{\lambda}$.

11.4 Conclusions

Spectral analysis of the weights of that follow the QHO model was based on previous studies of wavelets' energy spectrum. Spectral analysis of quantum associative memories has shown that: (a) The Gaussian basis functions of the weights express the distribution of the energy with respect to the weights' value. The smaller the spread of the basis functions is, the larger becomes the spectral (energy) content that can be captured therein. Narrow spread of the basis functions results in wide range of frequencies of the Fourier transformed pulse, (b) The stochastic weights satisfy an equation which is analogous to the principle of uncertainty. Through numerical and simulation tests it has been confirmed that the representation of the basis functions in the space and in the frequency domain cannot be, in both cases, two sharply localized.

Chapter 12
Neural Networks Based on the Eigenstates of the Quantum Harmonic Oscillator

Abstract The chapter introduces feed-forward neural networks where the hidden units employ orthogonal Hermite polynomials for their activation functions. These neural networks have some interesting properties: (a) the basis functions are invariant under the Fourier transform, subject only to a change of scale, (b) the basis functions are the eigenstates of the quantum harmonic oscillator (QHO), and stem from the solution of Schrödinger's harmonic equation. The proposed neural networks have performance equivalent to wavelet networks and belong to the general category of nonparametric estimators. They can be used for function approximation, system modelling, image processing and fault diagnosis. These neural networks demonstrate the particle-wave nature of information and give the incentive to analyze significant issues related to this dualism, such as the principle of uncertainty and Balian–Low's theorem.

12.1 Overview

Feed-forward neural networks (FNN) are the most popular artificial neural structures due to their structural flexibility, good representational capabilities, and availability of a large number of training algorithms. The hidden units in an FNN usually have the same activation functions and are often selected as sigmoidal functions or Gaussians. This chapter presents FNN that use orthogonal Hermite polynomials as basis functions. The proposed neural networks have some interesting properties: (a) the basis functions are invariant under the Fourier transform, subject only to a change of scale (b) the basis functions are the eigenstates of the quantum harmonic oscillator (QHO), and stem from the solution of Schrödinger's diffusion equation. The proposed neural networks belong to the general category of nonparametric estimators and are suitable for function approximation, image processing, and system fault diagnosis. Two-dimensional QHO-based neural networks can be also constructed by taking products of the one-dimensional basis functions.

G.G. Rigatos, *Advanced Models of Neural Networks*,
DOI 10.1007/978-3-662-43764-3_12, © Springer-Verlag Berlin Heidelberg 2015

FNN that use the eigenstates of the quantum harmonic oscillator (Hermite basis functions) demonstrate the particle-wave nature of information as described by Schrödinger's diffusion equation [34, 186]. Attempts to enhance connectionist neural models with quantum mechanics properties can be also found in [96, 135–137]. The proposed FNNs extend previous results on neural structures compatible with quantum mechanics postulates, given in [149, 163, 164, 168].

Through the analysis given in this chapter, it is shown that the input variable x of the neural network can be described not only by crisp values (particle equivalent) but also by the normal modes of a wave function (wave equivalent). Since the basis functions of the proposed FNN are the eigenstates of the quantum harmonic oscillator, the FNN's output will be the weighted sum $\psi(x) = \sum_{i=1}^{n} w_k \psi_k(x)$, where $|\psi(x)|^2$ is the probability that the input of the neural network (quantum particle equivalent) is found between x and $x + \Delta x$. Thus, the weight w_k provides a measure of the probability to find the input on the neural network in the region associated with the eigenfunction $\psi_k(x)$.

Furthermore, issues related to the uncertainty principle are examined in case of the QHO-based neural network. An expression of the uncertainty principle for Hermite basis functions is given. The uncertainty principle is a measure of the time-frequency localization of the activation functions in the QHO-based neural network and evaluates the degradation of localization when successive elements of these orthonormal basis functions are considered. It is shown that the Hermite basis functions as well as their Fourier transforms cannot be uniformly concentrated in the time-frequency plane [89, 141]. Simulation results support this argument.

12.2 Feed-Forward Neural Networks

FNN serve as powerful computational tools, in a diversity of applications including function approximation, image compression, and system fault diagnosis. When equipped with procedures for learning from measurement data they can generate models of unknown systems. FNN are the most popular artificial neural structures due to their structural flexibility, good representational capabilities, and availability of a large number of training algorithms.

The idea of function approximation with the use of FNN comes from generalized Fourier series. It is known that any function $\psi(x)$ in an L^2 space can be expanded in a generalized Fourier series in a given orthonormal basis, i.e.

$$\psi(x) = \sum_{k=1}^{\infty} c_k \psi_k(x), \; a \leq x \leq b \tag{12.1}$$

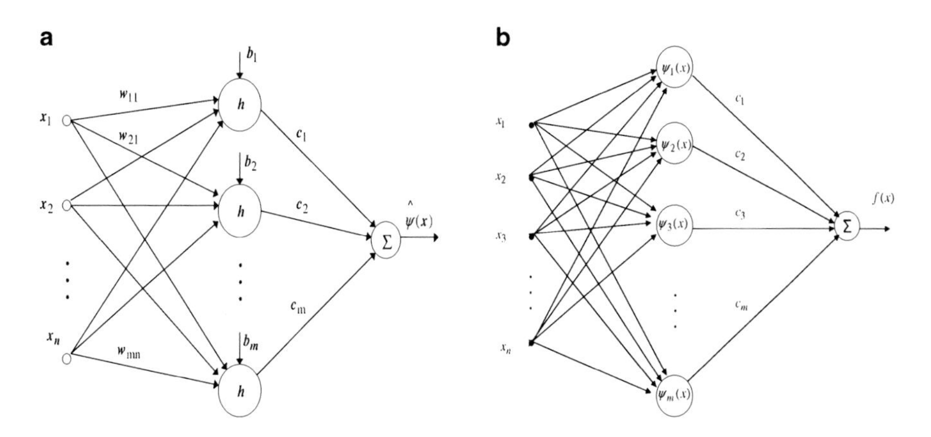

Fig. 12.1 (**a**) Feed-forward neural network (**b**) Quantum neural network with Hermite basis functions

Truncation of the series yields in the sum

$$S_M(x) = \sum_{k=1}^{M} a_k \psi_k(x) \tag{12.2}$$

If the coefficients a_k are taken to be equal to the generalized Fourier coefficients, i.e. when $a_k = c_k = \int_a^b \psi(x)\psi_k(x)dx$, then Eq. (12.2) is a mean square optimal approximation of $\psi(x)$.

Unlike generalized Fourier series, in FNN the basis functions are not necessarily orthogonal. The hidden units in an FNN usually have the same activation functions and are often selected as sigmoidal functions or gaussians. A typical feed-forward neural network consists of n inputs x_i, $i = 1, 2, \cdots, n$, a hidden layer of m neurons with activation function $h: R \to R$ and a single output unit (see Fig. 12.1a). The FNN's output is given by

$$\psi(x) = \sum_{j=1}^{n} c_j h\left(\sum_{i=1}^{n} w_{ji} x_i + b_j\right) \tag{12.3}$$

The root mean square error in the approximation of function $\psi(x)$ by the FNN is given by

$$E_{\text{RMS}} = \sqrt{\frac{1}{N}\sum_{k=1}^{N}(\psi(x^k) - \hat{\psi}(x^k))^2} \tag{12.4}$$

where $x^k = [x_1^k, x_2^k, \cdots, x_n^k]$ is the k-th input vector of the neural network. The activation function is usually a sigmoidal function $h(x) = \frac{1}{1+e^{-x}}$ while in the case of radial basis functions networks it is a Gaussian [78]. Several learning algorithms for neural networks have been studied. The objective of all these algorithms is to find numerical values for the network's weights so as to minimize the mean square error $E_{\rm RMS}$ of Eq. (12.4). The algorithms are usually based on first and second order gradient techniques. These algorithms belong to: (a) batch-mode learning, where to perform parameters update the outputs of a large training set are accumulated and the mean square error is calculated (back-propagation algorithm, Gauss–Newton method, Levenberg–Marquardt method, etc.), (b) pattern-mode learning, in which training examples are run in cycles and the parameters update is carried out each time a new datum appears (Extended Kalman Filter algorithm).

12.3 Eigenstates of the Quantum Harmonic Oscillator

FNN with Hermite basis functions (see Fig. 12.1b) show the particle-wave nature of information, as described by Schrödinger's diffusion equation, i.e.

$$i\hbar\frac{\partial\psi(x,t)}{\partial t} = -\frac{\hbar^2}{2m}\nabla^2\psi(x,t) + V(x)\psi(x,t) \Rightarrow i\hbar\frac{\partial\psi(x,t)}{\partial t} = H\psi(x,t) \qquad (12.5)$$

where $\hbar$ is Planck's constant, H is the Hamiltonian, i.e. the sum of the potential $V(x)$ and of the Laplacian $-\frac{\hbar^2}{2m}\nabla^2 = \frac{\hbar^2}{2m}\frac{\partial^2}{\partial x^2}$. The probability density function $|\psi(x,t)|^2$ gives the probability at time instant t the input x of the neural network (quantum particle equivalent) to have a value between x and $x + \Delta x$. The general solution of the quantum harmonic oscillator, i.e. of Eq. (12.5) with $V(x)$ being a parabolic potential, is [34, 186]:

$$\psi_k(x,t) = H_k(x)e^{-x^2/2}e^{-i(2k+1)t} \; k = 0, 1, 2, \cdots \qquad (12.6)$$

where $H_k(x)$ are the associated Hermite polynomials. In case of the quantum harmonic oscillator, the spatial component $X(x)$ of the solution $\psi(x,t) = X(x)T(t)$ is

$$X_k(x) = H_k(x)e^{-x^2/2} \; k = 0, 1, 2, \cdots \qquad (12.7)$$

Eq. (12.7) satisfies the boundary condition $\lim_{x\to\pm\infty} X(x) = 0$. From the above it can be noticed that the Hermite basis functions $H_k(x)e^{-\frac{x^2}{2}}$ are the eigenstates of the quantum harmonic oscillator (see Fig. 12.2). The general relation for the Hermite polynomials is

$$H_k(x) = (-1)^k e^{x^2}\frac{d^{(k)}}{dx^{(k)}}e^{-x^2} \qquad (12.8)$$

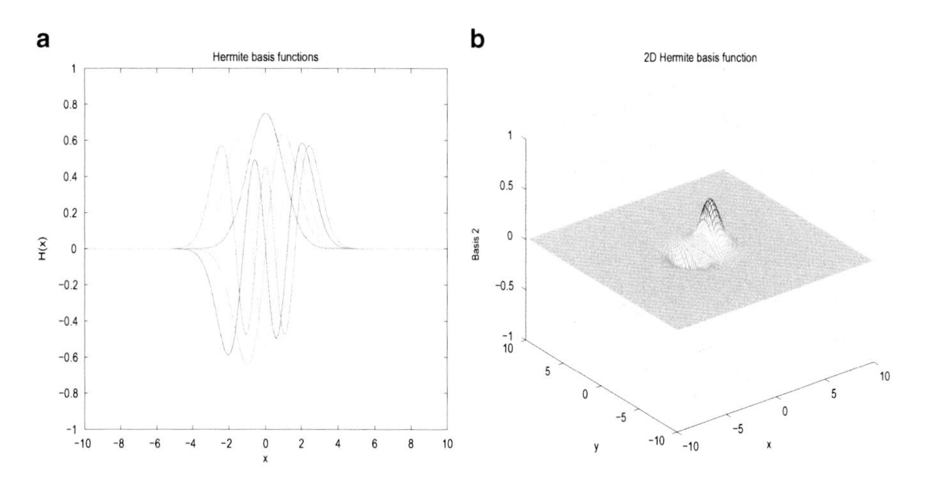

Fig. 12.2 (**a**) First five one-dimensional Hermite basis functions (**b**) 2D Neural Network based on the QHO eigenstates: basis function $B_{1,2}(x,\alpha)$

According to Eq. (12.8) the first five Hermite polynomials are:

$$H_0(x) = 1 \text{ for } \lambda = 1, a_1 = a_2 = 0, \quad H_1(x) = 2x \text{ for } \lambda = 3, a_0 = a_3 = 0$$
$$H_2(x) = 4x^2 - 2 \text{ for } \lambda = 5, a_1, a_4 = 0, \quad H_3(x) = 8x^3 - 12x \text{ for } \lambda = 7, a_0, a_5{=}0$$
$$H_4(x) = 16x^4 - 48x^2 + 12 \text{ for } \lambda = 9, a_1, a_6 = 0$$

It is known that Hermite polynomials are orthogonal [186]. Other polynomials with the property of orthogonality are: Legendre polynomials, Chebychev polynomials, and Laguerre polynomials [215, 217].

12.4 Neural Networks Based on the QHO Eigenstates

12.4.1 The Gauss–Hermite Series Expansion

The following normalized basis functions can now be defined [143]:

$$\psi_k(x) = [2^k \pi^{\frac{1}{2}} k!]^{-\frac{1}{2}} H_k(x) e^{-\frac{x^2}{2}}, \ k = 0, 1, 2, \cdots \tag{12.9}$$

where $H_k(x)$ is the associated Hermite polynomial. To succeed multi-resolution analysis Hermite basis functions of Eq. (12.9) are multiplied with the scale coefficient α. Thus the following basis functions are derived

$$\beta_k(x,\alpha) = \alpha^{-\frac{1}{2}} \psi_k(\alpha^{-1} x) \tag{12.10}$$

where α is a characteristic scale. The abovementioned basis functions have an input range from $-\infty$ to $+\infty$. The basis functions of Eq. (12.10) also satisfy orthogonality condition, i.e.

$$\int_{-\infty}^{+\infty} \beta_m(x,\alpha)\beta_k(x,\alpha)dx = \delta_{mk}, \tag{12.11}$$

where δ_{mk} is the Kronecker delta symbol [143]. Any continuous function $f(x),\ x \in R$ can be written as a weighted sum of the above orthogonal basis functions, i.e.

$$f(x) = \sum_{k=0}^{\infty} c_k \beta_k(x,\alpha) \tag{12.12}$$

The expansion of $f(x)$ using Eq. (12.12) is a Gauss–Hermite series. It holds that the Fourier transform of the basis function $\psi_k(x)$ of Eq. (12.9) satisfies the relation [143]

$$\Psi_k(s) = j^n \psi_k(s) \tag{12.13}$$

while for the basis functions $\beta_k(x,\alpha)$ of Eq. (12.10), it holds that the associated Fourier transform is

$$B_k(s,\alpha) = j^n \beta_k(s,\alpha^{-1}) \tag{12.14}$$

Therefore, it holds

$$f(x) = \sum_{k=0}^{\infty} c_k \beta_k(x,\alpha) \overset{F}{\rightarrow} F(s) = \sum_{k=0}^{\infty} c_k j^n \beta_k(s,\alpha^{-1}) \tag{12.15}$$

which means that the Fourier transform of Eq. (12.12) is the same as the initial function , subject only to a change of scale.

12.4.2 Neural Networks Based on the Eigenstates of the 2D Quantum Harmonic Oscillator

FNN with Hermite basis functions of two variables can be constructed by taking products of the one-dimensional basis functions $B_k(x,\alpha)$ [143]. Thus, setting $x = [x_1, x_2]^T$ one can define the following basis functions

$$B_{k_1,k_2}(x,\alpha) = \frac{1}{\alpha} B_{k_1}(x_1,\alpha) B_{k_2}(x_2,\alpha) \tag{12.16}$$

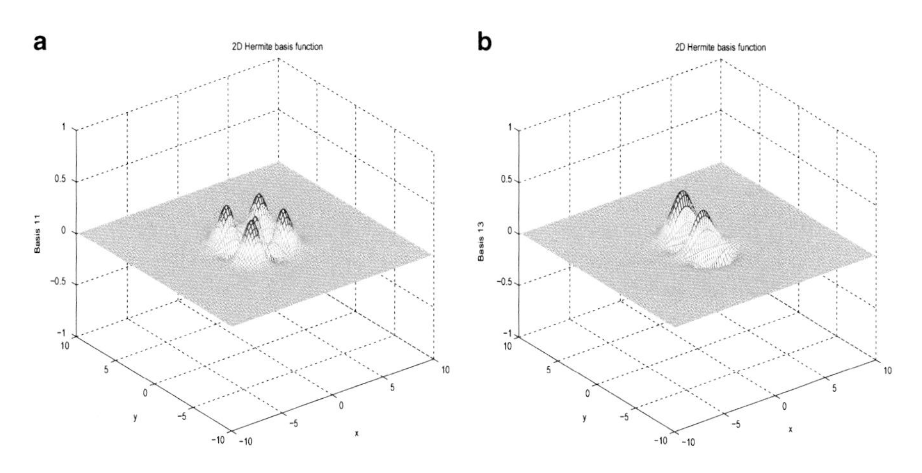

Fig. 12.3 2D Neural Network based on the QHO eigenstates: (**a**) basis function $B_{3,3}(x,\alpha)$ (**b**) basis function $B_{4,1}(x,\alpha)$

These two-dimensional basis functions are again orthonormal, i.e. it holds

$$\int d^2x\, B_n(x,\alpha) B_m(x,\alpha) = \delta_{n_1 m_1} \delta_{n_2 m_2} \tag{12.17}$$

The basis functions $B_{k_1,k_2}(x)$ are the eigenstates of the two-dimensional harmonic oscillator, which is a generalization of Eq. (12.5). These basis functions form a complete basis for integrable functions of two variables. A two-dimensional function $f(x)$ can thus be written in the series expansion:

$$f(x) = \sum_{k_1,k_2}^{\infty} c_k B_{k_1,k_2}(x,\alpha) \tag{12.18}$$

The choice of an appropriate scale coefficient α and of the maximum order $k_1^{\max}, k_2^{\max}$ is a practical issue. Indicative basis functions $B_{1,2}(x,\alpha)$, $B_{3,3}(x,\alpha)$ and $B_{4,1}(x,\alpha)$ of a 2D feed-forward quantum neural network are depicted in Fig. 12.2b, and Fig. 12.3a,b, respectively.

Remark. The significance of the results of Sect. 12.4 is summarized in the sequel:

(i) Orthogonality of the basis functions and invariance under the Fourier transform, (subject only to a change of scale): this means that the energy distribution in the proposed neural network can be estimated without moving to the frequency domain. The values of the weights of the neural network provide a measure of how energy is distributed in the various modes $\beta_k(x,\alpha)$ of the signal that is approximated by the neural network.

(ii) The basis functions of the neural network are the eigenstates of the quantum harmonic oscillator: this means that the proposed neural network can capture the particle-wave nature of information. The input variable x is viewed not only as a crisp value (particle equivalent) but is also distributed to the normal modes of a wave function (wave equivalent).

(iii) The basis functions of the neural network have different frequency characteristics, which means that the QHO-based neural network is suitable for multi-resolution analysis.

12.5 Uncertainty Principles for the QHO-Based Neural Networks

12.5.1 Uncertainty Principles for Bases and the Balian–Low Theorem

Overcoming limitations of existing image processing approaches is closely related to the harnessing of uncertainty principle [94]. A brief description of Heisenberg's principle of uncertainty is first given. The position of a quantum particle, i.e. of a particle the motion of which is subject to Schrödinger's equation is defined by x while its momentum is denoted by p. As explained in Chap. 11, it holds

$$\Delta x \Delta p \geq \hbar \tag{12.19}$$

where $\hbar$ is Planck's constant. The interpretation of Eq. (12.19) is as follows: it is impossible to define at a certain time instant both the position and the momentum of the particle with arbitrary precision. When the limit imposed by Eq. (12.19) is approached, the increase of the accuracy of the position measurement (decrease of Δx) implies a decrease in the accuracy of the momentum measurement (increase of Δp) and vice-versa.

The uncertainty principle in harmonic analysis is a class of theorems which state that a function and its Fourier transform cannot be both too sharply localized. The uncertainty principle can be seen not only as a statement about the time-frequency localization of a single function but also as a statement on the degradation of localization when one considers successive elements of a set of basis functions [71]. This means that the elements of a basis as well as their Fourier transforms cannot be uniformly concentrated in the time-frequency plane [24, 111].

First, a formulation of the uncertainly principle in the case of Gabor frames will be given, through the Balian–Low theorem. A Gabor function $g_{nm}, n, m \in Z$ is defined as (see, for instance, Fig. 11.2b):

$$g_{mn}(x) = e^{i \cdot m f_0 x} g(x - n b_0), \ n, m, \in Z \tag{12.20}$$

A Gabor frame is a set of functions $g_{mn}(x)$ that can be a basis of the L^2 space. The Gabor frame depends on two positive and bounded numbers b_0 and f_0 and on a function g, which has compact support is contained in an interval of length $\frac{2\pi}{f_0}$ [191]. If there exist two real constants C_1 and C_2 such that $0 < C_1 \leq \sum |g(x-nb_0)|^2 \leq C_2 < \infty$, then the family of Gabor functions $g_{mn}, n, m, \in Z$ is a frame in $L^2(R)$, with bounds $2\frac{\pi C_1}{f_0}$ and $2\frac{\pi C_2}{f_0}$.

The Balian–Low theorem describes the behavior of Gabor frames at the critical frequency f_0 defined by $b_0 f_0 = 2\pi$. The theorem is a key result in time-frequency analysis. It expresses the fact that time-frequency concentration and non-redundancy are incompatible properties for Gabor frames of Eq. (12.20). Specifically, if for some $\alpha > 0$ and $g \in L^2(R)$ the set $\{e^{2\pi lx/\alpha} g(x - k\alpha)\}_{k,l \in Z^2}$ is an orthonormal basis for $L^2(R)$, then

$$\left(\int_R |xg(x)|^2 dx\right)\left(\int_R |f\hat{g}(f)|^2 df\right) = \infty \tag{12.21}$$

This means that for a Gabor frame which is contracted from the function $g(x)$ one has necessarily $\int x^2 |g(x)|^2 dx = \infty$ (which means not fine localizaton on the space axis x) or $\int f^2 |\hat{g}(f)|^2 df = \infty$ (which means not fine localization on the frequency axis f).

In Sect. 11.3 a practical view of the application of the uncertainty principles to wavelets (which generalize Gabor functions) has been presented.

12.5.2 Uncertainty Principles for Hermite Series

For a function $\psi_k(x) \in L^2(R)$ the mean $\mu(\psi_k)$ and the variance $\Delta^2(\psi_k)$ are defined as follows [89]:

$$\mu(\psi_k) = \int x |\psi_k(x)|^2 dx, \quad \Delta^2(\psi_k) = \int |x - \mu(\psi_k)|^2 |\psi_k(x)|^2 dx \tag{12.22}$$

The Hermite basis functions $\psi_k(x)$ were defined in Eq. (12.6), while the Hermite polynomials $H_k(x)$ were given in Eq. (12.8). The Hermite functions are an orthonormal basis for $L^2(R)$. It was also shown that Hermite functions are the eigenfunctions of the solution of Schrödinger's equation, and remain invariant under Fourier transform, subject only to a change of scale as given in Eq. (12.13), i.e. $\Psi_k(s) = j^n \psi_k(s)$. Furthermore, the *mean-dispersion principle* provides an expression of the uncertainty principle for orthogonal series [89]. This states that there does not exist an infinite orthonormal sequence $\{\psi_k\}_{k=0}^{\infty} \subset L^2$, such that all four of $\mu(\psi_k)$, $\mu(\Psi_k)$, $\Delta(\psi_k)$ and $\Delta(\Psi_k)$ are uniformly bounded.

Using Eq. (12.22) for the Hermite basis functions one obtains $\mu(\psi_k) = \mu(\Psi_k) = 0$, and $\Delta^2(\psi_k) = \Delta^2(\Psi_k) = \frac{2k+1}{4\pi}$ which quantifies the results of the mean-dispersion principle and formulates an uncertainty principle for Gauss–Hermite basis functions (Fig. 12.4).

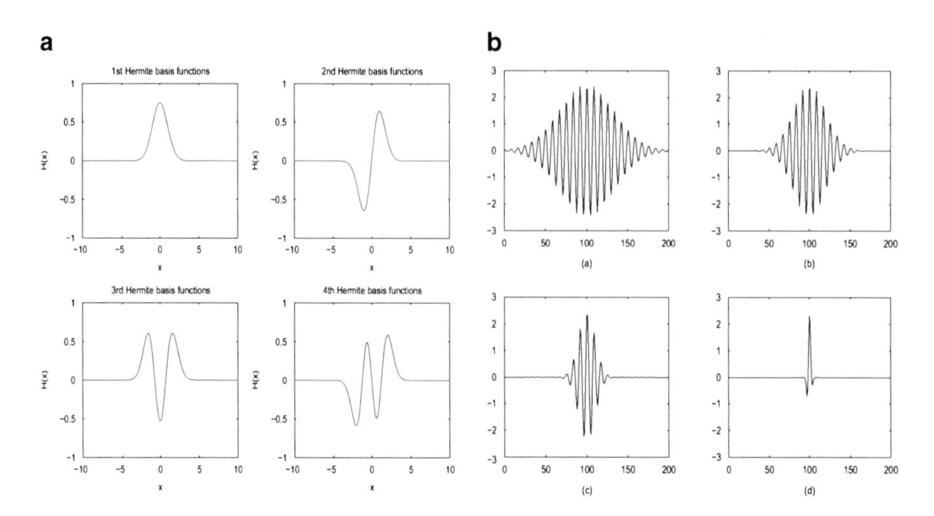

Fig. 12.4 (**a**) The first four modes of the Hermites series expansion (**b**) Fourier transform of the basic mode of the Hermite expansion assuming the following variances: (**a**) $3\sigma^2$, (**b**) $5\sigma^2$, (**c**) $10\sigma^2$, (**d**) $25\sigma^2$

12.6 Multiscale Modelling of Dynamical System

The performance of neural networks that use the eigenstates of the quantum harmonic oscillator as basis functions is compared to the performance of one hidden layer FNN with sigmoidal basis functions (OHL-FNN), as well as to Radial Basis Function (RBF) neural networks with Gaussian activation functions. It should be noted that the sigmoidal basis functions $\phi(x) = \frac{1}{1+\exp(-x)}$ in OHL-FNN and the Gaussian basis functions in RBF do not satisfy the property of orthogonality. Unlike this, neural networks with Hermite basis functions use the orthogonal eigenfunctions of the quantum harmonic oscillator which are given in Eq. (12.10).

In the sequel neural networks with Hermite basis functions, one hidden layer FNN with sigmoidal basis functions and RBF neural networks with Gaussian functions are used to approximate function $y_k = f(x_k) + v_k$, where v_k is a noise sequence, independent from x_k's. The obtained results are depicted in Fig. 12.5 to Fig. 12.14. The training pairs were x_k, y_k^d. Root mean square error, defined as $RMSE = \sqrt{\frac{1}{N}\sum_{k-1}^{N}(y_k - y_k^d)^2}$, gives a measure of the performance of the neural networks.

In the case of RBF and of neural networks with Hermite basis functions, training affects only the output weights, and can be performed with second order gradient algorithms. However, since the speed of convergence is not the primary objective of this study, the LMS (Least Mean Square) algorithm is sufficient for training. In the case of the OHL-FNN with sigmoidal basis functions, training concerns weights of both the hidden and output layer and is carried out using the back-propagation

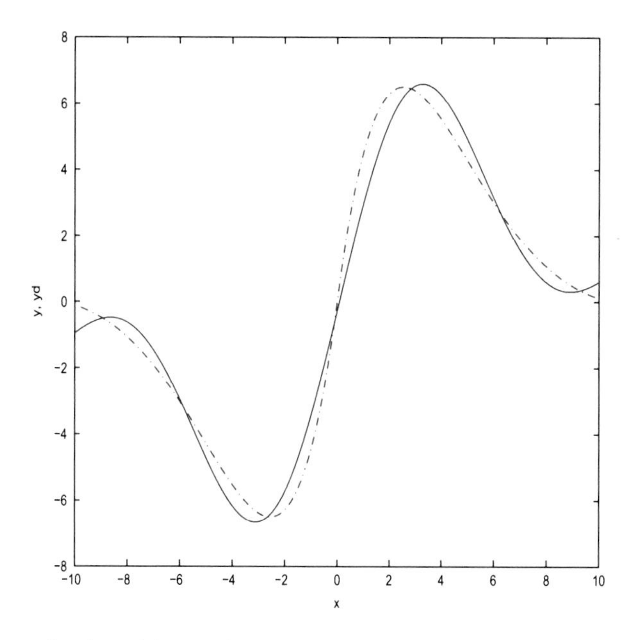

Fig. 12.5 Approximation of the test function of Eq. (12.23) (*dashed line*), using a feed-forward neural network with Hermite basis functions

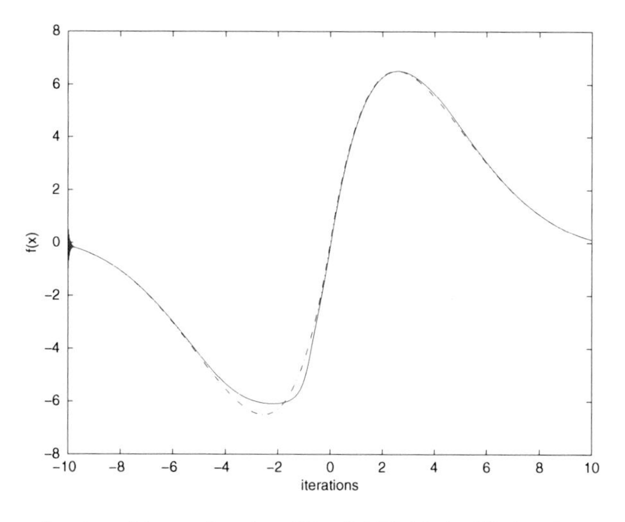

Fig. 12.6 Approximation of the test function of Eq. (12.23) (*dashed line*), using a one hidden layer FNN with sigmoidal basis functions

algorithm [78]. Alternatively, second order (Newton) training methods can be used [118].

The following test functions are examined (dashed red lines):

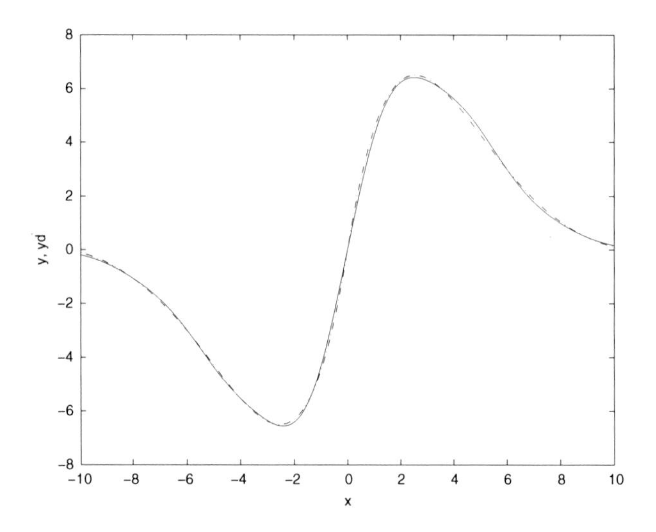

Fig. 12.7 Approximation of the test function of Eq. (12.23) (*dashed line*), using an RBF neural network with Gaussian basis functions

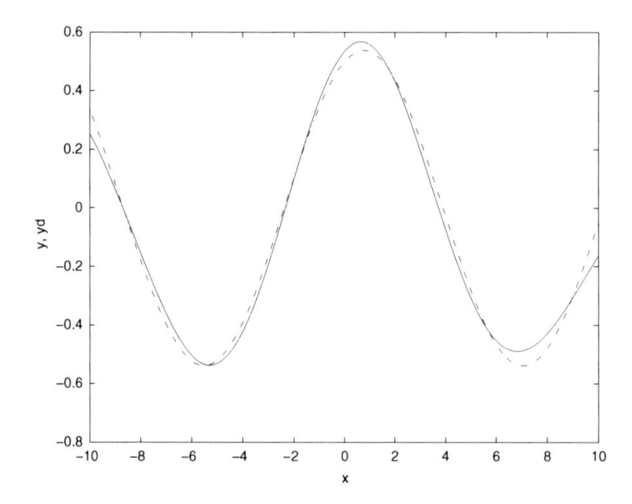

Fig. 12.8 Approximation of the test function of Eq. (12.24) (*dashed line*), using a feed-forward neural network with Hermite basis functions

1. Test function 1 given in Eq. (12.23) over the domain $D = [-10, 10]$:

$$f(x) = 0.5\cos(x/2) + 0.3\sin(x/2) \tag{12.23}$$

2. Test function 2 given in Eq. (12.24) over the domain $D = [-10, 10]$:

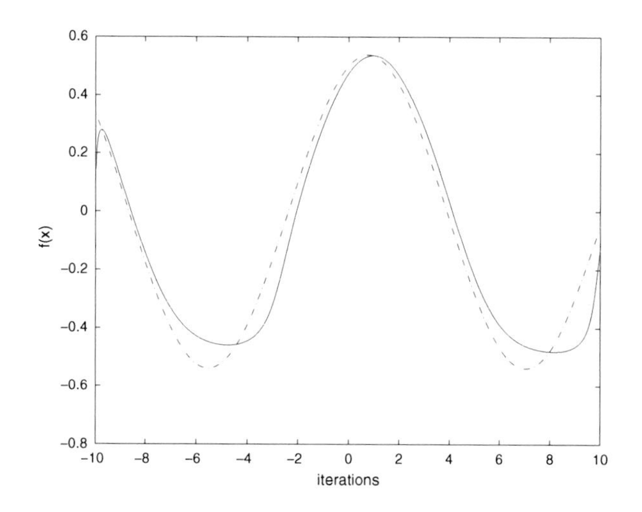

Fig. 12.9 Approximation of the test function of Eq. (12.24) (*dashed line*), using a one hidden layer FNN with sigmoidal basis functions

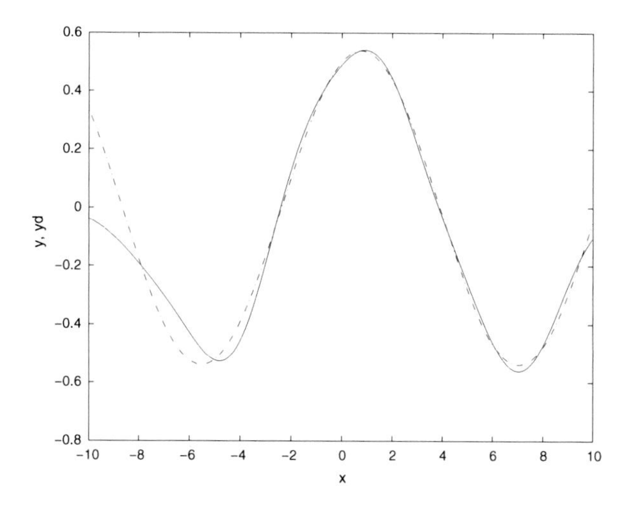

Fig. 12.10 Approximation of the test function of Eq. (12.24) (*dashed line*), using an RBF neural network with Gaussian basis functions

$$f(x) = \begin{array}{l} -2.186x - 12.864, \ -10 \leq x < -2 \\ 4.246x, \ -2 \leq x < 0 \\ 10e^{-(0.05x+0.5)} \sin[(0.001x + 0.05)x], \ 0 \leq x \leq 10 \end{array} \quad (12.24)$$

3. Test function 3 given in Eq. (12.25) over the domain $D = [-10, 10]$:

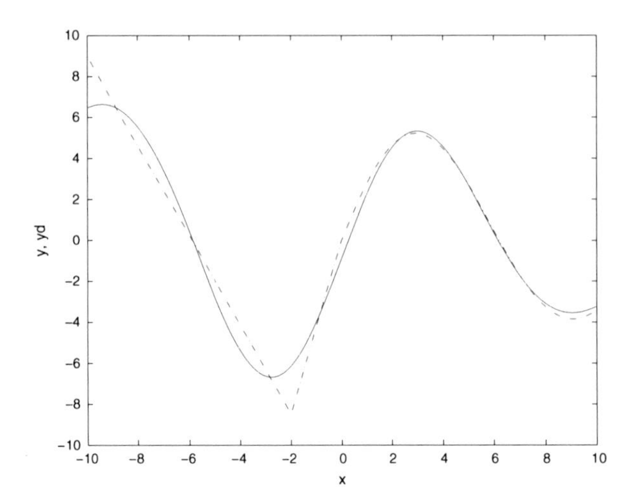

Fig. 12.11 Approximation of the test function of Eq. (12.25) (*dashed line*), using a feed-forward neural network with Hermite basis functions

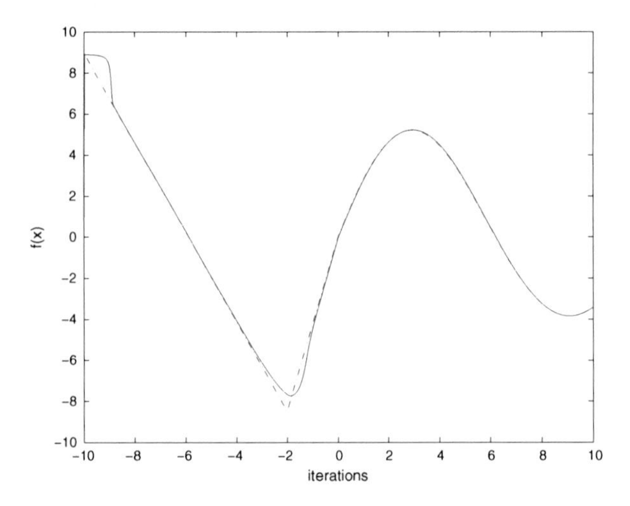

Fig. 12.12 Approximation of the test function of Eq. (12.25) (*dashed line*), using a one hidden layer FNN with sigmoidal basis functions

$$f(x) = 20.5e^{(-0.3|x|)} \sin(0.03x) \cos(0.7x) \tag{12.25}$$

4. Test function 4 given in Eq. (12.26) over the domain $D = [-10, 10] \times [-10, 10]$:

$$f(x, y) = \sin(0.3x) \sin(0.3y) e^{(-0.1|y|)} \tag{12.26}$$

The approximation results for the 2D function of Eq. (12.26), obtained by an NN with Hermite basis functions are shown in Fig. 12.15. The approximation

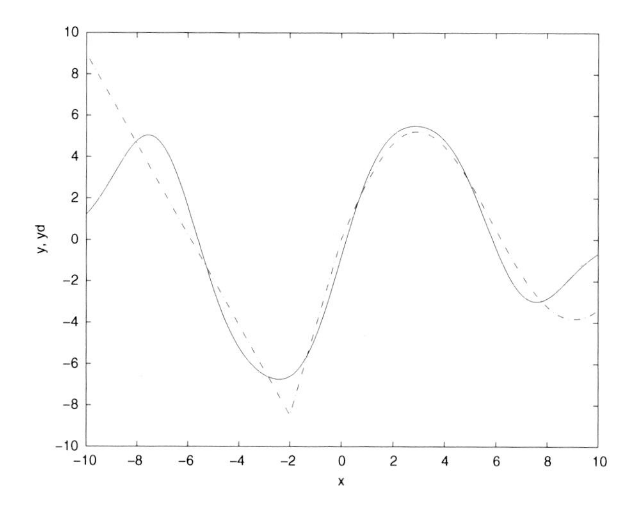

Fig. 12.13 Approximation of the test function of Eq. (12.25) (*dashed line*), using an RBF neural network with Gaussian basis functions

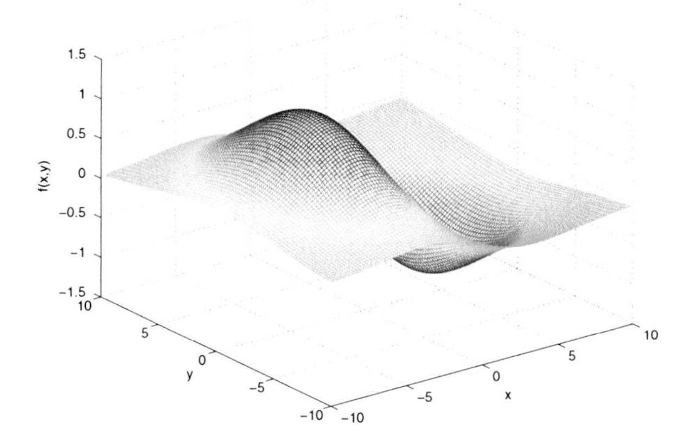

Fig. 12.14 Test function of Eq. (12.26)

of the test function obtained by a one hidden layer FNN with sigmoidal basis functions is shown in Fig. 12.16. Finally the approximation achieved by RBF neural network with Gaussian activation functions is depicted in Fig. 12.17.

5. Test function 5 given in Eq. (12.27) over the domain $D = [-10, 10] \times [-10, 10]$ (Fig. 12.18):

$$f(x, y) = 1.9e^{(-0.02x^2 - 0.02y^2)} \tanh(-0.3y) \tag{12.27}$$

The approximation results for the 2D function of Eq. (12.27), obtained by an NN with Hermite basis functions, are depicted in Fig. 12.19. The approximation

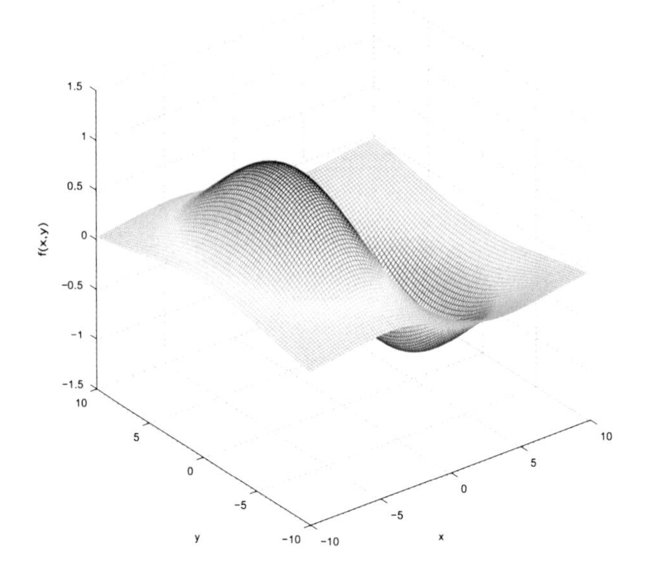

Fig. 12.15 Approximation of the test function of Eq. (12.26), using a feed-forward neural network with Hermite basis functions

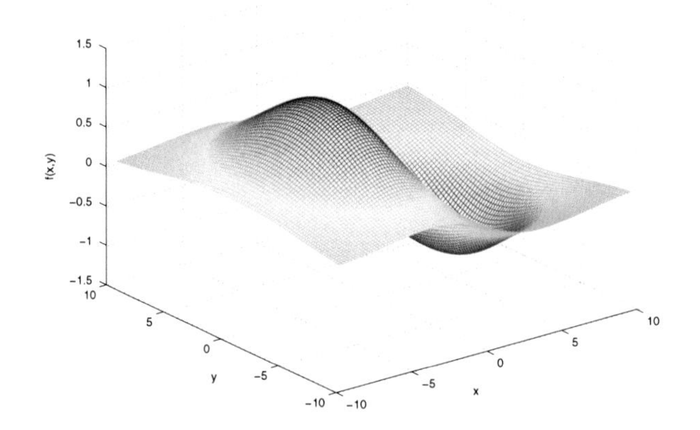

Fig. 12.16 Approximation of the test function of Eq. (12.26), using a one hidden layer FNN with sigmoidal basis functions

obtained by a one hidden layer FNN with sigmoidal basis functions is shown in Fig. 12.20. Finally the approximation achieved by an RBF neural network with Gaussian activation functions is given in Fig. 12.21.

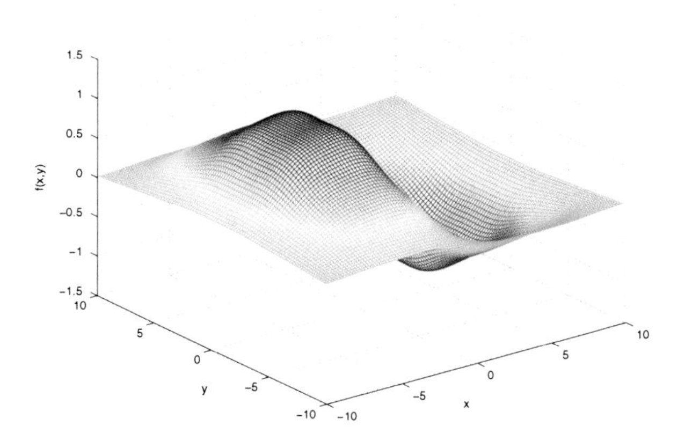

Fig. 12.17 Approximation of the test function of Eq. (12.26), using an RBF neural network with Gaussian basis functions

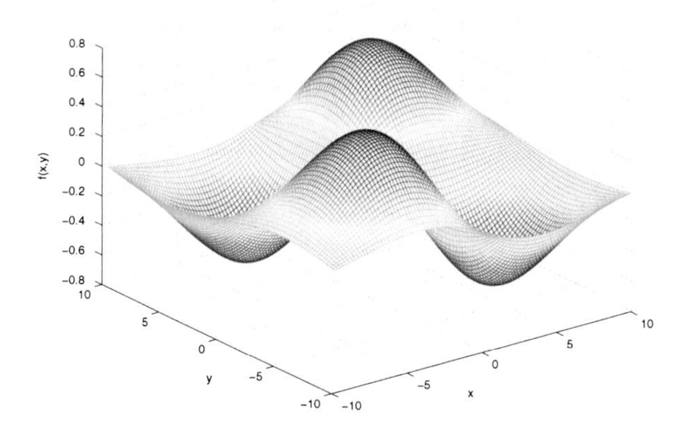

Fig. 12.18 Test function of Eq. (12.27)

Table 12.1 demonstrates the final RMSE values, succeeded by neural networks with Hermite basis functions (Hermite-FNN), the one hidden layer FNN with sigmoidal basis functions (OHL-FNN) and the RBF with Gaussian activation functions, after 50 epochs.

From Table 12.1, it is observed that in function approximation problems, neural networks with Hermite basis functions perform at least as well as one hidden layer FNN with sigmoidal activation functions (OHL-FNN) and RBF with Gaussian activation functions. As expected, the number of nodes of the hidden layer of the OHL-FNN and the variance of the Gaussian basis functions of the RBF affect the quality of function approximation. In the case of 1D functions Hermite-based neural

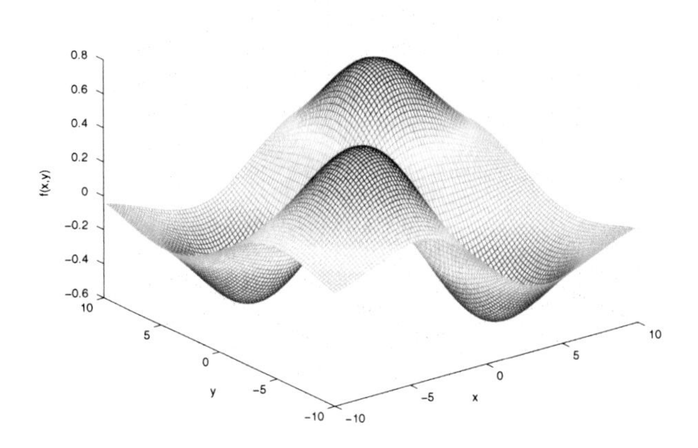

Fig. 12.19 Approximation of the test function of Eq. (12.27), using a feed-forward neural network with Hermite basis functions

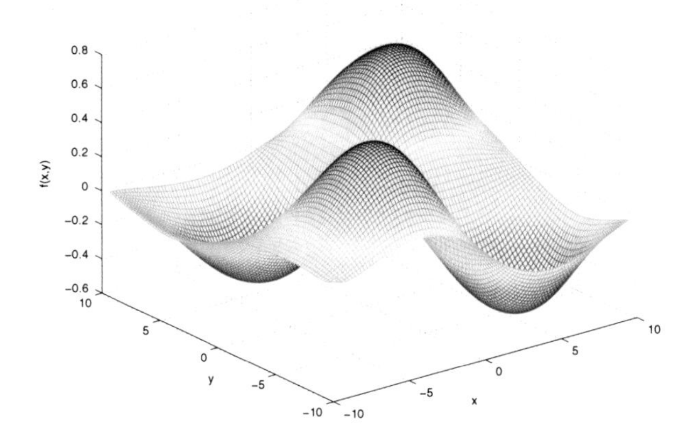

Fig. 12.20 Approximation of the test function of Eq. (12.27), using a one hidden layer FNN with sigmoidal basis functions

networks and OHL-FNN succeed better approximation than RBF. In the case of the 2D functions, the performance of OHL-FNN improved when more nodes were added to the hidden layer, while the performance of RBF improved when the spread of the gaussians was increased.

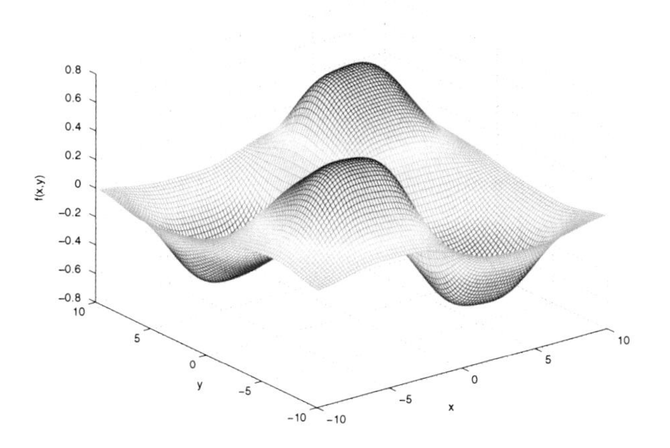

Fig. 12.21 Approximation of the test function of Eq. (12.27), using an RBF neural network with Gaussian basis functions

Table 12.1 Evaluation of NN with Hermite basis functions

Function	FNN type	Num nodes	RMSE
	Hermite	6	0.613
Eq. (12.23)	OHL-FNN	6	0.187
	RBF	6	0.223
	Hermite	6	0.034
Eq. (12.24)	OHL-FNN	6	0.059
	RBF	6	0.373
	Hermite	6	0.726
Eq. (12.25)	OHL-FNN	6	0.339
	RBF	6	1.565
	Hermite	16	0.047
Eq. (12.26)	OHL-FNN	16	0.121
	RBF	16	0.050
	Hermite	16	0.032
Eq. (12.27)	OHL-FNN	16	0.117
	RBF	16	0.028

12.7 Applications to Image Compression

The suitability of NN based on the QHO eigenstates for function approximation and system modelling has been evaluated in [164], using the benchmark tests given in [221]. Here, the image compression capabilities of FNN with Hermite basis functions will be evaluated. Regarding the structure of FNN used for image compression, the hidden layer has fewer nodes than the input layer, and the hidden layer output values are considered as the compressed image (see Fig. 12.22a). In order to reconstruct the compressed image, the weights between the nodes of the hidden and the output layer are also transmitted. If the number of the output values

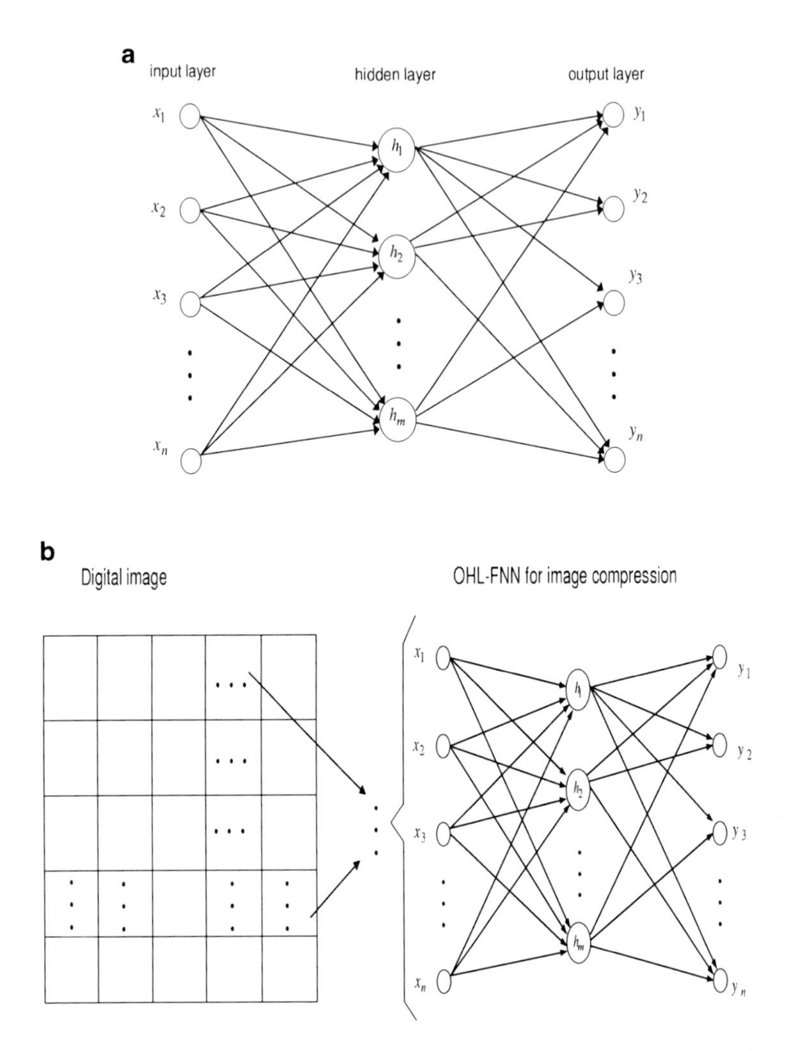

Fig. 12.22 (**a**) A one hidden layer FNN for image compression, having n neurons at the input and output layer and m neurons at the hidden layer ($m < n$) (**b**) Image compression using a one hidden layer FNN: the image is divided into square blocks that become input training patterns to the NN

for all the nodes of the hidden layer, plus the weights between the hidden and the output layer, is less than the total number of pixels of the image, compression is achieved. The fewer the number of units in the middle layer, the higher the degree of compression [117, 118].

In FNN-based image compression, an image of size $L \times L$ pixels is first divided into P square blocks of equal size $M_b \times M_b$. The total number of blocks is $P = \frac{L^2}{M_b^2}$. Each square block is then arranged into a vector of dimension $n \times 1$, i.e. $n = M_b^2$, that is fed to the neural network as an input training pattern. All such vectors are

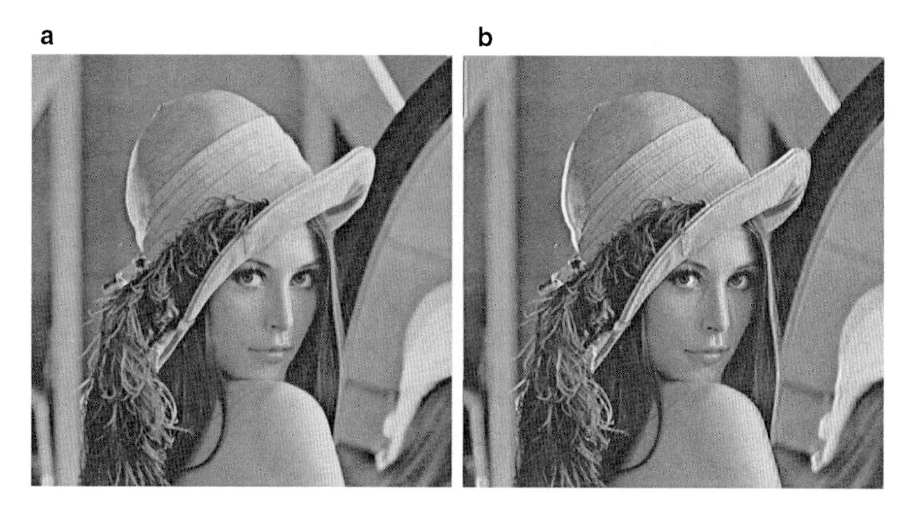

Fig. 12.23 (**a**) Lena image used in image compression tests (**b**) Reconstructed Lena image achieved by a one hidden layer FNN with sigmoidal basis functions

Fig. 12.24 (**a**) Reconstructed Lena image achieved by an FNN with Hermite basis functions, (**b**) Boats image used in image compression tests

put together to form a training matrix X of size $n \times P$. The target matrix D for the training of the neural net is considered to be the same as the input matrix X. The structure of an FNN for image compression is depicted in Fig. 12.22b. To quantify the quality of the reconstructed image, the compression ratio ρ can be used, which in a simple form is expressed as:

$$\rho = \frac{L \times L}{m \times P + m \times n} \tag{12.28}$$

Fig. 12.25 (**a**) Reconstructed Boats image achieved by a one hidden layer FNN with sigmoidal basis functions, (**b**) Reconstructed Boats image achieved by an FNN with Hermite basis functions

where the first term in the denominator is the number of outputs of the hidden layer with respect to all input patterns, and the second term is the number of the output weights.

In the sequel the image compression performed by FNN with Hermite basis functions is compared to the compression achieved by an FNN with sigmoidal activation functions. The hidden layer nodes of the FNN with Hermite basis functions contains the first 4 eigenstates of the quantum harmonic oscillator given in Eq. (12.10). Both neural structures consist of 16 nodes at the input layer, 4 nodes at the hidden layer, and 16 nodes at the output layer.

Tests are performed on two benchmark images, Lena and Boats. Both images are of size 512×512 pixels and the block size is selected as 4×4. Using Eq. (12.28), the compression ratio achieved by the examined neural structures is found to be $\rho \simeq 4$. The reconstructed Lena images are depicted in Fig 12.23a,b and in Fig. 12.24a.

The reconstructed Boats images are depicted in Fig 12.24b and in Fig. 12.25a,b, respectively.

12.8 Applications to Fault Diagnosis

12.8.1 Signals Power Spectrum and the Fourier Transform

It will be shown that neuronal networks using as activation functions the QHO eigenstates can be also used in fault diagnosis tasks. Apart from time-domain approaches to fault diagnosis, one can use frequency-domain methods to find out the existence of a fault in a system under monitoring [17,222,223]. Thus to conclude if a dynamical system has been subjected to fault of its components, information coming from the power spectrum of its outputs can be processed. To find the spectral density of a signal $\psi(t)$ with the use of its Fourier transform $\Psi(j\omega)$, the following definition is used:

$$\begin{aligned} E_\psi &= \int_{-\infty}^{+\infty}(\psi(t))^2dt = \frac{1}{2\pi}\int_{-\infty}^{+\infty}\psi(t)(\int_{-\infty}^{+\infty}\Psi(j\omega)e^{j\omega t}d\omega)dt \text{ i.e.} \\ E &= \frac{1}{2\pi}\int_{-\infty}^{+\infty}\Psi(j\omega)\Psi(-j\omega)d\omega \end{aligned} \tag{12.29}$$

Taking that $\psi(t)$ is a real signal it holds that $\Psi(-j\omega) = \Psi^*(j\omega)$ which is the signal's complex conjugate. Using this in Eq. (12.29) one obtains

$$\begin{aligned} E_\psi &= \frac{1}{2\pi}\int_{-\infty}^{+\infty}\Psi(j\omega)\Psi^*(j\omega)d\omega \text{ or} \\ E_\psi &= \frac{1}{2\pi}\int_{-\infty}^{+\infty}|\Psi(j\omega)|^2d\omega \end{aligned} \tag{12.30}$$

This means that the energy of the signal is equal to $\frac{1}{2\pi}$ times the integral over frequency of the square of the magnitude of the signal's Fourier transform. This is *Parseval's theorem*. The integrated term $|\Psi(j\omega)|^2$ is the energy density per unit of frequency and has units of magnitude squared per Hertz.

12.8.2 Power Spectrum of the Signal Using the Gauss–Hermite Expansion

As shown in Eqs. (12.11) and (12.17) the Gauss–Hermite basis functions satisfy the orthogonality property, i.e. for these functions it holds

$$\int_{-\infty}^{+\infty}\psi_m(x)\psi_k(x)dx = \begin{cases} 1 \text{ if } m = k \\ 0 \text{ if } m \neq k \end{cases}$$

Therefore, using the definition of the signal's energy one has

$$E = \int_{-\infty}^{+\infty}(\psi(t))^2dt = \int_{-\infty}^{+\infty}[\sum_{k=1}^{N}c_k\psi_k(t)]^2dt \tag{12.31}$$

and exploiting the orthogonality property one obtains

$$E = \sum_{k=1}^{N}c_k^2 \tag{12.32}$$

Therefore the square of the coefficients c_k provides an indication of the distribution of the signal's energy to the associated basis functions. One could arrive at the same results using the Fourier transformed description of the signal and Parseval's theorem. It has been shown that the Gauss–Hermite basis functions remain invariant under the Fourier transform subject only to a change of scale. Denoting by $\Psi(j\omega)$

the Fourier transformed signal of $\psi(t)$ and by $\Psi_k(j\omega)$ the Fourier transform of the k-th Gauss–Hermite basis function one obtains

$$\Psi(j\omega) = \sum_{k=1}^{N} c_k \Psi_k(j\omega) \tag{12.33}$$

and the energy of the signal is computed as

$$E_\psi = \frac{1}{2\pi}\int_{-\infty}^{+\infty} |\Psi(j\omega)|^2 d\omega \tag{12.34}$$

Substituting Eqs. (12.33) into (12.34) one obtains

$$E_\psi = \frac{1}{2\pi}\int_{-\infty}^{+\infty} |\sum_{k=1}^{N} c_k \Psi_k(j\omega)|^2 d\omega \tag{12.35}$$

and using the invariance of the Gauss–Hermite basis functions under Fourier transform one gets

$$E_\psi = \frac{1}{2\pi}\int_{-\infty}^{+\infty} |\sum_{k=1}^{N} c_k \alpha^{-\frac{1}{2}} \psi_k(\alpha^{-1} j\omega)|^2 d\omega \tag{12.36}$$

while performing the change of variable $\omega_1 = \alpha^{-1}\omega$ it holds that

$$E_\psi = \frac{1}{2\pi}\int_{-\infty}^{+\infty} |\sum_{k=1}^{N} c_k \alpha^{\frac{1}{2}} \psi_k(j\omega_1)|^2 d\omega_1 \tag{12.37}$$

Next, by exploiting the orthogonality property of the Gauss–Hermite basis functions one gets that the signal's energy is proportional to the sum of the squares of the coefficients c_k which are associated with the Gauss–Hermite basis functions, i.e. a relation of the form

$$E_\psi = \sum_{k=1}^{N} c_k^2 \tag{12.38}$$

12.8.3 Detection of Changes in the Spectral Content of the System's Output

The thermal model of a system, i.e. variations of the temperature output signal, has been identified considering the previously analyzed neural network with Gauss–Hermite basis functions. As shown in Figs. 12.26 and 12.27, thanks to the multi-frequency characteristics of the Gauss–Hermite basis functions, such a neural model can capture with increased accuracy spikes and abrupt changes in the temperature profile [164, 167, 169, 221]. The RMSE (Root Mean Square Error) of training the Gauss–Hermite neural model was of the order of 4×10^{-3}.

The update of the output layer weights of the neural network is given by a gradient equation (LMS-type) of the form

$$w^i(k+1) = w^i(k) - \eta e(k)\phi^T(k) \tag{12.39}$$

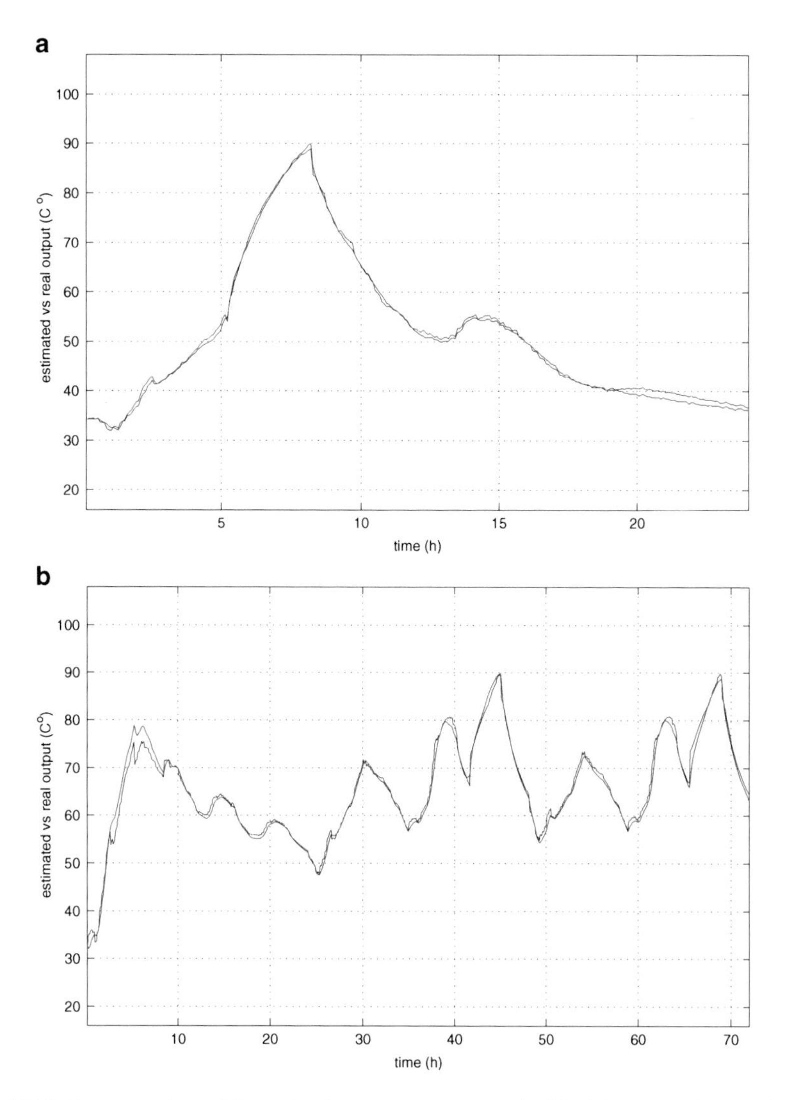

Fig. 12.26 Approximation of the system's temperature output (*red line*) by a neural network with Hermite polynomial basis functions (*blue-line*) (**a**) temperature's time variation—profile 1 (**b**) temperature's time variation—profile 2

where $e(k) = y(k) - y_d(k)$ is the output estimation error at time instant k and $\phi^T(k)$ is the regressor vector having as elements the values $\phi(x(k))$ of the Gauss–Hermite basis functions for input vector $x(k)$.

To approximate the temperature–signal's variations described in a data set consisting of 870 quadruplets of input–output data, a feed-forward neural network with 3-D Gauss–Hermite basis functions has been used. The neural network contained 64 nodes in its hidden layer.

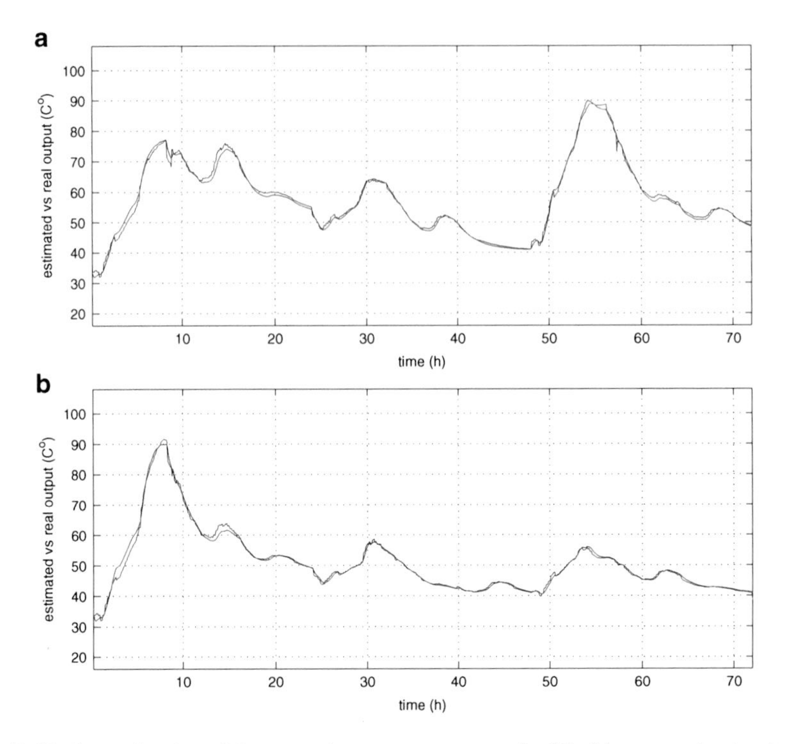

Fig. 12.27 Approximation of the system's temperature output (*red line*) by a neural network with Hermite polynomial basis functions (*blue-line*) (**a**) temperature's time variation—profile 3 (**b**) temperature's time variation—profile 4

The spectral components of the temperature signal for both the fault-free and the under-fault operation of the system have been shown in Figs. 12.28, 12.29, 12.30, 12.31. It can be noticed that after a fault has occurred the amplitude of the aforementioned spectral components changes and this can be a strong indication about failure of the monitored system.

Obviously, the proposed spectral decomposition of the monitored signal, with series expansion in Gauss–Hermite basis functions can be used for fault detection tasks. As it can be seen in Figs. 12.28, 12.29, 12.30, 12.31, in case of failure, the spectral components of the monitored signal differ from the ones which are obtained when the system is free of fault. Moreover, the fact that certain spectral components exhibit greater sensitivity to the fault and change value in a more abrupt manner is a feature which can be exploited for fault isolation. Specific failures can be associated with variations of specific spectral components. Therefore, they can provide indication about the appearance of specific types of failures and specific malfunctioning components.

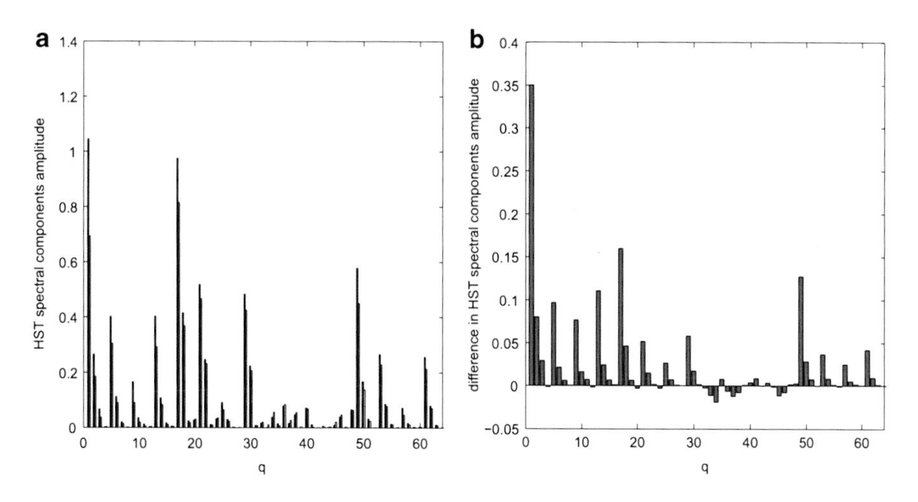

Fig. 12.28 Temperature's time variation—profile 1: (**a**) Amplitude of the spectral components of the temperature signal in the fault free case (*red bar line*) and when a fault had taken place (*yellow bar line*) (**b**) differences in the amplitudes of the spectral components between the fault-free and the faulty case (*green bar line*)

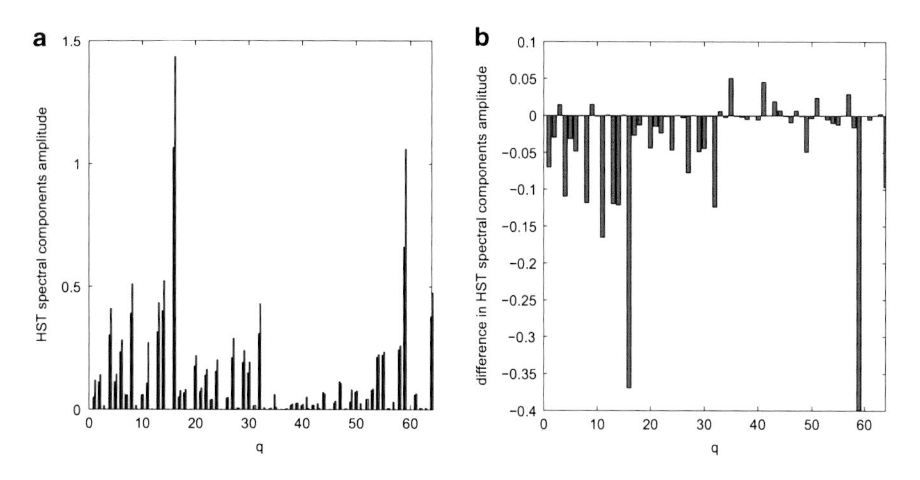

Fig. 12.29 Temperature time variation—profile 2: (**a**) Amplitude of the spectral components of the temperature signal in the fault free case (*red bar line*) and when a fault had taken place (*yellow bar line*) (**b**) differences in the amplitudes of the spectral components between the fault-free and the faulty case (*green bar line*)

12.9 Conclusions

In this chapter FNN that use the eigenstates of the quantum harmonic oscillator as basis functions have been studied. The proposed neural networks have some interesting properties: (a) the basis functions are invariant under the Fourier transform,

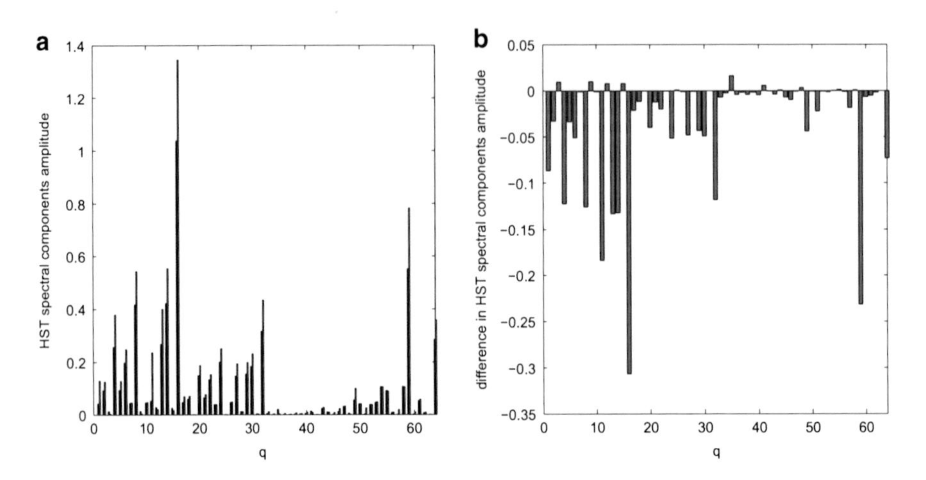

Fig. 12.30 Temperature time variation—profile 3: (**a**) Amplitude of the spectral components of the temperature signal in the fault free case (*red bar line*) and when a fault had taken place (*yellow bar line*) (**b**) differences in the amplitudes of the spectral components between the fault-free and the faulty case (*green bar line*)

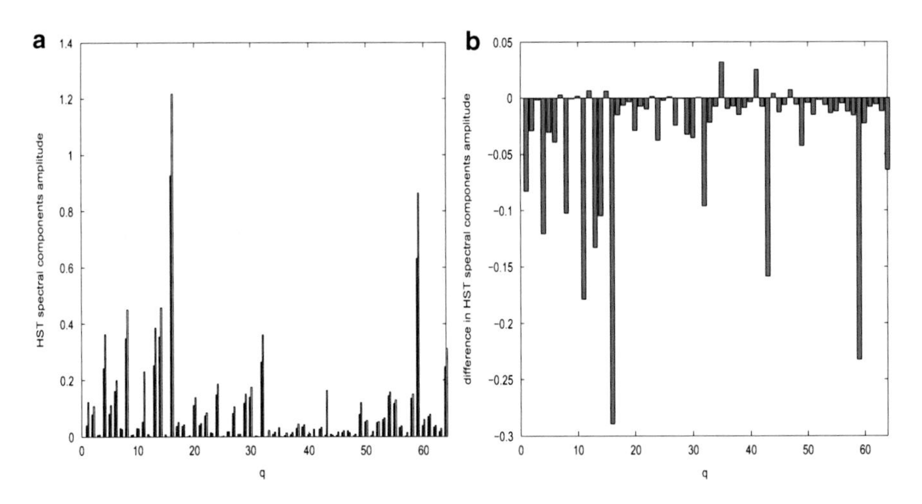

Fig. 12.31 Temperature time variation—profile 4: (**a**) Amplitude of the spectral components of the temperature signal in the fault free case (*red bar line*) and when a fault had taken place (*yellow bar line*) (**b**) differences in the amplitudes of the spectral components between the fault-free and the faulty case (*green bar line*)

subject only to a change of scale (b) the basis functions are the eigenstates of the quantum harmonic oscillator, and stem from the solution of Schrödinger's diffusion equation. The proposed FNN belong to the general category of nonparametric estimators and are suitable for function approximation, system modelling, and image processing.

FNN that use Hermite basis functions demonstrate the particle-wave nature of information as described by Schrödinger's diffusion equation. It is considered that the input variable x of the neural network can be described not only by crisp values (particle equivalent) but also by the normal modes of a wave function (wave equivalent). Since the basis functions are the eigenstates of the quantum harmonic oscillator, this means that the output of the proposed neural network is the weighted sum $\psi(x) = \sum_{k=1}^{m} w_k \psi_k(x)$ where $|\psi(x)|^2$ is the probability that the input of the neural network (quantum particle equivalent) is found between x and $x + \Delta x$. The weight w_k provides a measure of the probability to find the input on the neural network in the region of the patterns space associated with the eigenfunction $\psi_k(x)$.

Issues related to the uncertainty principle have been studied for the case of the QHO-based neural networks. The uncertainty principle was presented first through the Balian–Low theorem and next an expression of the uncertainty principle for Hermite basis functions has been given. It was shown that the Gauss–Hermite basis functions as well as their Fourier transforms cannot be uniformly concentrated in the time-frequency plane.

In the simulation tests, FNN with Hermite basis functions were evaluated in image compression problems. In this task the performance of FNN with Hermite basis functions was satisfactory and comparable to the compression succeeded by other neural structures, such as MLP or RBF networks. The fact that the basis functions of the QHO-based neural networks contain various frequencies makes them suitable for multi-resolution analysis.

Additionally, the use of FNN comprising as basis functions the eigenstates of the QHO was tested in the problem of fault diagnosis. Faults could be diagnosed by monitoring changes in the spectral content of a system and these in turn were associated with the values of the weights of the neural network.

Chapter 13
Quantum Control and Manipulation of Systems and Processes at Molecular Scale

Abstract Biological systems, at molecular level, exhibit quantum mechanical dynamics. A question that arises is if the state of such systems can be controlled by an external input such as an electromagnetic field or light emission (e.g., laser pulses). The chapter proposes a gradient method for feedback control and stabilization of quantum systems using Schrödinger's and Lindblad's descriptions. The eigenstates of the quantum system are defined by the spin model. First, a gradient-based control law is computed using Schrödinger's description. Next, an estimate of state of the quantum system is obtained using Lindblad's differential equation. In the latter case, by applying Lyapunov's stability theory and LaSalle's invariance principle one can compute a gradient control law which assures that the quantum system's state will track the desirable state within acceptable accuracy levels. The performance of the control loop is studied through simulation experiments for the case of a two-qubit quantum system.

13.1 Basics of Quantum Systems Control

Recently, evidence has emerged from several studies that quantum coherence is playing an important role in certain biological processes [95, 100, 107]. In [183] the quantization of systems in the quantum theory and their functional roles in biology has been analyzed. In [146] it has been shown that the electron clouds of nucleic acids in a single strand of DNA can be modeled as a chain of coupled quantum harmonic oscillators with dipole–dipole interaction between nearest neighbors. Finally it has been shown that the quantum state of a single base (A, C.G. or T) contains information about its neighbor, questioning the notion of treating individual DNA bases as independent bits of information. The next question that arises is if the state of such systems exhibiting quantum mechanical dynamics can be controlled by an external input, such as electromagnetic field or light emission (e.g., laser pulses). To this end this chapter analyzed the basics of quantum systems control and proposes a systematic method for assuring stability in quantum control feedback loops.

G.G. Rigatos, *Advanced Models of Neural Networks*,

DOI 10.1007/978-3-662-43764-3_13,

The main approaches to the control of quantum systems are: (a) open-loop control and (b) measurement-based feedback control [210]. In open-loop control, the control signal is obtained using prior knowledge about the quantum system dynamics and assuming a model that describes its evolution in time. Some open-loop control schemes for quantum systems have been studied in [153, 154]. Previous work on quantum open-loop control includes flatness-based control on a single qubit gate [41]. On the other hand, measurement-based quantum feedback control provides more robustness to noise and model uncertainty [30]. In measurement-based quantum feedback control, the overall system dynamics are described by the estimation equation called stochastic master equation or Belavkin's equation [20]. An equivalent approach can be obtained using Lindblad's differential equation [210]. Several researchers have presented results on measurement-based feedback control of quantum systems using the stochastic master equation or the Lindblad differential equation, while theoretical analysis of the stability for the associated control loop has been also attempted in several cases [18, 124, 203].

In this chapter, a gradient-based approach to the control of quantum systems will be examined. Previous results on control laws which are derived through the calculation of the gradient of an energy function of the quantum system can be found in [4, 9, 155, 163]. Convergence properties of gradient algorithms have been associated with Lyapunov stability theory in [22]. The chapter considers a quantum system confined in a cavity that is weakly coupled to a probe laser. The spin model is used to define the eigenstates of the quantum system. The dynamics of the quantum model are described by Lindblad's differential equation and thus an estimate of the system's state can be obtained. Using Lyapunov's stability theory a gradient-based control law is derived. Furthermore, by applying LaSalle's invariance principle it can be assured that under the proposed gradient-based control the quantum system's state will track the desirable state within acceptable accuracy levels. The performance of the control loop is studied through simulation experiments for the case of a two-qubit quantum system.

13.2 The Spin as a Two-Level Quantum System

13.2.1 Description of a Particle in Spin Coordinates

As explained in Chap. 7, the basic equation of quantum mechanics is *Schrödinger's equation*, i.e.

$$i\frac{\partial \psi}{\partial t} = H\psi(x,t) \tag{13.1}$$

where $|\psi(x,t)|^2$ is the probability density function of finding the particle at position x at time instant t, and H is the system's Hamiltonian, i.e. the sum of its kinetic and potential energy, which is given by $H = p^2/2m + V$, with p being the momentum of the particle, m the mass, and V an external potential. The solution of Eq. (13.1) is given by $\psi(x,t) = e^{-iHt}\psi(x,0)$ [34].

However, cartesian coordinates are not sufficient to describe the particle's behavior in a magnetic field and thus the spin variable taking values in SU(2) has been introduced. In that case the solution ψ of Schrödinger's equation can be represented in the basis $|r,\epsilon>$ where r is the position vector and ϵ is the spin's value which belongs in $\{-\frac{1}{2},\frac{1}{2}\}$ (fermion). Thus vector ψ which appears in Schrödinger's equation can be decomposed in the vector space $|r,\epsilon>$ according to $|\psi>=\sum_{\epsilon}\int d^3r|r,\epsilon>,<r,\epsilon|\psi>$. The projection of $|\psi>$ in the coordinates system r,ϵ is denoted as $<r,\epsilon|\psi>=\psi_{\epsilon}(r)$. Equivalently one has $\psi_{+}(r)=<r,+|\psi>$ and $\psi_{-}(r)=<r,-|\psi>$. Thus one can write $\psi(r)=[\psi_{+}(r),\psi_{-}(r)]^T$.

13.2.2 *Measurement Operators in the Spin State-Space*

It has been proven that the eigenvalues of the particle's magnetic moment are $\pm\frac{1}{2}$ or $\pm\hbar\frac{1}{2}$. The corresponding eigenvectors are denoted as $|+>$ and $|->$. Then the relation between eigenvectors and eigenvalues is given by

$$\begin{aligned}\sigma_z|+>&=+(\hbar/2)|+>\\ \sigma_z|->&=+(\hbar/2)|->\end{aligned} \tag{13.2}$$

which shows the two possible eigenvalues of the magnetic moment [34]. In general the particle's state, with reference to the spin eigenvectors, is described by

$$|\psi>=\alpha|+>+\beta|-> \tag{13.3}$$

with $|\alpha|^2+|\beta|^2=1$ while matrix σ_z has the eigenvectors $|+>=[1,0]$ and $|->=[0,1]$ and is given by

$$\sigma_z=\frac{\hbar}{2}\begin{pmatrix}1&0\\0&-1\end{pmatrix} \tag{13.4}$$

Similarly, if one assumes components of magnetic moment along axes x and z, one obtains the other two measurement (Pauli) operators

$$\sigma_x=\frac{\hbar}{2}\begin{pmatrix}0&1\\1&0\end{pmatrix},\quad \sigma_y=\frac{\hbar}{2}\begin{pmatrix}0&-i\\i&0\end{pmatrix} \tag{13.5}$$

13.2.3 The Spin Eigenstates Define a Two-Level Quantum System

The spin eigenstates correspond to two different energy levels. A neutral particle is considered in a magnetic field of intensity B_z. The particle's magnetic moment M and the associated kinetic moment Γ are collinear and are related to each other through the relation

$$M = \gamma \Gamma \tag{13.6}$$

The potential energy of the particle is

$$W = -M_z B_z = -\gamma B_z \Gamma_z \tag{13.7}$$

Variable $\omega_0 = -\gamma B_z$ is introduced, while parameter Γ_z is substituted by the spin's measurement operator S_z.

Thus the Hamiltonian H which describes the evolution of the spin of the particle due to field B_z becomes $H_0 = \omega_0 S_z$, and the following relations between eigenvectors and eigenvalues are introduced:

$$H|+>= +\tfrac{\hbar\omega_0}{2}|+>, \quad H|->= -\tfrac{\hbar\omega_0}{2}|-> \tag{13.8}$$

Therefore, one can distinguish 2 different energy levels (states of the quantum system)

$$\begin{aligned} E_+ &= +\tfrac{\hbar\omega_0}{2} \\ E_- &= -\tfrac{\hbar\omega_0}{2} \end{aligned} \tag{13.9}$$

By applying an external magnetic field the probability of finding the particle's magnetic moment at one of the two eigenstates (spin up or down) can be changed. This can be observed, for instance, in the Nuclear Magnetic Resonance (NMR) model and is the objective of quantum control [34].

13.3 The Lindblad and Belavkin Description of Quantum Systems

13.3.1 The Lindblad Description of Quantum Systems

It will be shown that the Lindblad and the Belavkin equation can be used in place of Schrödinger's equation to describe the dynamics of a quantum system. These equations use as state variable the probability density matrix $\rho = |\psi><\psi|$, associated with the probability of locating the particle at a certain eigenstate. The Lindblad and Belavkin equations are actually the quantum analogous of the Kushner–Stratonovich stochastic differential equation which denotes that the change of the probability of the state vector x to take a particular value depends on the difference (innovation) between the measurement $y(x)$ and the mean value of the estimation of the measurement $E[y(x)]$. It is also known that the Kushner–Stratonovich SDE can be written in the form of a Langevin SDE [155]

$$dx = \alpha(x)dt + b(x)dv \quad (13.10)$$

which finally means that the Lindblad and Belavkin description of a quantum system are a generalization of Langevin's SDE for quantum systems [210]. For a quantum system with state vector x and eigenvalues $\lambda(x) \in R$, the Lindblad equation is written as [27,210]

$$\hbar\dot{\rho} = -i[\hat{H}, \rho] + D[\hat{c}]\rho \quad (13.11)$$

where ρ is the associated probability density matrix for state x, i.e. it defines the probability to locate the particle at a certain eigenstate of the quantum system and the probabilities of transition to other eigenstates. The variable $\hat{H}$ is the system's Hamiltonian, operator $[A, B]$ is a Lie bracket defined as $[A, B] = AB - BA$, the vector $\hat{c} = (\hat{c}_1, \cdots, \hat{c}_L)^T$ is also a vector of operators, variable D is defined as $D[\hat{c}] = \sum_{l=1}^{L} D[\hat{c}_l]$, and finally $\hbar$ is Planck's constant.

13.3.2 The Belavkin Description of Quantum Systems

The Lindblad equation (also known as stochastic master equation), given in Eq. (13.11), is actually a differential equation which can be also written in the form of a stochastic differential equation that is known as *Belavkin equation*. The most general form of the Belavkin equation is:

$$\hbar d\rho_c = dt D[\hat{c}]\rho_c + H[-i\hat{H}dt + dz^+(t)\hat{c}]\rho_c \quad (13.12)$$

Variable H is an operator which is defined as follows:

$$H[\hat{r}]\rho = \hat{r}\rho + \rho\hat{r}^+ - Tr[\hat{r}\rho + \rho\hat{r}^+]\rho \quad (13.13)$$

Variable $\hat{H}$ stands for the Hamiltonian of the quantum system. Variable $\hat{c}$ is an arbitrary operator obeying $\hat{c}^+\hat{c} = \hat{R}$, where $\hat{R}$ is an hermitian operator. The infinite dimensional complex variables vector dz is defined as $dz = (dz_1, \cdots, dz_L)^T$, and in analogy to the innovation dv of the Langevin equation (see Eq. (13.10)), variable dz expresses innovation for the quantum case. Variable dz^+ denotes the conjugate-transpose $(dz^*)^T$. The statistical characteristics of dz are

$dzdz^{+} = \hbar H_c dt$, $dzdz^{T} = \hbar Y dt$. In the above equations matrix Y is a symmetric complex-valued matrix. Variable H_c is defined as $m_1 = \{H_c = diag(n_1, \cdots, n_L) : \forall l, n_l \in [0,1]\}$, where n_l can be interpreted as the possibility of monitoring the l-th output channel. There is also a requirement for matrix U to be positive semi-definite. As far as the *measured output* of the Belavkin equation is concerned one has an equation of complex currents $J^T dt =< \hat{c} H_c + \hat{c}^{+} Y >_c + dz^T$, where $<>$ stands for the mean value of the variable contained in it [210]. Thus, in the description of the quantum system according to Belavkin's formulation, the state equation and the output equation are given by Eq. (13.14).

$$\begin{aligned} \hbar d\rho_c &= dt D[\hat{c}]\rho_c + H[-i\hat{H}dt + dz^{+}(t)\hat{c}]\rho_c \\ J^T dt &=< \hat{c}^T H_c + \hat{c}^{+} Y_c > dt + dz^T \end{aligned} \tag{13.14}$$

where ρ_c is the probability density matrix (state variable) for remaining at one of the quantum system eigenstates, and J is the measured output (current).

13.3.3 Formulation of the Control Problem

The control loop consists of a cavity where the multi-particle quantum system is confined and of a laser probe which excites the quantum system. Measurements about the condition of the quantum system are collected through photodetectors and thus the projections of the probability density matrix ρ of the quantum system are turned into weak current. By processing this current measurement and the estimate of the quantum system's state which is provided by Lindblad's or Belavkin's equation, a control law is generated which modifies a magnetic field applied to the cavity. In that manner, the state of the quantum system is driven from the initial value $\rho(0)$ to the final desirable value $\rho_d(t)$ (see Fig. 13.1).

When Schrödinger's equation is used to describe the dynamics of the quantum system the objective is to move the quantum system from a state ψ, that is associated with a certain energy level, to a different eigenstate associated with the desirable energy level. When Lindblad's or Belavkin's equation is used to describe the dynamics of the quantum system, the control objective is to stabilize the probability density matrix $\rho(t)$ on some desirable quantum state $\rho_d(t) \in C^n$, by controlling the intensity of the magnetic field. The value of the control signal is determined by processing the measured output which in turn depends on the projection of $\rho(t)$ defined by $Tr\{P\rho(t)\}$.

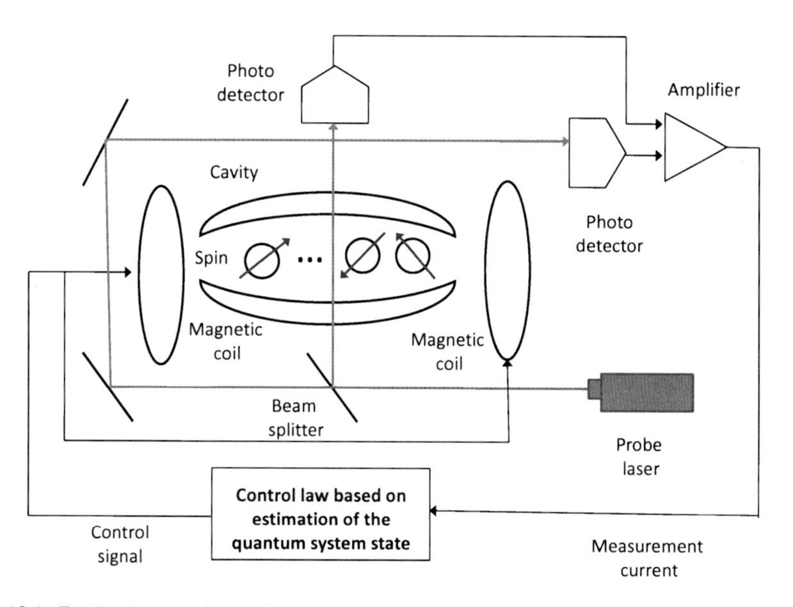

Fig. 13.1 Feedback control loop for quantum systems

13.4 A Feedback Control Approach for Quantum System Stabilization

13.4.1 Control Law Calculation Using Schrödinger's Equation

It is assumed that the dynamics of the controlled quantum system is described by a Schrödinger equation of the form

$$i\hbar\dot{\psi}(t) = [H_0 + f(t)H_1]\psi(t) \quad \psi(t) \in C^n \tag{13.15}$$

where H_0 is the system's Hamiltonian, H_1 is the control Hamiltonian, and $f(t)$ is the external control input. The following Lyapunov function is then introduced [9]

$$V(\psi) = (\psi^+ Z\psi - Z_d)^2 \tag{13.16}$$

where $+$ stands for the transposition and complex conjugation, Z is the quantum system's observable and is associated with the energy of the system. The term $\psi^+ Z\psi$ denotes the observed mean energy of the system at time instant t and Z_d is the desirable energy value. The first derivative of the Lyapunov function of Eq. (13.16) is

$$\dot{V}(\psi) = 2[\psi^+ Z\psi - Z_d][\dot{\psi}^+ Z\psi + \psi^+ Z\dot{\psi}] \tag{13.17}$$

while from Eq. (13.15) it also holds $\dot{\psi}(t) = -\frac{i}{\hbar}[H_0 + f(t)H_1]\psi(t)$, which results into

$$\dot{V}(\psi) = \tfrac{2i}{\hbar}(\psi^+ Z\psi - Z_d)\cdot\psi^+\{H_0 Z - ZH_0 + f(H_1 Z - ZH_1)\}\psi \quad (13.18)$$

Choosing the control signal $f(t)$ to be proportional to the gradient with respect to f of the first derivative of the Lyapunov function with respect to time (velocity gradient), i.e.

$$f(t) = k\nabla_f\{\dot{V}(\psi)\} \quad (13.19)$$

and for Z such that $\psi^+ H_0 Z\psi = \psi^+ ZH_0\psi$ (e.g. $Z = H_0$) one obtains

$$f(t) = \frac{2i}{\hbar}(\psi^+ Z\psi - Z_d)\psi^+(H_1 Z - ZH_1)\psi \quad (13.20)$$

Substituting Eqs. (13.20) into (13.18) provides

$$\dot{V}(\psi) = k\frac{2i}{\hbar}(\psi^+ Z\psi - Z_d)\psi^+[\frac{2i}{\hbar}(\psi^+ Z\psi - Z_d)\psi^+ \\ (H_1 Z - ZH_1)\psi(H_1 Z - ZH_1)]\psi \quad (13.21)$$

and finally results in the following form of the first derivative of the Lyapunov function

$$\dot{V}(\psi) = -k\frac{4}{\hbar^2}(\psi^* Z\psi - Z_d)^2\psi^{+^2}(H_1 Z - ZH_1)^2\psi^2 \leq 0 \quad (13.22)$$

which is non-positive along the system trajectories. This implies stability for the quantum system and in such a case La Salle's principle shows convergence not to an equilibrium but to an area round this equilibrium, which is known as *invariant set* [92].

Consequently, from Eq. (13.22) and LaSalle's theorem, any solution of the system $\psi(t)$ remains in the invariant set $M = \{\psi : \dot{V}(\psi) = 0\}$.

13.4.2 Control Law Calculation Using Lindblad's Equation

Next, it will be shown how a gradient-based control law can be formulated using the description of the quantum system according to Lindblad's equation. The following bilinear Hamiltonian system is considered (Lindblad equation)

$$\dot{\rho}(t) = -i[H_0 + f(t)H_1, \rho(t)] \quad (13.23)$$

where H_0 is the interaction Hamiltonian of the quantum system, H_1 is the control Hamiltonian of the quantum system, and $f(t)$ is the real-valued control field for the quantum system. The control problem consists of calculating the control function $f(t)$ such that the system's state (probability transition matrix $\rho(t)$) with initial conditions $\rho(0) = \rho_0$ converges to the desirable final state ρ_d for $t \to \infty$. It is considered that the initial state ρ_0 and the final state ρ_d have the same spectrum and this is a condition needed for reachability of the final state through unitary evolutions.

Because of the existence of the interaction Hamiltonian H_0 it is also considered that the desirable target state also evolves in time according to the Lindblad equation, i.e.

$$\dot{\rho}_d(t) = -i[H_0, \rho_d(t)] \tag{13.24}$$

The target state is considered to be stationary if it holds $[H_0, \rho_d(t)] = 0$, therefore in such a case it also holds $\dot{\rho}_d(t) = 0$. When $[H_0, \rho_d(t)] \neq 0$ then one has $\dot{\rho}_d \neq 0$ and the control problem of the quantum system is a tracking problem. The requirement $\rho(t) \to \rho_d(t)$ for $t \to \infty$ implies a trajectory tracking problem, while the requirement $\rho(t) \to O(\rho_d)(t)$ for $t \to \infty$ is an orbit tracking problem.

It will be shown that the calculation of the control function $f(t)$ which assures that $\rho(t)$ converges to $\rho_d(t)$ can be performed with the use of the Lyapunov method. To this end, a suitable Lyapunov function $V(\rho, \rho_d)$ will be chosen and it will be shown that there exists a gradient-based control law $f(t)$ such that $\dot{V}(\rho, \rho_d) \leq 0$.

The dynamics of the state of the quantum system, as well as the dynamics of the target state are jointly described by

$$\begin{aligned} \dot{\rho}(t) &= -i[H_0 + f(t)H_1, \rho(t)] \\ \dot{\rho}_d(t) &= -i[H_0, \rho_d(t)] \end{aligned} \tag{13.25}$$

A potential Lyapunov function for the considered quantum system is taken to be

$$V = 1 - Tr(\rho_d \rho) \tag{13.26}$$

It holds that $V > 0$ if $\rho \neq \rho_d$. The Lyapunov function given in Eq. (13.26) can be also considered as equivalent to the Lyapunov function $V(\psi, \psi_d) = 1 - |<\psi_d(t)|\psi(t)>|^2$, which results from the description of the quantum system with the use of Schrödinger's equation given in Eq. (13.1). The term $<\psi_d(t)|\psi(t)>$ expresses an internal product which takes value 1 if $\psi_d(t)$ and $\psi(t)$ are aligned. The first derivative of the Lyapunov function defined in Eq. (13.26) is

$$\dot{V} = -Tr(\dot{\rho}_d \rho) - Tr(\rho_d \dot{\rho}) \tag{13.27}$$

One can continue on the calculation of the first derivative of the Lyapunov function

$$\begin{aligned}
\dot{V} &= -Tr(\dot{\rho}_d\rho) - Tr(\rho_d\dot{\rho}) \Rightarrow \\
\dot{V} &= -Tr([-iH_0,\rho_d]\rho) - Tr(\rho_d[-iH_0,\rho]) - f(t)Tr(\rho_d[-iH_1,\rho]) \Rightarrow \\
\dot{V} &= -Tr([-iH_0\rho_d+\rho_d iH_0]\rho) - Tr(\rho_d[-iH_0\rho+\rho iH_0]) - f(t)Tr(\rho_d[-iH_1,\rho]) \Rightarrow \\
\dot{V} &= -Tr(-iH_0\rho_d\rho + \rho_d iH_0\rho - \rho_d iH_0\rho + \rho_d\rho iH_0) - f(t)Tr(\rho_d[-iH_1,\rho])
\end{aligned} \tag{13.28}$$

Using that $Tr(iH_0\rho_d\rho) = Tr(\rho_d\rho iH_0)$ one obtains

$$\dot{V} = -f(t)Tr(\rho_d[-iH_1,\rho]) \tag{13.29}$$

The control signal $f(t)$ is taken to be the gradient with respect to f of the first derivative of the Lyapunov function, i.e. $f(t) = -k\nabla_f\dot{V}(t)$, which gives

$$f(t) = kTr(\rho_d[-iH_1,\rho]) \ \ k > 0 \tag{13.30}$$

and which results in a negative semi-definite Lyapunov function $\dot{V} \leq 0$. Choosing the control signal $f(t)$ according to Eq. (13.30) assures that for the Lyapunov function of the quantum system given by Eq. (13.26) it holds

$$\begin{aligned}
V > 0 \ \forall \ (\rho,\rho_d) \neq 0 \\
\dot{V} \leq 0
\end{aligned} \tag{13.31}$$

and since a negative semi-definite Lyapunov function is examined, LaSalle's theorem is again applicable [92].

Again according to LaSalle's theorem, explained in Chap. 8, the state $(\rho(t),\rho_d(t))$ of the quantum system converges to the invariant set $M = \{(\rho,\rho_d)|\dot{V}(\rho(t),\rho_d(t)) = 0\}$. Attempts to define more precisely the convergence area for the trajectories of $\rho(t)$ when applying La Salle's theorem can be found in [203, 206].

13.5 Simulation Tests

Simulation tests about the performance of the gradient-based quantum control loop are given for the case of a two-qubit (four-level) quantum system. Indicative results from two different simulation experiments are presented, each one associated with different initial conditions of the target trajectory and different desirable final state.

The Hamiltonian of the quantum system was considered to be ideal, i.e. $H_0 \in C^4$ was taken to be strongly regular and $H_1 \in C^4$ contained non-zero non-diagonal elements. The two-qubit quantum system has four eigenstates which are denoted as $\psi_1 = (1000)$, $\psi_2 = (0100)$, $\psi_3 = (0010)$, and $\psi_4 = (0001)$. For the first case, the desirable values of elements ρ_{ii}^d, $i = 1,\cdots,4$ corresponding to quantum states ψ_1

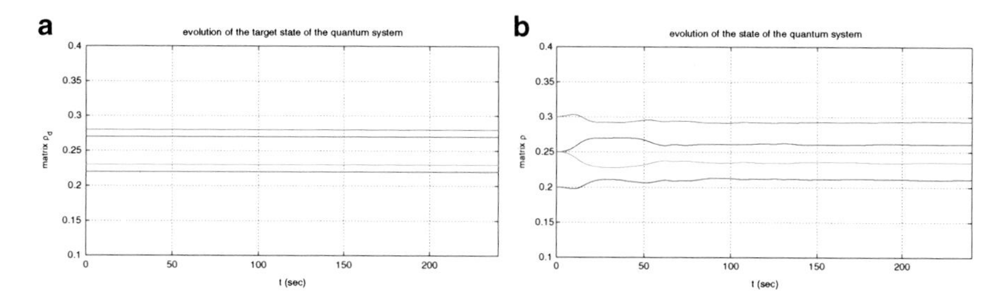

Fig. 13.2 (**a**) Desirable quantum states in the first test case, (**b**) Actual quantum states in the first test case

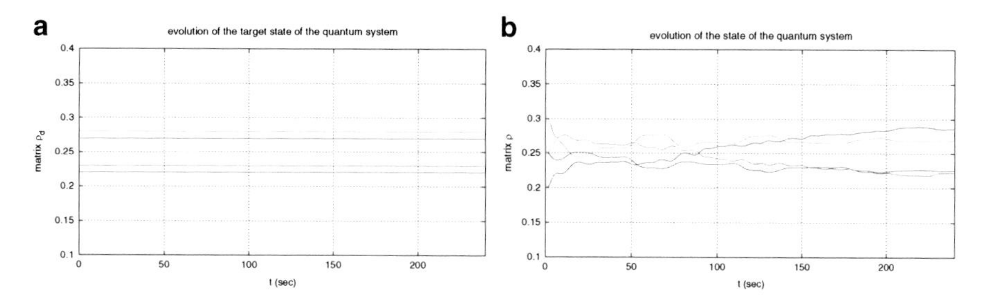

Fig. 13.3 (**a**) Desirable quantum states in the second test case, (**b**) Actual quantum states in the second test case

to ψ_4 are depicted in Fig. 13.2a, while the convergence of the actual values ρ_{ii}, $i = 1, \cdots, 4$ towards the associated desirable values is shown in Fig. 13.2b. Similarly, for the second case, the desirable values of elements ρ_{ii}^d, $i = 1, \cdots, 4$ are shown in Fig. 13.3a, while the associated actual values are depicted in Fig. 13.3b. It can be observed that the gradient-based control calculated according to Eq. (13.30) enabled convergence of ρ_{ii} to ρ_{ii}^d, within acceptable accuracy levels. Figure 13.4 presents the evolution in time of the Lyapunov function of the two simulated quantum control systems. It can be noticed that the Lyapunov function decreases, in accordance with the negative semi-definiteness proven in Eq. (13.31). Finally, in Fig. 13.5, the control signals for the two aforementioned simulation experiments are presented. The simulation tests verify the theoretically proven effectiveness of the proposed gradient-based quantum control scheme. The results can be extended to the case of control loops with multiple control inputs f_i and associated Hamiltonians H_i, $i = 1, \cdots, n$.

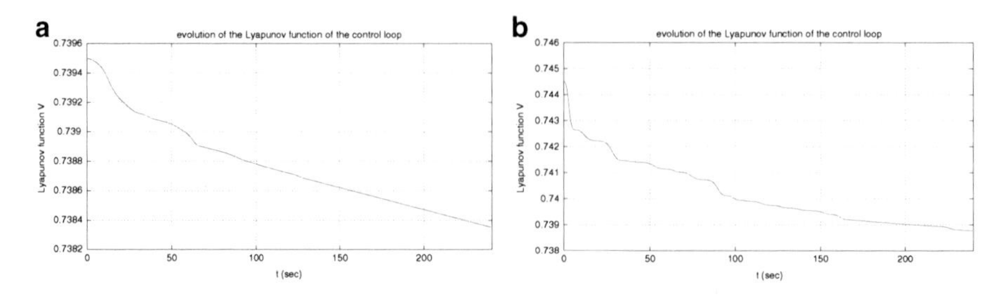

Fig. 13.4 (**a**) Lyapunov function of the first test case, (**b**) Lyapunov function of the second test case

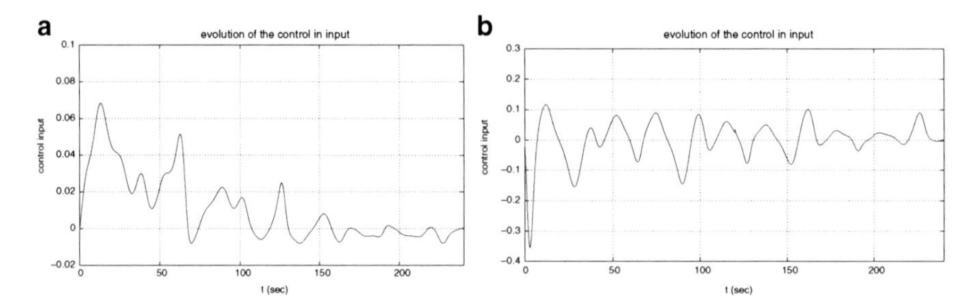

Fig. 13.5 (**a**) Control input of the first test case, (**b**) Control input of the second test case

13.6 Conclusions

The chapter has presented a gradient-based approach to feedback control of quantum systems. Different descriptions of the quantum system dynamics were formulated using Schrödinger's and Lindblad's differential equations, as well as Belavkin's stochastic differential equation. When Scrödinger's equation is used to describe the dynamics of the quantum system the objective is to move the quantum system from an eigenstate associated with a certain energy level to a different eigenstate associated with the desirable energy level. When Lindblad's or Belavkin's equations are used to describe the dynamics of the quantum system the control objective is to stabilize the probability density matrix ρ on some desirable quantum state $\rho_d \in C^n$ by controlling the intensity of the magnetic field. The control input is calculated by processing the measured output, which in turn depends on the projection of the probability density matrix ρ, as well as on processing of the estimate of ρ provided by Lindblad's or Belavkin's equation. It was shown that using either the Schrödinger or the Lindblad description of the quantum system a gradient-based control law can be formulated which assures tracking of the desirable quantum state within acceptable accuracy levels. The convergence properties of the gradient-based control scheme were proven using Lyapunov stability theory and LaSalle's invariance principle. Finally, simulation experiments for the case of a two-qubit (four-level) quantum system verified the theoretically established efficiency of the proposed gradient-based control approach.

References

1. E. Abdalla, B. Maroufi, B. Cuadros Melgar, M.B. Sedra, Information transport by sine-Gordon solitons in microtubules. Physica A **301**, 169–173 (2001)
2. P.A. Addison, *The Illustrated Wavelet Transform Handbook: Introductory Theory and Applications in Science, Engineering, Medicine and Finance* (Institute of Physics Publishing, Bristol, 2002)
3. H. Adeli, S. Sumonway Ghosh-Dastidar, *Automated EEG-Based Diagnosis of Neurological Disorders: Inventing the Future of Neurology* (CRC Press, Boca Raton, 2010)
4. E. Aero, A. Fradkov, B. Andrievsky, S. Vakulenko, Dynamics and control of oscillations in a complex crystalline lattice. Phys. Lett. A **353**, 24–29 (2006)
5. M. Aqil, K.S. Hong, M.Y. Jeong, Synchronization of coupled chaotic FitzHugh–Nagumo systems. Commun. Nonlinear Sci. Numer. Simul. **17**, 1615–1627 (2012)
6. E. Alfinito, G. Vitiello, Formation and life-time of memory domains in the dissipative quantum model of the brain. Int. J. Mod. Phys. B **14**, 853–868 (2000)
7. S. Aiyer, M. Niranjan, F. Fallside, Theoretical investigation into the performance of the Hopfield model. IEEE Trans. Neural Netw. **15**(1), 204–215 (1990)
8. B. Ambrosio, M.A. Aziz-Alaoui, Synchronization and control of coupled reaction diffusion systems of the FitzHugh–Nagumo type. Comput. Math. Appl. **64**, 934–943 (2012)
9. M.S. Ananievskii, A.L. Fradkov, Control of the observables in the finite-level quantum systems. Autom. Rem. Control. **66**(5), 734–745 (2005)
10. R. Appali, U. van Rienen, T. Heimburg, A comparison of the Hodgkin-Huxley model the soliton theory for the action potential in nerves. Adv. Planar Lipid Bilayers Liposomes **16**, 275–298 (2012)
11. P. Arena, L. Fortuna, M. Frasca, G. Sicurella, An adaptive self-organizing dynamical system for hierarchical control of bio-inspired locomotion. IEEE Trans. Syst. Man Cybern. B **34**(4), 1823–1837 (2004)
12. M.D. Armani, S.V. Chaudhary, P. Probst, B. Shapiro, Using feedback control of microflows to independently steer multiple particles. IEEE J. Microelectromech. Syst. **15**(4), 945–956 (2006)
13. K.J. Astrom, *Introduction to Stochastic Control Theory* (Dover, New York, 2006)
14. N.Bagheri, J. Stelling, F.J. Doyle, Optimal Phase-Tracking of the Nonlinear Circadian Oscillator, in *2005 American Control Conference*, Portland, 2005
15. L. Bahry, M. Pace, S. Saighi, Global parameter estimation of a Hdgkin-Huxley formalism using membrane voltage recordings: application to neuro-mimetic analog integrated circuits. Neurocomputing **81**, 75–85 (2012)
16. G. Baird Emertrout, D.H. Terman, *Mathematical Foundations of Neuroscience*. Interdisciplinary Applied Mathematics, vol. 35 (Springer, New York, 2009)

G.G. Rigatos, *Advanced Models of Neural Networks*,
DOI 10.1007/978-3-662-43764-3,

17. M. Basseville, I. Nikiforov, *Detection of Abrupt Changes: Theory and Applications* (Prentice-Hall, Englewood Cliffs, 1993)
18. K. Beauchard, J.M. Coron, M. Mirrahimi, P. Rouchon, Implicit Lyapunov control of finite dimensional Schrödinger equations. Syst. Control Lett. **56**, 388–395 (2007)
19. E.C. Behrman, L.R. Nash, J.E. Steck, V.G. Chandrashekar, S.R. Skinner, Simulations of Quantum neural networks. Inf. Sci. **128**, 257–269 (2002)
20. V.P. Belavkin, On the theory of controlling observable quantum systems. Autom. Rem. Control **44**(2), 178–188 (1983)
21. J. Bell, G. Crociun, A distributed parameter identification problem in neuronal cable theory models. Math. Biosci. **194**, 1–19 (2005)
22. A. Benvensite, P.Metivier, P. Priouret, *Adaptive Algorithms and Stochastic Approximations*. Applications of Mathematics Series, vol. 22 (Springer, Heidelberg, 1990)
23. C. Bertoglio, D. Chapelle, M.A. Fernandez, J.F. Gerbeau, P. Moireau, State observers of a vascular fluid-structure interaction model through measurements in the solid. INRIA research report no 8177, Dec. 2012
24. A. Bonami, B. Demange, P. Jaming, Hermite functions and uncertainty principles for the Fourier and Windowed Fourier transforms. Rev. Matematica Iberoamericana **19**(1), 23–55 (2003)
25. S. Bououden, D. Boutat, G. Zheng, J.P. Barbot, F. Kratz, A triangular canonical form for a class of 0-flat nonlinear systems. Int. J. Control **84**(2), 261–269 (2011)
26. P.C. Bresslof, J.M. Newby, Stochastic models of intracellular transport. Rev. Mod. Phys. **85**, 135–196 (2013)
27. H.P. Breuer, F. Petruccione, *The Theory of Open Quantum Systems* (Clarendon Press, Oxford, 2006)
28. C. Chaudhary, B. Shapiro, Arbitrary steering of multiple particles independently in an electro-osmotically driven microfluidic system. IEEE J. Control Syst. Technol. **14**(4), 669–680 (2006)
29. J. Chauvin, Observer design for a class of wave equations driven by an unknown periodic input, in *18th World Congress*, Milano, 2011
30. C. Chen, G. Rigatos, D. Dong, J. Lam, Partial feedback control of quantum systems using probabilistic fuzzy estimator, in *IEEE CDC 2009, Joint 48th IEEE Conference on Decision and Control and 28th Chinese Control Conference*, Sanghai, 2009
31. J.R. Chumbley, R.J. Dolan, K.J. Friston, Attractor models of working memory and their modulation by reward. Biol. Cybern. **98**, 11–18 (2008)
32. J. Clairambault, Physiologically based modelling of circadian control on cell proliferation, in *28th IEEE EMBS Annual International Conference*, New York, 2006
33. J. Clairambault, A step toward optimization of cancer therapeutics: physiologically based modeling of circadian control on cell proliferation. IEEE Eng. Med. Biol. Mag. (2008)
34. C. Cohen-Tannoudji, B. Diu, F. Laloë, in *Mécanique Quantique I*, Hermann, 1998
35. F. Comet, T. Meyre, *Calcul stochastique et modèles de diffusion* (Dunod, Paris, 2006)
36. G. Chen, J.L. Moiola, H.O. Wang, Bifurcation control: theories, methods and applications. Int. J. Bifurcat. Chaos, **10**(3), 511–548 (2000)
37. K.L. Chung, *Green, Brown and Probability & Brownian Motion on the Line* (World Scientific, Singapore, 2002)
38. S. Coombes, P.C. Bressloff, Solitary waves in a model of dendritic cable with active spines. SIAM J. Appl. Math. **61**(2), 432–453 (2000)
39. D. Csercsik, I. Farkas, G. Szederkenyi, E. Hrabovszky, Z. Liposits, K.M. Hangas, Hodgkin-Huxley type modelling and parameter estimation of the GnRH neurons. Biosystems **100**, 198–207 (2010)
40. A. Crespi, A. Ijspeert, Online optimization of swimming and crawling in an amphibious snake robot. IEEE Trans. Robot. **24**(1), 75–87 (2008)
41. P.S.P. da Silva, P. Rouchon, Flatness-based control of a single qubit-gate. IEEE Trans. Automat. Control **53**(3), 775–779 (2008)
42. I. Debauchies, The wavelet transform, time-frequency localization and signal processing. IEEE Trans. Inf. Theory **36**, 961–1005 (1990)

43. G. Deco, E.T. Rolls, Attention, short-memory and action selection: a unifying theory. Prog. Neurobiol. **76**, 236–256 (2005)
44. G. Deco, D. Marti, A. Ledberg, R. Reig, M.V. Sanchez Vives, Effective reduced diffusion-models: A data driven approach to the analysis of neuronal dynamics. PLoS Comput. Biol. **5**(12), e1000587 (2009)
45. G. Deco, E.T. Rolls, R. Romo, Stochastic dynamics as a principle of brain function. Prog. Neurobiol. **88**, 1–16 (2009)
46. M.A. Demetriou, Design of consensus and adaptive consensus filters for distributed parameter systems. Automatica **46**, 300–311 (2010)
47. M.J. Denham, Characterising neural representation in terms of the dynamics of cell membrane potential activity: a control theoretic approach using differential geometry. Biosystems **79**, 53–60 (2005)
48. D. Deutsch, Quantum computational networks. Proc. R. Soc. Lond. A **425**, 73 (1989)
49. R.O. Dorak, Control of repetitive firing in Hodgkin-Huxley nerve fibers using electric fields. Chaos Solitons Fract. **52**, 66–72 (2013)
50. H. Duan, C. Cai, C. Han, Y. Che, Bifurcation control in Morris-Lecar neuron model via washout filter with a linear term based on filter-aided dynamic feedback. Adv. Mater. Res. **485**, 600–603 (2012)
51. D. Dubois, L. Foulloy, G. Mauris and H. Prade, Probability-possibility transformations, triangular fuzzy sets and probabilistic inequalities. Reliab. Comput. **10**(4), 273–297 (2004)
52. M. Duflo, *Algorithmes Stochastiques*. Mathématiques et Applications, vol. 23 (Springer, Berlin, 1996)
53. K.L. Eckel-Mahan, Cirdadian oscillations within the hippocampus support memory formation and persistence. Front. Mol. Neurosci. **5**, 46 (2012)
54. D.V. Efimov, A.L. Fradkov, Adaptive tuning to bifurcation for time-varying nonlinear systems. Automatica **42**, 417–425 (2006)
55. R.T. Faghih, K. Savla, M.A. Dahleh, E.N. Brown, The FitzHugh–Nagumo model: Firing modes with time-varying parameters and parameter estimation, in *IEEE 32nd International Conference of the Engineering in Biology and Medicine Society*, Buenos Aires, 2010
56. W.G. Faris, *Diffusion, Quantum Theory, and Radically Elementary Mathematics* (Princeton University Press, Princeton, 2006)
57. M. Fliess, Probabilités et fluctuations quantiques. Comp. Rendus Math. (Phys. Math.), C. R. Acad. Sci. Paris **344**, 663–668 (2007)
58. M. Fliess, H. Mounier, Tracking control and π-freeness of infinite dimensional linear systems, in *Dynamical Systems, Control, Coding and Computer Vision*, vol. 258, ed. by G. Picci, D.S. Gilliam (Birkhaüser, Boston, 1999), pp. 41–68
59. R. Feynman, Quantum mechanical computers. Found. Phys. **16**, 507–531 (1986)
60. R.F. Fox, Y. Lu, Emergent collective behavior in large numbers of globally coupled independently stochastic ion channels. Phys. Rev. E **49**(4), 3421–3431 (1994)
61. P. Francois, A model for the neurospora circadian clock. Biophys. J. **88**, 2369–2383 (2005)
62. Y. Fukuoka, H. Kimura, A. Cohen, Adaptive dynamic walking of a quadruped robot on irregular terrain based on biological concepts. Int. J. Robot. Res. **3–4**, 187–202 (2003)
63. H. Fujii and I. Tsuda, Neocortical gap junction coupled interneuron systems may induce chaotic behavior itinerant among quasi-attractors exhibiting transient synchrony. Neurocomputing **58–60**, 151–157 (2004)
64. V. Gazi, K. Passino, Stability analysis of social foraging swarms. IEEE Trans. Syst. Man Cybern. B Cybern. **34**, 539–557 (2004)
65. W. Gerstner, W.M. Kistler, *Spiking Neuron Models: Simple Neurons, Populations, Plasticity* (Cambridge University Press, Cambridge, 2002)
66. M. Gitterman, *The Noisy Oscillator: The First Hundred Years, From Einstein Until Now* (World Scientific, Singapore, 2005)
67. D. Gonze, J.C. Leloup, A. Goldbeter, Theoretical models for circadian rhythms in Neurospora and Drosophila. C. R. Acad. Sci. III **323**, 57–67 (2000)

68. D. Gonze, J. Halloy, J.C. Leloup, A. Goldbeter, Stochastic models for circadian rhythmes: effect of molecular noise on periodic and chaotic behaviour. C. R. Biol. **326**, 189–203 (2003)
69. B.F. Grewe, D. Langer, H. Kasper, B.M. Kampa, F. Helmchen, High-speed in vivo calcium imaging reveals neuronal network activity with near-millisecond precision. Nat. Methods **7**(5), 399–405 (2010)
70. S. Grillner, Biological pattern generation: the cellular and computational logic of networks in motion. Neuron **52**, 751–766 (2006)
71. K. Gröchenig, D. Han, C. Heil, G. Kutyniak, The Balian-Low theorem for symplectic lattices in higher dimensions. Appl. Comput. Harmon. Anal. **13**, 169–176 (2002)
72. B.Z. Guo, C.Z. Xu, H. Hammouri, Output feedback stabilization of a one-dimensional wave equation with an arbitrary time-delay in boundary observation, in *ESAIM: Control, Optimization and Calculus of Variations*, vol. 18, pp. 22–25, 2012
73. S. Hagan, S.R. Hameroff, J.A. Tuzyinski, *Quantum Computation in Brain Microtubules: Decoherence and Biological Feasibility*. Phys. Rev. E **65**, 1–11 (2002)
74. G. Haine, Observateurs en dimension infinie. Application à l étude de quelques problèmes inverses, Thèse de doctorat, Institut Elie Cartan Nancy, 2012.
75. L.M. Harisson, K. David, K.J. Friston, Stochastic models of neuronal dynamics. Phil. Tran. R. Soc. B **360**, 1075–1091 (2005)
76. R. Haschke, J.J. Steil, Input-space bifurcation manifolds of recurrent neural networks. Neurocomputing **64**, 25–38 (2005)
77. M. Havlicek, K.J. Friston, J. Jan, M. Brazdil, V.D. Calhoun. Dynamic modeling of neuronal responses in fMRI using cubature Kalman Filtering. Neuroimage **56**, 2109–2128 (2011)
78. S. Haykin, *Neural Networks: A Comprehensive Foundation* (McMillan, New York, 1994)
79. R. Héliot, B. Espiau, Multisensor input for CPG-based sensory-motor coordination. IEEE Trans. Robot. **24**, 191–195 (2008)
80. T. Heimburg, A.D. Jackson, On soliton propagation in biomembranes and nerves. Proc. Natl. Acad. Sci. **102**(28), 9790–9795 (2005)
81. Z. Hidayat, R. Babuska, B. de Schutter, A. Nunez, Decentralized Kalman Filter comparison for distributed parameter systems: a case study for a 1D heat conduction process, in *Proceedings of the 16th IEEE International Conference on Emerging Technologies and Factory Automatio (ETFA 2011)*, Toulouse, 2011
82. J.J. Hopfield, Neural networks as physical systems with emergent computational abilities. Proc. Natl. Acad. Sci. USA **79**, 2444–2558 (1982)
83. S. Hu, X. Liao, X. Mao, Stochastic Hopfield neural networks. J. Phys. A Math. Gen. **35**, 1–15 (2003)
84. J.H. Huggins, L. Paninski, Optimal experimental design for sampling voltage on dendritic trees in the low-SNR regime. J. Comput. Neurosci. **32**(2), 347–66 (2012)
85. A.J. Ijspeert, Central pattern generator for locomotion control in animals and robots: a review. Neural Netw. **21**, 642–653 (2008)
86. V.G. Ivancevic, T.T. Ivancevic, *Quantum Neural Computation* (Springer, Netherlands, 2010)
87. I. Iwasaki, H. Nakasima, T. Shimizu, Interacting Brownian particles as a model of neural network. Int. J. Bifurcat. Chaos **8**(4), 791–797 (1998)
88. B.S. Jackson, Including long-range dependence in integrate-and-fire models of the high interspike-interval variability of the cortical neurons. Neural Comput. **16**, 2125–2195 (2004)
89. P. Jaming, A.M. Powell, Uncertainty principles for orthonormal sequences. J. Funct. Anal. **243**, 611–630 (2007)
90. F. Jedrzejewski, *Modèles aléatoires et physique probabiliste* (Springer, Paris, 2009)
91. Y. Katory, Y. Otsubo, M. Okada, K. Aihara, Stability analysis of associative memory network composed of stochastic neurons and dynamic synapses. Front. Comput. Neurosci. **7**, 6 (2013)
92. H. Khalil, *Nonlinear Systems* (Prentice Hall, Englewood Cliffs, 1996)
93. F.C. Klebaner, *Introduction to Stochastic Calculus with Applications* (Imperial College Press, London, 2005)
94. M. Kolobov, *Quantum Imaging* (Springer, New York, 2007)

95. I.K. Kominis, Quantum Zeno effect explains magnetic-sensitive radical-ion-pair reactions. Phys. Rev. E **80**, 056115 (2009)
96. B. Kosko, *Neural Networks and Fuzzy Systems: A Dynamical Systems Approach to Machine Intelligence* (Prentice Hall, Englewood Cliffs, 1992)
97. T. Kurikawa, K. Kaneko, Learning to memorize input-output mapping as bifurcation in neural dynamics: relevance of multiple time-scales for synaptic changes. Neural Comput. Appl. **31**, 725–734 (2012)
98. K. Kushima, Y. Kawamura, J. Imura, Oscillation analysis of coupled piecewise affine systems: Application to spatio-temporal neuron dynamics. Automatica **47**, 1249–1254 (2011)
99. M. Lankarany, W.P. Zhu, M.N.S. Swamy, Parameter estimation of Hodgkin-Huxley neuronal model using the dual Extended Kalman Filter, in *IEEE ISCAS 2013 International Symposium on Circuits and Systems*, Beijing, 2013
100. N. Lambert, Y.N. Chen, Y.C. Cheng, C.M. Li, G.Y. Chen, Franco Nori Nat. Phys. **9**, 10–18 (2013)
101. B. Laroche, P. Martin, N. Petit, *Commande par platitude: Equations différentielles ordinaires et aux derivées partielles* (Ecole Nationale Supérieure des Techniques Avancées, Paris, 2007)
102. B. Laroche, D. Claude, Flatness-based control of PER protein oscillations in a Drosophila model. IEEE Trans. Automat. Contr. **49**(2), 175–183 (2004)
103. B. Lautrup, R. Appali, A.D. Jackson, T. Heimburg, The stability of solitons in biomembranes and nerves. Eur. Phys. J. E **34**(57), 1–9 (2011)
104. J.C. Leloup, D. Gonze, A. Goldbeter, Computational models for circadian rythms: deterministic versus stochastic approaches, in *Computational Systems Biology* ed. by A. Kriete, R. Eils (Elsevier, San Diego, 2006)
105. F.A. Lévi, The Circadian timing system: a coordinator of life processes. IEEE Eng. Med. Biol. Mag. (2008)
106. J.H. Lee, A. Sancar, Circadian clock disruption improves the efficacy of chemotherapy through p73-mediated apoptosis, in *Proceedings of the National Academy of Sciences*, 2011
107. V.S. Letokhov, *Laser Control of Atoms and Molecules* (Oxford University Press, Oxford, 2007)
108. H. Levine, W.J. Rappel, Self-organization in systems of self-propelled particles. Phys. Rev. E **63**, 2000
109. J. Lévine, On necessary and sufficient conditions for differential flatness. Appl. Algebra Eng. Commun. Comput. **22**(1), 47–90 (2011)
110. J. Lévine, D.V. Nguyen, Flat output characterization for linear systems using polynomial matrices. Syst. Control Lett. **48**, 69–75 (2003)
111. Z. Li, L. Liu, Uncertainty principles for Sturm-Liouville operators. Constr. Approx. **21**, 195–205 (2005)
112. H.Y. Li, Y.K. Wang, W.L. Chan, The asymptotic structure of the Morris-Lecar models. Neurocomputing **74**, 2108–2113 (2011)
113. X. Liao, K.W. Wong, Z. Wu, Bifurcation analysis of a two-neuron system with distributed delays. Physica D **149**, 123–141 (2001)
114. M. Loh, E.T. Rolls, G. Deco, Statistical fluctuation in attractor networks related to Schizophrenia. Pharmacopsychiatry **40**, 78–84 (2007)
115. M. Loh, E.T. Rolls, G. Deco, A dynamical systems hypothesis of Schizophrenia. PLoS Comput. Biol. **3**(11), 1–11 (2007)
116. M. Lysetskiy, J.M. Zurada, Bifurcating neuron: computation and learning. Neural Netw. **17**, 225–232 (2004)
117. L. Ma, K. Khorasani, Application of adaptive constructive neural networks to image compression. IEEE Trans. Neural Netw. **13**(5), 1112–1125 (2002)
118. L. Ma, K. Khorasani, Constructive feedforward neural networks using hermite polynomial activation functions. IEEE Trans. Neural Netw. **16**(4), 821–833 (2005)
119. S. Mallat, *A Wavelet Tour of Signal Processing* (Academic, San Diego, 1998)
120. G. Mahler, V.A. Weberuss, *Quantum Networks: Dynamics of Open Nanostructures* (Springer, New York , 1998)

121. T. Manninen, M.L. Liane, K. Ruohonen, Developing Itô stochastic differential equation models for neuronal signal transduction pathways. Comput. Biol. Chem. **30**, 280–291 (2006)
122. R. Marino, P. Tomei, Global asymptotic observers for nonlinear systems via filtered transformations. IEEE Trans. Automat. Control **37**(8), 1239–1245 (1992)
123. J.D. Marshall, M.J. Schnitzer, ACS Nano (Am. Chem. Soc.) **7**(5), 4601–4609 (2013)
124. M. Mirrahimi, P. Rouchon, Controllability of quantum harmonic oscillators. IEEE Trans. Automat. Control. **45**(5), 745–747 (2004)
125. Ph. Martin, P. Rouchon, Systèmes plats: planification et suivi des trajectoires. Journées X-UPS, École des Mines de Paris, Centre Automatique et Systèmes, Mai (1999)
126. H. Mounier, J. Rudolph, Trajectory tracking for π-flat nonlinear dealy systems with a motor example, in *Nonlinear Control in the Year 2000*, vol. 1, ed. by A. Isidori, F. Lamnabhi-Lagarrigue, W. Respondek. Lecture Notes in Control and Information Sciences, vol. 258 (Springer, London, 2001), pp. 339–352
127. G. Müller, *Quantum Mechanics: Symmetries*, 2nd edn. (Springer, New York, 1998)
128. B. Nagy, Comparison of the bifurcation curves of a two-variable and a three-variable circadian rhythm model. Appl. Math. Model. **38**, 1587–1598 (2008)
129. B. Nagy, Analysis of the biological clock of Neurospora. J. Comput. Appl. Math. **226**, 298–305 (2009)
130. L.H. Nguyen, K.S. Hong, Synchronization of coupled chaotic FitzHugh–Nagumo neurons via Lyapunov functions. Math. Comput. Simul. **82**, 590–603 (2011)
131. L.H. Nguyen, K.S. Hong, Hopf bifurcation control via a dynamic state-feedback control. Phys. Lett. A **376**, 442–446 (2012)
132. L.H. Nguyen, K.S. Hong, S. Park, Bifurcation control of the Morris-Lecar neuron model via a dynamic state feedback ontrol. Biol. Cybern. **106**, 587–594 (2012)
133. M. Nielsen, A. Chuang, *Quantum Computation and Quantum Information* (Cambridge University Press, Cambridge, 2000)
134. L. Paninski, Fast Kalman filtering on quasilinear dendritic trees. J. Comput. Neurosci. **28**(2), 211–228 (2010)
135. M. Perus, Neural networks as a basis for quantum associative networks. Neural Netw. World **10**, 1001–1013 (2000)
136. M. Perus, Multi-level synergetic computation in brain. Nonlinear Phenom. Complex Syst. **4**(2), 157–193 (2001)
137. M. Perus, H. Bischof, J. Caulfield, C.K. Loo, Quantum implementable selective reconstruction of high resolution images. Appl. Opt. **43**, 6134–6138 (2004)
138. D.S. Peterka, H. Takahashi, R. Yuste, Imaging voltage in neurons. Neuron Cell Press **69**, 9–21 (2011)
139. M. Pinsky, *Partial Differential Equations and Boundary Value Problems* (Prentice-Hall, Englewood Cliffs, 1991)
140. M. Pospischil, M. Toledo-Rodriguez, C. Monier, Z. Piwkowska, T. Bal, Y. Frégnac, H. Markram, A. Destexhe, Minimal Hodgkin–Huxley type models for different classes of cortical and thalamic neurons. J. Biol. Cybern. **99**, 427–441 (2008)
141. A.M. Powell, Time-frequency mean and variance sequences of orthonormal bases. J. Fourier Anal. Appl. **11**, 375–387 (2005)
142. H. Puebla, M. Ortiz-Vargas, R. Aguilar-Lopez, E. Hernandez-Martinez, Control of coupled circadian oscillators, in *10th IFAC Symposium on Computer Applications in Biotechnology*, Cancun, 2007
143. A. Refregier, Shapelets - I. A method for image analysis. Mon. Not. R. Astron. Soc. **338**, 35–47 (2003)
144. M. Rehan, K.S. Hong, M. Aqil, Synchronization of multiple chaotic FitzHugh–Nagumo neurons with gap junctions under external electrical stimulation. Neurocomputing **74**, 3296–3304 (2011)
145. G. Resconi, A.J. van der Wal, Morphogenic neural networks encode abstract rules by data. Inf. Sci. **142**, 249–273 (2002)

146. E. Rieper, Quantum coherence in biological systems. Ph.D. Thesis, Centre for Quantum Technologies, National University of Singapore, 2011
147. G.G. Rigatos, Distributed gradient for multi-robot motion planning, in *ICINCO'05, 2nd International Conference on Informatics in Control, Automation and Robotics*, Barcelona, 2005
148. G.G. Rigatos, Energy spectrum of quantum associative memory. *Proceedings of the IEEE WCCI'06 Conference*, Vancouver, 2006
149. G.G. Rigatos, Quantum wave-packets in fuzzy automata and neural associative memories. Int. J. Mod. Phys. C **18**(9), 1551 (2007)
150. G.G. Rigatos, Coordinated motion of autonomous vehicles with the use of a distributed gradient algorithm. Appl. Math. Comput. **199**, 494–503 (2008)
151. G.G. Rigatos, Stochastic Processes in Machine Intelligence: neural structures based on the model of the quantum harmonic oscillator. Opt. Memories Neural Netw. (Inf. Opt.) **17**(2), 101–110 (2008)
152. G.G. Rigatos, Distributed gradient and particle swarm optimization for multi-robot motion planning. Robotica **26**(3), 357–370 (2008)
153. G.G. Rigatos, Open-loop control of particle systems based on a model of coupled stochastic oscillators, in *ICQNM 2009 International Conference on Quantum Micro and Nano Technologies*, Mexico, 2009
154. G.G. Rigatos, Spin-orbit interaction in particles' motion as a model of quantum computation, in *PhysCon 2009, 4th International Conference on Physics and Control*, Catania, 2009
155. G.G. Rigatos, Cooperative behavior of nano-robots as an analogous of the quantum harmonic oscillator. Ann. Math. Artif. Intell. **55**(3–4), 277–294 (2009)
156. G.G. Rigatos, Stochastic processes and neuronal modelling: quantum harmonic oscillator dynamics in neural structures. Neural Process. Lett. **32**(2), 167–199 (2010)
157. G.G. Rigatos, Modelling and Control for Intelligent Industrial Systems: Adaptive Algorithms in Robotcs and Industrial Engineering (Springer, Heidelberg, 2011)
158. G.G. Rigatos, A derivative-free Kalman Filtering approach to state estimation-based control of nonlinear dynamical systems. IEEE Trans. Ind. Electron. **59**(10), 3987–3997 (2012)
159. G.G. Rigatos, Nonlinear Kalman Filters and Particle Filters for integrated navigation of unmanned aerial vehicles. Robot. Auton. Syst. **60**, 978–995 (2012)
160. G. Rigatos, E. Rigatou, A Derivative-free nonlinear Kalman Filtering approach to estimation of wave-type dynamics in neurons' membrane, in *ICNAAM 2013, 11th International Conference of Numerical Analysis and Applied Mathematics*, Rhodes, 2013
161. G. Rigatos, E. Rigatou, A Kalman Filtering approach to robust synchronization of coupled neural oscilltors, in *ICNAAM 2013, 11th International Conference on Numerical Analysis and Applied Mathematics*, Rhodes, 2013
162. G.G. Rigatos, P. Siano, Design of robust electric power system stabilizers using Kharitonov's theorem. Math. Comput. Simul. **82**(1), 181–191 (2011)
163. G.G. Rigatos, S.G. Tzafestas, Parallelization of a fuzzy control algorithm using quantum computation. IEEE Trans. Fuzzy Syst. **10**, 451–460 (2002)
164. G.G. Rigatos, S.G. Tzafestas, Neural structures using the eigenstates of the quantum harmonic oscillator. Open Syst. Inf. Dyn. **13**, 27–41 (2006)
165. G.G. Rigatos, S.G. Tzafestas, Quantum learning for neural associative memories. Fuzzy Sets Syst. **157**(13), 1797–1813 (2006)
166. G.G. Rigatos, S.G. Tzafestas, Extended Kalman filtering for fuzzy modelling and multi-sensor fusion. Math. Comput. Model. Dyn. Syst. **13**, 251–266 (2007)
167. G.G. Rigatos, S.G. Tzafestas, Neurodynamics and attractors in quantum associative memories. J. Integr. Comput. Aid. Eng. **14**(3), 225–242 (2007)
168. G.G. Rigatos, S.G. Tzafestas, Attractors and energy spectrum of neural structures based on the model of the quantum harmonic oscillator, in *Complex-Valued Neural Networks: Utilizing High-Dimensional Parameters*, ed. by T. Nitta (IGI Publications, Hershey, 2008)
169. G. Rigatos, Q. Zhang, Fuzzy model validation using the local statistical approach. Fuzzy Sets Syst. **60**(7), 882–904 (2009)

170. G. Rigatos, P. Siano, N. Zervos, PMSG sensorless control with the use of the Derivative-free nonlinear Kalman Filter, in *IEEE ICCEP 2013, IEEE International Conference on Clean Electrical Power*, Alghero, 2013
171. E.T. Rolls, M. Loh, G. Deco, An attractor hypothesis of obsessive compulsive disorder. Eur. J. Neurosci. **28**, 782–793 (2008)
172. P. Rouchon, Flatness-based control of oscillators. ZAMM Zeitschr. Angew. Math. Mech. **85**(6), 411–421 (2005)
173. J. Rudolph, *Flatness Based Control of Distributed Parameter Systems*. Steuerungs- und Regelungstechnik (Shaker, Aachen, 2003)
174. S.A. Salberg, P.S. Maybeck, M.E. Oxley, Infinite-dimensional sampled-data Kalman Filtering and stochastic heat equation, in *49th IEEE Conference on Decision and Control*, Atlanta, 2010
175. A.K. Schierwagen, Identification problems in distributed parameter neuron models. Automatica **26**(4), 739–755 (1990)
176. S.J. Schiff, Kalman meets neuron: the emerging intersection of control theory with neuroscience, in *IEEE 31st Annual International Conference of the Engineering in Biology and Medicine Society*, Minneapolis, 2006
177. H. Sira-Ramirez, S. Agrawal, *Differentially Flat Systems* (Marcel Dekker, New York, 2004)
178. Y. Song, M. Han, J. Wei, Stability and Hopf bifurcation analysis on a simplified BAM network with delays. Physica D **200** 185–204 (2005)
179. T.T. Soong, M. Grigoriou, *Random Vibration of Mechanical and Structural Systems* (Prentice Hall, Upper Saddle River, 1992)
180. R.G. Spencer, Bipolar spectral associative memories. IEEE Trans. Neural Netw. **12**, 463–475 (2001)
181. R. Srebro, The Duffing oscillator: a model for the dynamics of the neuronal groups comprising the transient evoked potential. Electroengephalongr. Clin. Neurophysiol. **96**, 561–573 (1995)
182. G.B. Stan, A. Hamadeh, R. Sepulchre, J. Gonçalves, Output synchronization in networks of cyclic biochemical oscillators, in *2007 American Control Conference*, New York, 2007
183. R.H. Steele, Harmonic oscillators: the quantization of simple systems in the old quantum theory and their functional roles in biology. Mol. Cell. Biochem. **310**, 19–42 (2008)
184. N.J. Stevenson, M. Mesbah, G.B. Poylan , P.B. Colditz, B. Boushash, A nonlinear model of newborn EEG with non-stationary inputs. Ann. Biomed. Eng. **38**(9), 3010–3021 (2010)
185. S. Still, G. Le Masson, Travelling waves in a ring of three inhibitory coupled model neurons. Neurocomputing **26–27**, 533–539 (1999)
186. W.A. Strauss, *Partial Differential Equations: An Introduction* (Wiley, New York, 1992)
187. N. Sureshbabu, J. Farell, Wavelet-based system identification for nonlinear control. IEEE Trans. Automat Control **44**, 412–417 (1999)
188. T. Takeuchi, T. Hinohara, K. Uchida, S. Shibata, Control theoretic views on circadian rhythms, in *2006 IEEE International Conference on Control Applications*, Munich, 2006
189. L.T. That, Z. Ding, Reduced-order observer design of multi-output nonlinear systems with application to a circadian model. Trans. Inst. Meas. Control **35**(4), 417425 (2012)
190. L. Ton That, Z. Ding, Circadian phase resetting using nonlinear output-feedback control. J. Biol. Syst. **20**, 1–19 (2013)
191. B. Torrésani, *Analyse Continue Par Ondelettes* (CNRS Editions, Paris, 1995)
192. R. Toscano, P. Lyonnet, Robust static output feedback controller synthesis usingKharitonov's theoremand evolutionary algorithms. Inf. Sci. **180**, 2023–2028 (2010)
193. K. Tsumoto, T. Yoshinaga, H. Iida, H. Kawakami, K. Ashara, Bifurcations in a mathematical model for circadian oscillations of clock genes. J. Theor. Biol. **239**, 101–122 (2006)
194. H.C. Tuckwell, Stochastic partial differential equations in neurobiology: linear and nonlinear models for spiking neurons, *Stochastic Biomathematical Models*. Lecture Notes in Mathematics (Springer, Heidelberg, 2013), pp. 149–179
195. S.G. Tzafestas, G.G. Rigatos, Stability analysis of an adaptive fuzzy control system using Petri Nets and learning automata. Math. Comput. Simul. **51**, 315–339 (2000)
196. F. Ventriglia, V. di Maio, A Brownian model of glutamate diffusion in excitatory synapses of hippocampus. BioSystems **58**, 67–74 (2000)

197. F. Ventiglia, V. di Maio, Stochastic fluctuations of the synaptic function. Biosystems **67**, 287–294 (2002)
198. D. Ventura, T. Martinez, Quantum associative memory. Inf. Sci. **124**, 273–296 (2000)
199. A. Vidal, Q. Zhang, C. Médigue, S. Fabre, F. Clément, DynPeak: an algorithm for pulse detection and frequency analysis in hormonal time series. PLoS One **7**(7), e39001 (2012)
200. J. Villagra, B. d'Andrea-Novel, H. Mounier, M. Pengov, Flatness-based vehicle steering control strategy with SDRE feedback gains tuned via a sensitivity approach. IEEE Trans. Control Syst. Technol. **15**, 554–565 (2007)
201. E. Villagran Vargas, A. Ludu, R. Hustert, P. Grumrich, A.D. Jackson, T. Heimburg, Periodic solutions and refractory periods in the soliton theory for nerves and the locust femoral nerve. Biophys. Chem. **153**, 159–167 (2011)
202. D. Vucinic, T.J. Sejnowski, A compact multiphoton 3d imaging system for recording fast neuronal activity. PLoS One **2**(8), e699 (2007)
203. X. Wang, S. Schirmer, Analysis of Lyapunov control for Hamiltonian quantum systems, in *ENOC 2008*, St. Petersburg, 2008
204. H. Wang, Y. Yu, R. Zhu, S. Wang, Two-parameter bifurcation in a two-dimensional simplified Hodgkin-Huxley model. Commun. Nonlinear Sci. Numer. Simul. **18**, 184–193 (2013)
205. J. Wang, L. Chen, X. Fei, Bifurcation control of the Hodgkin-Huxley equations. Chaos Solitons Fract. **33**, 217–224 (2004)
206. W. Wang, L.C. Wang, X.X. Yi, Lyapunov control on quantum open systems in decoherence-free subspaces. Phys. Rev. A **82**, 034308 (2010)
207. D.Q. Wei, X.S. Luo, B. Zhang, Y.H. Qin, Controlling chaos in space-clamped FitzHugh–Nagumo neuron by adaptive passive method. Nonlinear Anal. Real World Appl. **11**, 1752–1759 (2010)
208. Y. Wei, G. Ullah, R. Parekh, J. Ziburkus, S.J. Schiff, Kalman Filter tracking of Intracellular Neuronal Voltage and Current, in *2011-50ty IEEE Conference on Decision and Control and European Control Conference (CDC-ECC 11)*, Orlando, 2011
209. S. Wiggins, *Introduction to Applied Nonlinear Dynamical Systems and Chaos*. Texts in Applied Mathematics, vol. 2 (Springer, New York, 2003)
210. H.M. Wiseman, G.J. Milburn, *Quantum Measurement and Control* (Cambridge University Press, Cambridge, 2010)
211. F. Woittennek, J. Rudolph, Controller canonical forms and flatness-based state feedback for 1D hyperbolic systems, in *7th Vienna International Conference on Mathematical Modelling, MATHMOD* (2012)
212. Q.D. Wu, C.J. Liu, J.Q. Zhang, Q.J. Chen, Survey of locomotion control of legged robots inspired by biological concept. Sci China Ser F **52**(10), 1715–1729 (2009)
213. H.N. Wu, J.W. Wang, H.K. Li, Design of distributed H_∞ fuzzy controllers with constraint for nonlinear hyperbolic PDE systems. Automatica **48**, 2535–2543 (2012)
214. X. Xu, H.Y. Hu, H.L. Wang, Stability switches, Hopf bifurcations and chaos of a neuron model with delay dependent parameters. Phys. Lett. A **354**, 126–136 (2006)
215. S. Yang, C. Cheng, An orthogonal neural network for function approximation. IEEE Trans. Syst. Man Cybern. B Cybern. **26**, 779–784 (1996)
216. X.S. Yang, Y. Huang, Complex dynamics in simple Hopfield networks. AIP Chaos **16**, 033114 (2006)
217. H. Yu, J. Wang, B. Deng, X. Wei, Y. Che, Y.K. Wong, W.L. Chan, K.M. Tsang, Adaptive backstepping sliding mode control for chaos synchronization of two coupled neurons in the external electrical stimulation. Commun. Nonlinear Sci. Numer. Simul. **17**, 1344–1354 (2012)
218. D. Yu, S. Chakravotry, A randomly perturbed iterative proper orthogonal decomposition technique for filtering distributed parameter systems, in *American Control Conference*, Montreal, 2012
219. J. Zhang, A. Bierman, J.T. Wen, A. Julius, M. Figueiro, Circadian System Modeling and Phase Control, in *49th IEEE Conference on Decision and Control*, Atlanta, 2010
220. T. Zhang, J. Wang, X. Fei, B. Deng, Synchronization of coupled FitzHugh–Nagumo systems via MIMO feedback linearization control. Chaos Solitons Fract. **33**(1), 194–202 (2007)

221. Q. Zhang, A. Benveniste, Wavelet networks. IEEE Trans. Neural Netw. **3**(6), 869–898 (1993)
222. Q. Zhang, M. Basseville, A. Benveniste, Early warning of slight changes in systems. Automatica **30**, 95–113 (special issue on Statistical Methods in Signal Processing and Control) (1994)
223. Q. Zhang, M. Basseville, A. Benveniste, Fault detection and isolation in nonlinear dynamic systems: a combined input–output and local approach. Automatica **34**, 1359–1373 (1998)
224. P. Zheng, W. Tang, J. Zhang, Some novel double-scroll chaotic attractors in Hopfield networks. Neurocomputing **73**, 2280–2285 (2010)
225. W. Zuo, Y. Zhu, L. Cai , Fourier-Neural-Network-Based Learning control for a class of nonlinear systems with flexible components. IEEE Trans. Neural Netw. **20**, 139–151 (2009)

Index

G.G. Rigatos, *Advanced Models of Neural Networks*,
DOI 10.1007/978-3-662-43764-3, © Springer-Verlag Berlin Heidelberg 2015